T0140605

Predicting the Dynamics of Research Impact

Yannis Manolopoulos • Thanasis Vergoulis
Editors

Predicting the Dynamics of Research Impact

 Springer

Editors
Yannis Manolopoulos
Open University of Cyprus
Nicosia, Cyprus

Thanasis Vergoulis
Information Management Systems Institute
Marousi, Greece

ISBN 978-3-030-86670-9   ISBN 978-3-030-86668-6   (eBook)
https://doi.org/10.1007/978-3-030-86668-6

This Springer imprint is published by the registered company Springer Nature Switzerland AG
The registered company address is: Gewerbestrasse 11, 6330 Cham, Switzerland

Table of Contents

Preface

In the past, many scientists from various disciplines have attempted to study science itself and its evolution in a quantitative fashion. Bringing science in the focal point was so common that it gave birth to a dedicated research field named *Science of Science*[1] [6, 18]. Although the access to fresh and extensive scholarly data is the catalyst for successful investigations of this kind, during the previous decades, such data remained, in their vast majority, fragmented and restricted in the data silos of scientific publishers. Luckily, nowadays, the wide adoption of the *Open Science* principles paves the way for the field of Science of Science to flourish by making large amounts of scholarly data openly available.

The investigation of temporal aspects of science, has always been interesting for researchers and has historically attracted much attention. Not surprisingly, the field of *Science dynamics* [12, 17] lies in the heart of Science of Science. Indicatively, longitudinal studies on scholarly data dominate most journal volumes in the respective scientific domain. Such studies attempt to reveal hidden patterns or useful insights for scientists, articles, journals and other involved entities. But even more interesting are approaches that attempt to project the revealed dynamics in the future to estimate, as accurately as possible, the future state and/or the evolution of the respective data. *Prediction* in the context of science of science [2, 5] usually focuses on the forecasting of future performance (or *impact*) of an entity, either this is a research article or a scientist (e.g., [1, 10]), and also the prediction of future links in collaboration networks or identifying missing links in citation networks (e.g., [8, 13, 21]).

The aforementioned subjects have received significant attention in the past. This is because, there are many real-world applications that may significantly benefit from accurate predicting science dynamics. For instance, consider a system that recommends articles to scientists according to their research interests. Due to the large growth rate in the number of published articles [7], a large number of suggestions will be given for almost any subject of interest. The recommendation system could leverage estimations for the expected impact of articles to prioritise the list of the

---

[1] In fact, this field is known with numerous names like *Meta-Science, Meta-Research, Research on Research*, etc.

suggested articles, giving high-priority to those with higher estimated impact. Of course, similar benefits would be applicable to other use cases like expert finding, collaboration recommendation, etc. It is important to highlight that, despite the wide spectrum of potential applications, relying on predictions should always be done with extreme caution; there are many pitfalls in this approach and attention should be paid to avoid misconceptions.

The objective of this edited book is not to give a birds-eye view of the field of prediction and science dynamics in Science of Science. Instead, we preferred to follow a more focused approach, including chapters that attempt to bridge important gaps in the current literature. Each chapter can be read independently, since it includes a detailed description of the problem being investigated along with a thorough discussion and study of the respective state-of-the-art. In addition, some of the chapters, go even beyond the state-of-the-art, introducing and evaluating completely new approaches that notably outperform the respective rivals.

Chapter 1 provides a useful contribution in the theoretical foundations of the field of Scientometircs and Science of Science. In particular, Leo Egghe and Ronald Rousseau study under which circumstances the production functions that map observed data assets (e.g., an author's publications, the respective citations) to measures (e.g., a bibliometric indicator) are able to encode the whole information included in the assets. They introduce a new class of general function systems, called *Complete Evaluation Systems* (CESes), that have this property and they examine a set of impact measures on whether they can derive CESes or not.

Next, Chapters 2-4 turn the focal point to the study of factors that affect research impact and its dynamics. In Chapter 2, Natsuo Onodera and Fuyuki Yoshikane investigate the factors that affect citation rates of research articles, focusing on those factors that are not related to the merit or the content of the articles (i.e., the *extrinsic* factors). The authors survey the respective literature examining the current trends and provide a useful classification of the most important extrinsic factors. Finally, they conduct their own experiments measuring the ability of various extrinsic factors to predict citations. Next, in Chapter 3, Nikolay K. Vitanov and Zlatinka I. Dimitrova investigate the dynamics of research production of individual researchers, organisations, and groups of organisations. The authors mathematically formulate various of the factors that affect these dynamics and use the proposed model to study the effect they have in research production, in each of the aforementioned cases. Finally, in Chapter 4, Manolis Antonoyiannakis focusses on how the attention a scientific article receives in the press affects its citations. The study considers articles published in Physical Review Letters (PRL) of the American Physics Society (APS) and examines their dissemination in eight platforms, which are available for the respective discipline. After investigating the contribution of each platform in the citations attracted by the corresponding articles (performing a multiple linear regression analysis on the underlying data) and analysing the percentage of highlighted papers in the top 1% cited articles of the field, the author identifies that the level of attention an article receives in press is, to an extent, correlated with its future citations.

Chapters 5-7 focus on article-level measures that quantify the current and future impact of scientific articles. In Chapter 5, Ilias Kanellos et al. study the expected short-term impact of scientific articles, i.e., their current popularity and the volume of citations they are expected to receive in the near future. More specifically, they formally define the related problem of ranking articles by their expected short-term impact and they survey several methods that are capable of solving this problem. They also discuss experimental findings on the effectiveness, the strengths, and weaknesses of these methods. Next, in Chapter 6, Natsuo Onodera and Fuyuki Yoshikane study the citation durability (i.e., citation *aging*) property of research articles and its relationship with other properties of research articles. They also experimentally study the characteristics of a particular measure of citation durability called *Citation Delay*. Finally, In Chapter 7, Tanmoy Chakraborty et al. study how the current and future influence of an article is reflected into the volume and the connectivity (according to cross-citations) of its citing articles. They introduce the notion of the *Influence Dispersion Tree (IDT)*, which captures this type of information, and investigate the relationship between an article's influence and its IDT's depth and breadth. Finally, they experimentally evaluate the effectiveness of impact measures that leverage IDT against the one of raw citation counts, in terms of estimating an article's current and future influence showing that IDT-based measures outperform their competitors.

Chapters 8-10 investigate subjects relevant to predicting the future impact of individual researchers. First, in Chapter 8, Giannis Nikolentzos et al. investigate whether the collaboration patterns of an author are good predictors of their future h-index. The authors leverage advanced graph neural network techniques in an attempt to predict the h-index of researchers relying solely on their collaboration and the contextual content of a subset of their papers. Their experiments indicate that there is indeed some relationship between the future h-index of researchers and their structural role in the co-authorship network. Next, in Chapter 9, Antônio de Abreu Batista-Jr et al. experiment with researcher-level impact measures alternative metrics. They argue that traditional measures (e.g., the h-index), which attempt to quantify the current impact of a researcher, are biased against young researchers, something that results in problems in various applications (e.g., selecting members of editorial boards or program committees). To alleviate this issue, the authors investigate the effectiveness of alternative measures that attempt to estimate a researcher's potential for future achievements. Their experiments indicate that this type of measures are reliable in estimating a researcher's future impact, thus, they can be useful in applications that benefit from the reduction of the bias against young researchers. Finally, in Chapter 10, Gangan Prathap et al. introduce two citation-based measures, *Citation Currency Ratio* (CCR) and *Citation Currency Exergy* (CCX), to measure current citation performance, aiming to help the identification of scientists who are at different stages of their career (rising, steady, and fading attention). The authors perform a preliminary evaluation study that indicates that these measures are able to identify scientists who are at different stages of their career.

Finally, Chapters 11-13 focus more on science evolution and dynamics, leveraging heterogeneous and interconnected data. First of all, the analysis of research topic trends and their evolution has always played a key role in impact prediction approaches and quantitative analyses in the field of bibliometrics. In Chapter 11,

Angelo Salatino et al. present a framework for detecting, analysing, and forecasting research topics, which leverages large-scale Scientific Knowledge Graphs. Such publicly available graphs are rich data sources for scholarly communication and have become very popular in the last few years. The authors discuss the advantages of the presented framework and describe how it has been applied to produce bibliometric studies and tools for analysing and predicting research dynamics. Moreover, predicting links in heterogeneous bibliographic networks, can be useful in the context of providing added-value services to researchers and other stakeholders, such as collaboration recommendation and assisting the curation of scholarly knowledge bases. Next, in Chapter 12, Xiaoli Chen and Tao Han introduce a framework for detecting emerging technologies at an early stage based on temporal data from citation networks and social media. The authors combine various machine learning approaches to leverage these data and train the respective models, including graph embeddings, outlier detection, and classification approaches. Finally, in Chapter 13 Pantelis Chronis et al. experimentally study models from two distinct approaches for link prediction on bibliographic networks: path counting models and embeddings. Their findings indicate that models of the former approach, although conceptually simpler, slightly outperform models of the latter one.

Although there is a variety of recent books on Science of Science and science dynamics [3, 4, 9, 11, 12, 14, 15, 16, 17, 18, 19, 20], to the best of our knowledge, there are no authored or edited books to discuss science dynamics subjects with the main focus on prediction in the Science of Science realm; the existing books either include a very limited number of relevant chapters or no such chapters at all. We believe that this book is a valuable addition to the respective literature, covering very interesting subjects that have not previously been discussed in an adequate detail.

We hope that this edited book provides the reader with an enjoyable introduction to interesting prediction and science dynamics problems in the field of Science of Science. The chapters are written in a way that can help the reader gain a detailed technical understanding on the corresponding subjects, the strength and weaknesses of the state-of-the-art approaches for each described problem, as well as the currently open challenges.

Due to the interdisciplinarity of the Science of Science field, the book may be useful to interested readers from a variety of disciplines like information science, information retrieval, network science, informetrics, scientometrics, and machine learning, to name a few. The profiles of the readers may also be diverse ranging from researchers and professors in the respective fields to students and developers being curious about the covered subjects.

Acknowledgements

A lot of people helped us in the production of this edited volume and we would like to take this opportunity to thank them. First of all, we are grateful to our chapter authors for their hard work throughout all the phases of the production of this book. Each of them has been invited due to their significant expertise in the respective areas, while the editorial team gave them the liberty to focus on the subject they considered to be the most interesting and to organise the corresponding chapter in the way they believe that better fits their needs.

Of course, thanks are due to the reviewers for their constructive and insightful comments, which have helped our authors to considerably improve the quality of the respective chapters:

Manolis Antonoyiannakis, Columbia University, USA
Alberto Baccini, University of Siena, Italy
Panayiotis Bozanis, International Hellenic University, Greece
Xiaoli Chen, Beijing Jiaotong University, China
Dimitris Dervos, International Hellenic University, Greece
Georgios Evangelidis, University of Macedonia, Greece
Ilias Kanellos, Athena Research Center, Greece
Dimitrios Katsaros, University of Thessaly, Greece
Sameer Kumar, University of Malaya, Malaysia
Francesco Osborne, The Open University, UK
Gangan Prathap, APJ Abdul Kalam Technological University, India
Dimitris Sacharidis, Free University of Brussels, Belgium
Angelo Antonio Salatino, The Open University, UK
Dimitrios Skoutas, Athena Research Center, Greece
Antonis Sidiropoulos, International Hellenic University, Greece
Christos Tryfonopoulos, University of the Peloponnese, Greece
Theodora Tsikrika, Information Technologies Institute, CERTH, Greece
Theodoros Tzouramanis, University of Thessaly, Greece
Nikolay Vitanov, Bulgarian Academy of Sciences, Bulgaria

Last but not least, we would like to thank Ralf Gestner and the technical staff of Springer Nature for the excellent guidance he gave us throughout the whole process and for the promptness in responding to our questions.

Yannis Manolopoulos, Nicosia, Cyprus
Thanasis Vergoulis, Marousi, Greece
July 2021

References

1. Bai, X., Zhang, F., Lee, I.: Predicting the citations of scholarly paper. Journal of Informetrics **13**(1), 407–418 (2019)
2. Clauset, A., Larremore, D., Sinatra, R.: Data-driven predictions in the science of science. Science **355**(6324), 477–480 (2017)
3. Ding, Y., Rousseau, R., Wolfram, D.: Measuring Scholarly Impact: Methods and Practice. Springer (2014)
4. Gingras, Y.: Bibliometrics and Research Evaluation: Uses and Abuses. The MIT Press (2014)
5. Hou, J., Pan, H., Guo, T., Lee, I., Kong, X., Xia, F.: Prediction methods and applications in the science of science: A survey. Computer Science Review **34**, 1–12 (2019)
6. Ioannidis, J.P., Fanelli, D., Dunne, D.D., Goodman, S.N.: Meta-research: Evaluation and improvement of research methods and practices. PLoS Biology **13**(10), e1002264 (2015)
7. Larsen, P., Von Ins, M.: The rate of growth in scientific publication and the decline in coverage provided by science citation index. Scientometrics **84**(3), 575–603 (2010)

8. Liu, H., Kou, H., Yan, C., Qi, L.: Link prediction in paper citation network to construct paper correlation graph. EURASIP Journal on Wireless Communications and Networking (2019)
9. Moed, H.F. (ed.): Applied Evaluative Informetrics (Qualitative and Quantitative Analysis of Scientific and Scholarly Communication). Springer (2018)
10. Nezhadbiglari, M., Goncalves, M.A., Almeida, J.A.: Early prediction of scholar popularity. In: Proceedings of the 17th ACM/IEEE Joint Conference on Digital Libraries (JCDL), pp. 181–190 (2016)
11. Qiu, J., Zhao, R., Yang, S., Dong, K.: Informetrics: Theory, Methods and Applications. Springer (2017)
12. Scharnhorst, A., Börner, K., Van den Besselaar, P.: Models of Science Dynamics: Encounters between Complexity Theory and Information Sciences. Springer Science & Business Media (2012)
13. Shibata, N., Kajikawa, Y., Sakata, I.: Link prediction in citation networks. Journal of the American Society for Information Science & Technology 63(1), 78–85 (2012)
14. Sugimoto, C.R. (ed.): Theories of Informetrics and Scholarly Communication. De Gruyter (2017)
15. Thelwall, M.: Web Indicators for Research Evaluation: A Practical Guide. Synthesis Lectures on Information Concepts, Retrieval, and Services. Morgan & Claypool Publishers (2017)
16. Todeschini, R., Baccini, A.: Handbook of Bibliometric Indicators. Wiley (2016)
17. Vitanov, N.K.: Science Dynamics and Research Production. Springer (2016)
18. Wang, D., Barabási, A.L.: The Science of Science. Cambridge University Press (2021)
19. Wani, Z.A., Zainab, T.: Scholarly Content and its Evolution by Scientometric Indicators: Emerging Research and Opportunities. IGI Global (2018)
20. Zhao, D., Strotmann, A.: Analysis and Visualization of Citation Networks. Synthesis Lectures on Information Concepts, Retrieval, and Services. Morgan & Claypool Publishers (2015)
21. Zhou, W., Gu, J., Jia, Y.: h-index-based link prediction methods in citation network. Scientometrics 117(1), 381–390 (2018)

List of Authors

Manolis Antonoyiannakis is an Associate Editor and Bibliostatistics Analyst at the American Physical Society, and an Adjunct Associate Research Scientist at Columbia University. He received his MSc from the University of Illinois at Urbana Champaign, and his PhD from Imperial College London, both in Physics. He has 15+ years of editorial experience on scholarly publishing, having handled >10,000 manuscripts in the Physical Review family of journals, and delivered >50 invited talks. He served as scientific advisor to the President of the European Research Council (2008-2010). Since 2020, he is an Editorial Board Member of the Metrics Toolkit. He is interested in the science of science, scientometrics, research assessment, and peer review. More information is available at `bibliostatistics.org`.

Spiros Athanasiou is a Research Associate and Project Manager at the Information Management Systems Institute of "Athena" Research Center. He received his Diploma in Electrical Engineering from the National Technical University of Athens, and has worked as a researcher and project manager in R&I projects of the public and private sector. His research interests include (among others) big data and semantic web infrastructures. More information is available at `users.uop.gr/~spiros`.

Antonio de Abreu Batista-Jr is a PhD student at the Center for Mathematics, Computing, and Cognition of the Federal University of ABC, Brazil. He obtained a BSc from the Estácio de Sá University and an MSc from the Federal University of Maranhão, all in Computer Science. His ongoing research focuses on finding statistical regularities in the scientific output of junior researchers that may explain their research performances in the future. More information is available at `deinf.ufma.br/~antonio`.

Sumit Bhatia is a Senior Machine Learning Scientist at Media and Data Science Research Lab at Adobe Systems. His research interests are in information retrieval, knowledge graphs and reasoning, and scholarly data mining and he has authored >50 journal and conference papers. Previously, at IBM Research, he served as lead search scientist for IBM Watson Knowledge Graph and developed algorithms for entity-oriented search and exploration of knowledge graphs. He was the lead architect of IBM Expressive Reasoning Graph Store. He also teaches at IIIT Delhi, where he was awarded the Teaching Excellence award based on student feedback. He regularly serves as a reviewer for several conferences and journals including ACL, NAACL, ECIR, EMNLP, WWW, CIKM, TKDE, IJCAI, and AAAI. More information is available at sumitbhatia.net.

Tanmoy Chakraborty is an Assistant Professor and Ramanujan Fellow at IIIT Delhi. Prior to this, he was a post-doc at the University of Maryland, College Park. He completed his PhD in 2015 as a Google India PhD scholar at IIT Kharagpur. His research group, LCS2, broadly works in the areas of social network analysis and natural language processing. He received several prestigious awards including faculty awards/fellowships from Google, IBM, Accenture and LinkedIn. He has authored two books: "Data Science for Fake News Detection: Surveys and Perspectives" and "Social Network Analysis: Concepts and Applications". He has been appointed as a project director of the "Technology Innovation Hub", a massive initiative taken by the Government of India. More information is available at faculty.iiitd.ac.in/~tanmoy.

Xiaoli Chen is a Librarian of the National Science Library, Chinese Academy of Sciences since 2013. Her research interests include hierarchical knowledge organization and representation, science and technology assessment and forecasting, and data mining of scientific publications. She received her PhD in Information Science from the University of Chinese Academy of Sciences in 2021, and her MSc in Telecommunication and Information Systems from Beijing Jiaotong University in 2013. She has authored >20 papers at reputed conference proceedings, including ISSI, JCDL and GTM. She has translated >10 books including "Python Data Science Handbook", and "Mastering Social Media Mining with Python".

Pantelis Chronis is a PhD student of Data Science at the Department of Informatics and Telecommunications, University of the Peloponnese. He received his Diploma in Computer Engineering at the University of Patras and has worked at the Information Management Systems Institute of "Athena" Research Center as a Research Assistant. He has worked in various fields of machine learning, including link prediction and clustering.

Zlatinka I. Dimitrova is an Associate Professor at the Institute of Mechanics of the Bulgarian Academy of Sciences in Sofia. Her research interests are in the area of mathematical modeling of nonlinear systems from the area of biology, social sciences and research networks. She is specialist in the area of application of statistical methods and construction of indexes for assessment the state and evolution of complex systems as well as in the area of use of nonlinear differential equations for modeling evolution of systems from fluid mechanics and population biology.

Leo Egghe was born in 1952 and has 3 children. He has a Doctorate in Mathematics (University of Antwerp, 1978) and a PhD in Information Science (City University London, 1989). He was the Chief Librarian of the University of Hasselt (1979-2017), and Professor at the University of Antwerp (1983-2016), where he taught courses on informetrics and information retrieval. He was Visiting Professor in several scientific institutes in the world. He conducted several development projects in Africa. He (together with R. Rousseau) organized the first ISSI conference in 1987 (the name ISSI was given later). In 2001, he (together with R. Rousseau) received the Derek De Solla Price Award. He is the Founding Editor-in-Chief of the Journal of Informetrics, Elsevier (2007). He authored >300 scientific publications (most of them in JCR source journals), among which 5 books. His main interests are: the mathematical development of the theory of power laws (Zipf, Lotka, which resulted in 2005 in a book, published by Elsevier) and of impact measures. In 2006 he invented the well-known g-index, described in a Scientometrics paper with >1000 citations.

Iakovos Evdaimon is a BSc student in Informatics at Athens University of Economics and Business. He is passionate about machine learning, data mining and graph algorithms. His target is to pursue a scientific path in machine learning, solving complex problems and to combine its applications in biology and chemistry.

Fábio Castro Gouveia is a Public Health Technologist at Oswaldo Cruz Foundation, Brazil. Biologist, MSc in Microbiology and DSc in Biological Chemistry at the Federal University of Rio de Janeiro, with a short post-doc at the Katolieke Universiteit Leuven, Belgium. He develops research in Information Science with emphasis on metric information studies (scientometrics, webometrics, altmetrics and science, technology and innovation indicators), digital methods, data science and science and health communication, with an emphasis on studies on internet and social media. More information is available at orcid.org/0000-0002-0082-2392.

Annu Joshi is a Data Engineer building analytics solutions at QuantumBlack, a McKinsey company. Her experience involves building distributed data pipelines on complex data systems across diverse problem domains. Previously, she worked as a student researcher at IIIT Delhi, working at the intersection of graphs and machine learning, specializing in academic networks. She has an overall experience of 3+ years in the big data domain after graduating with an MSc in Computer Science from Delhi University. She has a passion for enabling others to learn and can be often found facilitating a workshop to encourage women to enter the world of technology.

Ilias Kanellos is a Scientific Associate at the Information Management Systems Institute of "Athena" Research Center. He received his Diploma in Electrical and Computer Engineering from the National Technical University of Athens (2012). He then completed his PhD at the same institution, under the supervision of Prof. Yannis Vassiliou (2020). He has been involved in several EU and national R&D projects and his research interests include research analytics, scientific data management, cloud computing, bioinformatics, and data mining. He has authored 17 journal and conference papers.

Dimitrios Katsaros is an Associate Professor with the Department of Electrical and Computer Engineering of the University of Thessaly, Greece. He received a PhD in Informatics from the Aristotle University of Thessaloniki, Greece (2004), and since 2009 he is with the University of Thessaly. He has spent semester terms as a vising scientist/professor in Electrical Engineering Department of Yale University and in Yale Institute for Network Science (2015 and 2017), and also in KIOS Research and Innovation Centre of Excellence at the University of Cyprus (2019). His research interests lie in the area of distributed algorithms and systems and in network science. In 2006 he co-invented the popular contemporary and trend h-index scientometric indicators described in a Scientometrics paper. More information is available at dana.e-ce.uth.gr.

Andrea Mannocci is a Research Fellow at the Institute of Information Science and Technologies of the Italian Research Council. He holds a PhD in Information Engineering from the University of Pisa. He has been Research Associate at the Knowledge Media Institute, The Open University, UK, where he joined the SKM3 group and worked on data science applied to scholarly big data and research analytics. He currently works for EU projects like OpenAIRE Nexus and EOSC. His research interests span from the development of Open Science enabling technologies to Science of Science and the analysis of research as a global-scale phenomenon with geopolitical and socioeconomic implications. More information is available at andremann.github.io.

Yannis Manolopoulos is a Professor and Vice-Rector at the Open University of Cyprus as well as a Professor Emeritus at Aristotle University of Thessaloniki. He has been with the University of Toronto, University of Maryland at College Park, University of Cyprus and Hellenic Open University. He has also served as President of the Board of the University of Western Macedonia in Greece and Vice-President of the Greek Computer Society. He has authored 6 monographs and 10 textbooks in Greek, as well as >350 journal and conference papers. He has received 5 best paper awards, >16000 citations from >2300 distinct academic institutions from >100 countries (h-index=57). Through his research interests in Data Management, he contributed in Scientometrics with a number of indices, such as, the contemporary h-index, the trend h-index, the perfectionism index and the fractal dimension of a citation curve. More information is available at yannismanolopoulos.eu.

Jesús P. Mena-Chalco is an Associate Professor at the Center for Mathematics, Computing, and Cognition of the Federal University of ABC, Brazil. He received an MSc and a PhD in Computer Science from the University of São Paulo. His research interests include Pattern recognition, Graph mining, and Scientometrics. More information is available at professor.ufabc.edu.br/~jesus.mena.

Giannis Nikolentzos received a Diploma in Electrical and Computer Engineering from the University of Patras, an MSc in AI from the University of Southampton, and a PhD in graph mining from Athens University of Economics and Business (2017). He is currently a post-doc at École Polytechnique, Paris. He has authored >25 journal and conference papers. He is recipient of the distinguished paper award of IJCAI'2018. His research interests are in the field of machine learning on graphs. He has been teaching postgraduate courses at Athens University of Economics and Business, while he has also been involved in several research projects.

Natsuo Onodera is an Affiliated Fellow at National Institute of Science and Technology Policy, Japan. Until 2009, he worked as Professor at the Graduate School of Library and Information Science, the University of Tsukuba, Japan (now, Professor Emeritus). He received a BA and an MA in Science from Osaka University in 1966 and 1968, respectively, and a PhD in Education from the University of Tokyo in 2018. He served as the President of the Information Science and Technology Association, Japan from 2009 to 2014. His current work focuses on citation analysis of academic articles and analysis of skewed bibliometric distributions.

Francesco Osborne is a Research Fellow at the Knowledge Media Institute, The Open University, UK. He has an MSc and a PhD in Computer Science from the University of Turin. He has authored >80 papers in the fields of AI, information extraction, knowledge graphs, science of science, semantic web, research analytics, and semantic publishing. He developed many open resources such as the Computer Science Ontology, which is currently the largest taxonomy of research areas in the field, and the Artificial Intelligence Knowledge Graph, which describes 850K research entities extracted from the most cited articles in AI. He organized several workshops, the most recent ones being: the Scientific Knowledge (Sci-K) workshop at TheWebConf 2021 and the DL4KG workshop at ISWC 2021. More information is available at `people.kmi.open.ac.uk/francesco`.

George Panagopoulos is a PhD candidate at École Polytechnique, Paris. He received his MSc from the University of Houston (2018), for which he received the best MSc thesis award from the Department of Computer Science. At that time, he was Research Assistant at the Computational Physiology Lab, and prior to this, at the Software Knowledge and Engineering Lab, NCSR "Demokritos" in Athens. His research interests lie in machine learning and its applications to network science and graph algorithms. He has >10 papers in reputed fora, including conferences such as AAAI, ICWSM, and journals such as IEEE Transactions of Knowledge and Data Engineering and Transactions of Intelligent Transportation Systems.

Partha Sarathi Paul received an MSc in Pure Mathematics from the University of Calcutta, an MTech in Computer Science from the Indian Statistical Institute, Kolkata, and a PhD in Computer Science and Engineering from the National Institute of Technology Durgapur. He is currently working as Senior Project Officer at the Department of Computer Science and Engineering, IIT Kharagpur. His research interests include networking systems and applications, mobile and pervasive systems, human-computer interaction, scholarly network analysis, and geographical information systems. He authored papers in reputed conferences, including: ACM MobileHCI, IEEE COMSNETS, ACM MobiCom CHANTS, and IEEE PerCom WiP, as well as in reputed journals, including: Journal of Network and Computer Applications, Ad Hoc Networks, and Scientometrics. More information is available at `sites.google.com/view/personal-site-partha-sarathi`.

Gangan Prathap was trained as an aerospace engineer and specialized in mathematical modeling and computer simulation of complex problems in aerospace engineering. For 30+ years, he has pursued a parallel interest in research evaluation, bibliometrics and scientometrics and the application of physical and mathematical insights for research assessment. Recently, he proposed a thermodynamic basis for bibliometric sequences which can lead to better indicators for research evaluation like the p-index and the EEE-sequences. He has served at many premier institutions in India like the CSIR-NAL, CSIR-C-MMACS (now CSIR 4PI), CUSAT, CSIR-NISCAIR, CSIR-RAB, Saarc Documentation Centre, CSIR-NIIST and APJ Abdul Kalam Technological University.

Ronald Rousseau was born in 1949 in Antwerp, Belgium. He obtained his PhD in Mathematics (University of Leuven, 1977), a Habilitation degree in Mathematics (University of Leuven, 1983) and a Phd in Library and Information Science (University of Antwerp, 1992). For his mathematical work he received the Prize of the Belgian Academy of Sciences (1979). He was a Mathematics Professor at the Engineering College KHBO (Oostende) and taught different courses for the Education in Library and Information Science at the University of Antwerp. Together with Leo Egghe he wrote "Introduction to Informetrics" (1990) and "Becoming Metric-wise" (2018), with Raf Guns as third author. In 2001 he and Leo Egghe received the Derek De Solla Price Award for their work in scientometrics. He became an Honorary Professor at Zhejiang University and Henan Normal University. He is the former President of the International Society of Scientometrics and Informetrics (ISSI). His main interest is citation analysis and diversity measurement. More information is available at `en.wikipedia.org/wiki/Ronald_Rousseau`.

Dimitris Sacharidis is an Assistant Professor at the Université Libre de Bruxelles. Prior to this, he was an Assistant Professor at the Technical University of Vienna, and a Marie Skłodowska Curie fellow at "Athena" Research Center and Hong Kong University of Science and Technology. He received his Diploma and Phd in Computer Engineering from the National Technical University of Athens, and in between an MSc in Computer Science from the University of Southern California. He has served as PC member and has been involved in the organization of top data science related conferences, as well as has acted as editorial member of associated journals. His research interests include databases, data mining, and recommender systems. More information is available at `dsachar.github.io`.

Angelo Salatino is a Research Associate at the Intelligence Systems and Data Science (ISDS) group, at the Knowledge Media Institute (KMi) of The Open University, UK. He obtained a PhD studying methods for the early detection of research trends. In particular, his project aimed at identifying the emergence of new research topics at their embryonic stage (i.e., before being recognized by the research community). His research interests are in the areas of semantic web, network science and knowledge discovery technologies, with focus on the structures and evolution of science. More information is available at `salatino.org`.

Spiros Skiadopoulos is a Professor at the Department of Informatics and Telecommunications at the University of the Peloponnese and Director of an MSc program in Data Science at NCSR "Demokritos" in Athens. He has been Department Head on the first year of its establishment (ac. year 2013-14). He has served as PC member of several venues and participated in various research and development projects. His current research interests focus on big data management. His contribution is internationally acknowledged and includes journal and conference publications and a large number of citations. He received his Diploma and his PhD from the National Technical University of Athens and his MPhil from the University of Manchester Institute of Science and Technology. More information is available at uop.gr/~spiros.

Dimitrios Skoutas is a Principal Researcher at the Information Management Systems Institute of "Athena" Research Center. He received his Diploma and PhD in Electrical and Computer Engineering from the National Technical University of Athens, and he has worked as a post-doc at the L3S Research Center, Leibniz Universität Hannover. His research interests include spatial and temporal data integration and mining, similarity search, and heterogeneous information network analysis, having authored >70 papers. He has been the Principal Investigator in several projects and has served as PC member in several international conferences and workshops. More information is available at web.imsi.athenarc.gr/~dskoutas.

Han Tao, PhD in information Science, is a Professor at the University of the Chinese Academy of Sciences. He has carried out research and practice on information analysis technology of science and technology for many years, focusing on the integration of complex science, such as complex network, link prediction and evolutionary dynamics, with scientometrics and information science. Relevant research has been presented at reputed conferences such as ASIS&T, ISSI, ÜIM, and JCDL. He has presided over or participated in >10 research projects such as National Natural Science Foundation of China, Chinese Academy of Sciences, Ministry of Science and Technology, China Association for Science and Technology, authored >30 papers, and authored 8 monographs.

Michalis Vazirgiannis is a Distinguished Professor at École Polytechnique, Paris. He has been intensively involved in data science and AI related research. He has authored >200 journal and conference papers, whereas his h-index equals 50. He has supervised 22 PhD theses. Finally, he has attracted significant funding for research from national and international research agencies as well as industrial partners (Google, Airbus, Huawei, Deezer, BNP, LVMH). He lead(s) academic research chairs (DIGITEO 2013-15, ANR/HELAS 2020-24) and an industrial one (AXA, 2015-2018). He has received a Marie Curie Intra European Fellowship (2006-08), a "Rhino-Bird International Academic Expert Award" in recognition of his academic/professional work @ Tencent (2017), and best paper awards (IJCAI'2018, CIKM'2013). More information is available at: www.lix.polytechnique.fr/labo/michalis.vazirgiannis.

Thanasis Vergoulis is a Scientific Associate at the Information Management Systems Institute of "Athena" Research Center. He received his Diploma in Computer Engineering and Informatics from the University of Patras and his PhD in Computer Science from the National Technical University of Athens, under the supervision of Prof. Timos Sellis. He has been involved in EU and national ICT projects related to big data, scientific data management, open science, and linked data. His research interests also span bioinformatics, text mining and information retrieval for scientific publications, scientometrics, and research analytics. Finally, he has been teaching undergrad and postgrad courses in academic institutions in Greece and Cyprus. More information is available at: thanasis-vergoulis.com/.

Nikolay K. Vitanov is a Professor and Department Head at the Institute of Mechanics of the Bulgarian Academy of Sciences in Sofia. Part of his research interests are in the area of mathematical modeling of complex social and economic systems. He authored the book "Science Dynamics and Research Production. Indicators, Indexes, Statistical Laws and Mathematical Models", published by Springer (2016). Since 2016, he is Vice-chair of the Commission of the Ministry of Education and Science of the Republic of Bulgaria for the observation and assessment of the research activities and production of the universities and research organizations in Bulgaria. He is one of the leading specialists in Bulgaria in the area of development of national and regional research networks and in the area of assessment of research production of universities and research organizations.

Fuyuki Yoshikane, PhD, works as Professor at the University of Tsukuba, Japan. Until 2009, he worked as Assistant Professor at the Department of Research for University Evaluation, National Institution for Academic Degrees and University Evaluation, Japan. He received a BA, MA, and PhD in Education from the University of Tokyo in 1994, 2000, and 2011, respectively. He serves on the research committee of Japan Society of Library and Information Science. His current work focuses on citation analysis of academic articles and patents, including science linkage. More information is available at slis.tsukuba.ac.jp/~yoshikane.fuyuki. gt/index-e.html.

Chapter 1
On Complete Evaluation Systems

Leo Egghe and Ronald Rousseau

Abstract Parametrized systems of evaluation measures derived from the h-index, the g-index and other informetric measures are applied to sets of functions, denoted as **Z**. This results in a new class of general function systems, referred to as Complete Evaluation Systems (CESes). Two different functions in a CES can never have the same value for every value of the parameter. Examples are given of systems that are CES, mainly those derived from well-known measures such as the h-index, the g-index and the average, and of other systems that are not.

1.1 Introduction

1.1.1 Informetrics and Research Evaluation

While consensus exists that research and research outcomes are essential for mankind, a lot of debate exists about how research activities and outcomes can be compared and evaluated and hence to make the best use of scarce funding. To answer such questions, modern science policy has created an applied field as part of the science of science, a term itself used since the 1930s [14]. This application-oriented field was introduced in [12] under the name of evaluative bibliometrics. Within this field indicators such as the journal impact factor, the total number of received citations, the h-index, and many more have been proposed and applied in

Leo Egghe
University of Hasselt, Belgium,
e-mail: leo.egghe@uhasselt.be

Ronald Rousseau [1,2]
[1] University of Antwerp, 2020 Antwerpen, Belgium
[2] Centre for R&D Monitoring (ECOOM) and MSI Department, KU Leuven, 3000 Leuven, Belgium,
e-mail: ronald.rousseau@uantwerpen.be, e-mail: ronald.rousseau@kuleuven.be

© The Author(s), under exclusive license to Springer Nature Switzerland AG 2021
Y. Manolopoulos, T. Vergoulis (eds.), *Predicting the Dynamics of Research Impact*,
https://doi.org/10.1007/978-3-030-86668-6_1

practical research assessment exercises. For a general overview of these so-called bibliometric indicators, we refer to [17, 18].

One of the most popular indicators in use nowadays is the h-index introduced by Hirsch [10]. Slightly adapting the original formulation, the h-index for authors is defined as follows [17]:

> Consider the list of articles [co-] authored by scientist S, ranked decreasingly according to the number of citations each of these articles has received. Articles with the same number of citations are given different rankings (the exact order does not matter). Then the h-index of scientist S is h if the first h articles received each at least h citations, while the article ranked $h+1$ received strictly less than $h+1$ citations. Stated otherwise: scientist S' h-index is h if the number h is the largest natural number such that the first h publications received each at least h citations.

Nowadays, informetric (bibliometric, scientometrics, altmetric) techniques are often considered as a part of the toolbox of research evaluation. This, however, is a very narrow interpretation of the original meaning of the word informetrics, and, of the aim of the field. In our book [17] (page 3) we defined informetrics as:

> The study of the quantitative aspects of information in any form and any social group

and placed it on the intersection between applied mathematics and the social sciences. Bibliometrics, scientometrics, webmetrics, and altmetrics are all subfields of informetrics. In the authors' mind research evaluation and theoretical, mathematical studies both have a place in the field of informetrics.

1.1.2 Some Mathematical Notation

By a function Z we mean the following triple: the domain set D; a set C, referred to as the codomain of Z; and a prescription which realizes a correspondence between every $d \in D$ and a unique point, denoted $Z(d) \in C$. The set $\{Z(d); d \in D\} \subset C$, is called the *range of Z*. All functions Z_1, Z_2 with the same domain are said to be the *same* (are equal) if their prescriptions always lead to the same image, i.e., $\forall d \in D$, $Z_1(d) = Z_2(d)$. Functions with a different domain are considered to be different functions. Yet, considered as functions restricted to the intersection of their domains they may be said to be *equal*.

Functions $Z(x)$ used further on are to be considered as abstract representations of observed data, such as an author's publications and the corresponding received citations. Then parametrized systems of measures derived from the h-index, the g-index [4, 6] and other informetric measures will be applied to sets $\mathbf{Z}$ of functions Z.

1.1.3 Motivation of this Work and an Overview of its Contents

Applying a measure associates one number to a set of data, modeled through a production function Z. This act entails a strong reduction of the information included in Z. Yet, such a reduction is inevitable in an evaluation exercise. We next wondered if it was possible that all information contained in Z could be saved in some way or other. We found a solution to this problem by introducing measures m_θ. One could see this as replacing the variable $x \in Z(x)$ with the variable θ. Yet, it is not always true that $m(Z_1) = m(Z_2)$ implies that $Z_1 = Z_2$. This observation makes it interesting to name and study mapping m for which this property holds. These are the Complete Evaluation Systems, in short CESes. In further work these reflections have led to a classification of impact measures [5].

In the next section, we will formally introduce CESes. Examples will be given of systems that are CESes, and of those who are not. We even construct an infinite number of non-CES examples. The theoretical part ends with an example that uses the Laplace transform, as an example of an integral transform, regularly used in engineering applications. Before coming to the conclusion we inserted a short section on the importance of mathematics in the sciences and the information sciences in particular.

1.2 Complete Evaluation Systems

1.2.1 Introducing a New Concept

Let $\mathbf{Z}$ be a set of positive functions Z defined on a subset $dom(Z)$ of $[0, +\infty[$. We associate with every $Z \in \mathbf{Z}$ a function $m(Z)$ as follows:

$$m(Z) : dom(m(Z)) \subset [0, +\infty[\to R^+ : \theta \to m_\theta(Z)$$

Hence m is a mapping from $\mathbf{Z}$ to $\mathbf{F}$, the set of positive, hence real-valued, functions defined on a subset of $[0, +\infty[$.

Definition 1 The mapping m on $\mathbf{Z}$ is a Complete Evaluation System, in short a CES, if:

$$m(Z_1) = m(Z_2) \Rightarrow Z_1 = Z_2 \tag{1.1}$$

In other words, m on $\mathbf{Z}$ is a CES if the mapping m from $\mathbf{Z}$ to $\mathbf{F}$ is injective and hence invertible on its range. We point out that $m(Z_1) = m(Z_2)$ means that these functions are defined for the same values of θ (they must have the same domain) and for each of these values θ, the real number $m_\theta(Z_1)$ must be equal to the real number $m_\theta(Z_2)$.

Functions m can be considered as parametrized measures (via the parameter θ). When used in abstract evaluation systems applied to objects $\mathbf{Z}$ they are 'complete' in the sense that no information about $\mathbf{Z}$ is lost. We note that if $\mathbf{Z}_0 \subset \mathbf{Z}$ and m is a CES on $\mathbf{Z}$, then m is also a CES on $\mathbf{Z}_0$.

A trivial example: let $\mathbf{Z}$ consists of functions defined on the same subset S of $[0, +\infty[$, and let m_0 be defined as follows:

$$m_0(Z) : dom(m_0(Z)) \subset [0, +\infty[\to R : \theta \to Z(\theta) \tag{1.2}$$

Here we see that $dom(m_0(Z))$ is equal to S. If now $m_0(Z_1) = m_0(Z_2)$, then we have that $\forall \theta \in S : Z_1(\theta) = Z_2(\theta)$, and hence $Z_1 = Z_2$. This shows that m_0 is a CES on $\mathbf{Z}$.

The following examples illustrate why we use the term "evaluation" in the expression "complete evaluation system". Indeed, we show that indicators such as the h-index, the g-index and averages all lead to CESes.

1.2.2 Examples

A CES derived from the generalized h-index as introduced in [7].
Let $\mathbf{Z}_h$ be a set of positive, decreasing, continuous functions defined on a fixed interval $[b, T]$, with $0 < b < T$, or on $]b, T]$ if $b = 0$. We note that $\mathbf{Z}_h$ may include constant functions. A function $Z(x)$ is said to be *positive* if $\forall x \in [b, T]$, $Z(x) \geq 0$. In [7] we introduced the continuous analog of the (discrete) van Eck-Waltman h-index [3] as follows. For $Z \in \mathbf{Z}_h$, we define the function $\theta \to h_\theta(Z)$, where $h_\theta(Z)$ is the unique solution of the equation:

$$Z\big(h_\theta(Z)\big) = \theta\, h_\theta(Z) \tag{1.3}$$

Denoting the function m (from the CES-framework) as m_h (the m-function related to the generalized h-index,) we see that for each Z, $m_h(Z)$ is the function $\theta \to h_\theta(Z)$, defined for $\theta \geq Z(T)/T$, i.e. $dom(m_h(Z)) = [Z(T)/T, Z(b)/b]$, if $b > 0$. If $b = 0$, then $dom(m_h(Z)) = [Z(T)/T, +\infty[$.

We now show that m_h determines a CES on the set $\mathbf{Z}_h$ of positive, decreasing, continuous functions as defined above. We first recall the following easy lemma.

Lemma 1. For every $Z \in \mathbf{Z}_h$, $\{h_\theta(Z);\ \theta \in [Z(T)/T, Z(b)/b]\} = [b, T]$.

Proof. For every $\theta \in [Z(T)/T, Z(b)/b]$ the corresponding $h_\theta(Z)$ belongs to the closed interval $[b, T]$, while $\forall x_0 \in [b, T]$ if suffices to take $\theta = Z(x_0)/x_0$, as then $Z(x_0) = \theta x_0$, or $x_0 = h_\theta(Z)$. $\qquad\square$

Theorem 1. The function m_h as defined above is a CES on the set $\mathbf{Z}_h$.

Proof. Assume that $m_h(Z_1) = m_h(Z_2)$, where Z_1 and Z_2 are functions defined on $[b, T]$, $0 < b < T$. Now, if $x = h_\theta(Z_1)$, then $Z_1(x) = \theta x$, and hence, as $h_\theta(Z_1) = h_\theta(Z_2)$, also $Z_2(x) = \theta x$, hence $Z_1(x) = Z_2(x)$. As, by Lemma 1, $\forall Z \in \mathbf{Z}$, $\{h_\theta(Z);\ \theta \in [Z(T)/T, Z(b)/b]\} = [b, T]$ this proves that m_h is a CES on the set of all positive, decreasing, continuous functions, defined on $[b, T]$. This proof is also valid (with small adaptations) if $b = 0$. $\qquad\square$

Corollary. Using the same set $\mathbf{Z}_h$, define the function $m_h^{(r)}$ through $m_h^{(r)}(Z) : \theta \to 1/h_\theta(Z)$. Clearly $m_h^{(r)}(Z_1) = m_h^{(r)}(Z_2)$ implies $Z_1 = Z_2$, and hence also $m_h^{(r)}$ is a CES on $\mathbf{Z}_h$. This special case will play a role (as a factor) in other functions, e.g. $m^{(1/2)}$ (see Theorem 10). This function $m^{(1/2)}$ is constructed as a product of two CESes, but is not a CES itself.

Next, we show how a similar reasoning can be applied for functions defined on $\mathbf{R}_0^+$.

Let $\mathbf{Z}_h^*$ be a set of positive, decreasing, continuous functions defined on $]0, +\infty[= \mathbf{R}_0^+$. Also here $\mathbf{Z}_h^*$ may include constant functions and a function $Z(x)$ defined on $\mathbf{R}_0^+$ is said to be *positive* if $\forall x \in \ ]0, +\infty[\ Z(x) \geq 0$. For $Z \in \mathbf{Z}_h^*$, we define the function $\theta \to h_\theta(Z)$ in the same way as above, namely as the unique solution of the equation:

$$Z(h_\theta(Z)) = \theta h_\theta(Z)$$

Denoting the function m as m_h^* we see that for each Z, $m_h^*(Z)$ is the function $\theta \to h_\theta(Z)$, defined for $0 < \theta < +\infty$. We now show that m_h^* determines a CES on the set $\mathbf{Z}_h*$.

We first mention the following lemma.

Lemma 1bis. For every $Z \in \mathbf{Z}_h^*$, $\{h_\theta(Z);\ 0 < \theta < +\infty\} = \mathbf{R}_0^+$.

Proof. For every θ, $0 < \theta < +\infty$ the corresponding $h_\theta(Z)$ belongs to $\mathbf{R}_0^+$, while $\forall x_0 \in \mathbf{R}_0^+$ if suffices to take $\theta = Z(x_0)/x_0$, as then $Z(x_0) = \theta x_0$, or $x_0 = h_\theta(Z)$. □

Theorem 1bis. The function m_h^* as defined above is a CES on the set $\mathbf{Z}_h^*$.

Proof. Assume that $m_h^*(Z_1) = m_h^*(Z_2)$, where Z_1 and Z_2 are functions defined on $\mathbf{R}_0^+$. Now, if $x = h_\theta(Z1)$, then $Z_1(x) = \theta x$, and hence, as $h_\theta(Z_1) = h_\theta(Z_2)$, also $Z_2(x) = \theta x$, hence $Z_1(x) = Z_2(x)$. As, by Lemma 1bis, $\forall Z \in \mathbf{Z}_h^*$, $\{h_\theta(Z); 0 < \theta < +\infty\} = \mathbf{R}_0^+$, this proves that m_h^* is a CES on the set $\mathbf{Z}_h^*$ of all positive, decreasing, continuous functions, defined on $\mathbf{R}_0^+$. □

A CES derived from the generalized g-index as introduced in [7]

Besides the continuous analog of the discrete van Eck-Waltman h-index, we similarly introduced a generalized g-index as follows [7]. Let $\mathbf{Z}_g$ be a set of positive, decreasing, continuous functions defined on $\mathbf{R}_0^+$ for which the improper integrals $Y_Z(x) = \int_0^x Z(s)ds$, $0 \leq x < +\infty$, exist. For $Z \in \mathbf{Z}_g$, we define the function $\theta \to g_\theta(Z)$ as the unique solution of the equation (where it exists):

$$Y_Z(g_\theta(Z)) = \int_0^{g_\theta(Z)} Z(s)ds = \theta(g_\theta(Z))^2 \tag{1.4}$$

Denoting the function m here as m_g (the m-function related to the generalized g-index) we see that $m_g(Z)$ is the function $\theta \to g_\theta(Z)$, defined for $0 < \theta < +\infty$.

Lemma 2. For every $Z \in \mathbf{Z}_g$, $\{g_\theta(Z); 0 < \theta < +\infty\} = \mathbf{R}_0^+$.

Proof. For every θ, $0 < \theta < +\infty$, the corresponding $g_\theta(Z)$ belongs to the interval $]0, +\infty[$, while $\forall x_0 \in \ ]0, +\infty[$ if suffices to take $\theta = \frac{1}{x_0^2} \int_0^{x_0} Z(s) ds$ as then $\int_0^{x_0} Z(s) ds = \theta x_0^2$ or $x_0 = g_\theta(Z)$. $\qquad\square$

We next show that m_g is a CES.

Theorem 2. The function m_g as defined above is a CES on the set $\mathbf{Z_g}$ of positive, decreasing, continuous functions defined on $\mathbf{R_0^+}$ for which the improper integrals $Y_Z(x) = \int_0^x Z(s) ds$, $0 \le x < +\infty$, exist.

Proof. Assume that $\forall \theta > 0$ and $\forall Z_1, Z_2 \in \mathbf{Z_g} : g_\theta(Z_1) = g_\theta(Z_2)$. Now, if $x = g_\theta(Z_1)$, then $Y_{Z_1}(x) = \theta x^2$, and hence, as $g_\theta(Z_1) = g_\theta(Z_2)$ also $Y_{Z_2}(x) = \theta x^2$. From this we derive that $Y_{Z_1}(x) = Y_{Z_2}(x)$ or $\int_0^x Z_1(s) ds = \int_0^x Z_2(s) ds$. As $\forall Z \in \mathbf{Z_g}, \{g_\theta(Z); 0 < \theta < +\infty\} = \mathbf{R_0^+}$ (Lemma 2), this equality holds $\forall x \in \ ]0, +\infty[$. Taking the derivative of both sides of this equality, we obtain $Z_1(x) = Z_2(x)$ or $Z = Y$ on $]0, +\infty[$. $\qquad\square$

A similar result can be formulated for functions defined on a fixed interval $]0, T]$, where $m_g(Z)$ is the function $\theta \to g_\theta(Z)$, defined for $\theta \ge Y_Z(T)/T^2$, i.e. $dom(m_g) \subset [Y_Z(T)/T^2, +\infty[$ [7]. In this case we have that $\forall Z \in \mathbf{Z_g}, \{g_{\theta(Z)}; \theta \in [Y_Z(T)/T^2, +\infty[\ \} = \ ]0, T]$, which can be shown in the same way as Lemma 2.

Averages

The next case uses average functions. Let $\mathbf{Z_\mu}$ be a set of continuous functions Z defined on $[0, +\infty[$. Now we associate with every $Z \in \mathbf{Z_\mu}$ a function $m_\mu(Z)$ as follows:

$$m_\mu(Z) : \ ]0, +\infty[\to \mathbf{R} : \theta \to \mu_\theta(Z) \tag{1.5}$$

with $\mu_\theta(Z) = \frac{1}{\theta} \int_0^\theta Z(s) ds$.

Theorem 3. The averaging measure m_μ determines a CES on the set $\mathbf{Z_\mu}$, of continuous functions Z defined on $[0, +\infty[$.

Proof. Assume that $\forall \theta > 0$ and $\forall Z_1, Z_2 \in \mathbf{Z} : \mu_\theta(Z_1) = \mu_\theta(Z_2)$. This means that $\forall \theta > 0 : \frac{1}{\theta} \int_0^\theta Z_1(s) ds = \frac{1}{\theta} \int_0^\theta Z_2(s) ds$. Consequently: $\forall \theta > 0 : \int_0^\theta Z_1(s) ds = \int_0^\theta Z_2(s) ds$ and taking the derivative with respect to θ: $\forall \theta > 0 : Z_1(\theta) = Z_2(\theta)$. Hence we find that $Z_1 = Z_2$ on $[0, +\infty[$. $\qquad\square$

For further use we define on the same set $\mathbf{Z_\mu}$ as used in Theorem 3, a function $m_\lambda(Z)$ as follows:

$$m_\lambda(Z) : \ ]0, +\infty[\to \mathbf{R} : \theta \to \lambda_\theta(Z) \tag{1.6}$$

with $\lambda_\theta(Z) = \frac{1}{\theta} \sqrt{\int_0^\theta Z(s) ds}$.

Proposition 1. The measure m_λ determines a CES on the set $\mathbf{Z_\mu}$ of continuous functions Z on $[0, +\infty[$.

Proof. Assume that $\forall \theta > 0$ and $\forall Z_1, Z_2 \in \mathbf{Z} : \lambda_\theta(Z_1) = \lambda_\theta(Z_2)$. This means that $\forall \theta > 0 : \frac{1}{\theta} \sqrt{\int_0^\theta Z_1(s) ds} = \frac{1}{\theta} \sqrt{\int_0^\theta Z_2(s) ds}$. Consequently: $\forall \theta > 0 : \int_0^\theta Z_1(s) ds =$

$\int_0^\theta Z_2(s)ds$ and taking derivatives: $\forall \theta > 0 : Z_1(\theta) = Z_2(\theta)$. Hence we find that $Z_1 = Z_2$ on $[0, +\infty[$. □

Percentiles

Let $\mathbf{Z_p}$ be a set of positive, continuous functions defined on a fixed interval $[0, T]$. Now we associate with every $Z \in \mathbf{Z_p}$ a function $m_p(Z)$ as follows:

$$m_p(Z) : [0, 1] \to R : \theta \to Z(\theta T)$$

Note that here θ is defined on the closed interval $[0, 1]$.

Theorem 4. The measure m_p determines a CES on the set $\mathbf{Z_p}$ of positive, continuous functions defined on a fixed interval $[0, T]$.

Proof. Assume that for $Z_1, Z_2 \in \mathbf{Z_p} : m_p(Z_1) = m_p(Z_2)$. This means that $\forall \theta \in [0, 1] : Z_1(\theta T) = Z_2(\theta T)$ and hence $Z_1 = Z_2$, on $[0, T]$. This shows that m_p is a CES on $\mathbf{Z_p}$. □

Theoretical cases

Next we consider a theoretical case involving an injective function.

Theorem 5. Consider a given positive, injective function i defined on $\mathbf{R}$. Let m be a CES on a set $\mathbf{Z}$, then m_i, defined as $m_i(Z) : dom(m_i(Z)) \subset]0, +\infty[\to \mathbf{R} : \theta \to (m_i(Z))(\theta) = i(m_\theta(Z))$ is also a CES on $\mathbf{Z}$.

Proof. Assume that $\forall \theta \in dom(m_i(Z))$ and for $Z_1, Z_2 \in \mathbf{Z} : i(m_\theta(Z_1)) = i(m_\theta(Z_2))$. Because i is injective, this means that $\forall \theta \in dom(m_i(Z)) : m_\theta(Z_1) = m_\theta(Z_2)$ and because m is a CES this implies that $Z_1 = Z_2$. □

Another case involving an injective function.

Let f be an injective function defined on $\mathbf{R}$; let $\mathbf{Z}$ be a set of functions such that $\forall Z \in \mathbf{Z}$, also $f \circ Z \in \mathbf{Z}$. Then we have the following proposition.

Proposition 2. Let m be a CES on a set $\mathbf{Z}$, then m_f, defined as $m_f(Z) : dom(m_f(Z)) \subset]0, +\infty[\to R : \theta \to (m_f(Z))(\theta) = m_\theta(f \circ Z)$ is also a CES on $\mathbf{Z}$.

Proof. Assume that $\forall \theta \in dom(m_f(Z))$ and for $Z_1, Z_2 \in \mathbf{Z} : f(Z_1(\theta)) = f(Z_2(\theta))$. Because f is injective, this means that $\forall \theta \in dom(m_f(Z)) : Z_1(\theta) = Z_2(\theta)$, and hence $Z_1 = Z_2$. □

Besides the classical h- and g-indices, also an R-index has been introduced [11]. Let now $\mathbf{Z_R}$ be a set of positive, decreasing continuous functions defined on $[0, +\infty[$, then a continuous version of this R-index can be parametrized, as follows:

$$R_\theta^2(Z) = \int_0^{h_\theta(Z)} Z(s)ds \tag{1.7}$$

defined on the domain of $\theta \to h_\theta(Z)$ (and we note that because Z is continuous these integrals exists). If $\theta = 1$, this is the continuous version of the discrete R-index.

Now we associate with every $Z \in \mathbf{Z_R}$ a function $m_R(Z)$ as follows:

$$m_R(Z) : dom(m_R(Z)) \subset]0, +\infty[\to \mathbf{R} : \theta \to m_R(Z) = R_\theta(Z)$$

Theorem 6. The function m_R as defined above is a CES on the set $\mathbf{Z_R}$ of positive, decreasing, continuous functions defined on $\mathbf{R}^+$.

Proof. Assume that for $Z_1, Z_2 \in \mathbf{Z_h}$ and $\forall \theta \in dom(m_R(Z_1)) \cap dom(m_R(Z_2))$: $R_\theta(Z_1) = R_\theta(Z_2)$. Then (for these θ): $\int_0^{h_\theta(Z_1)} Z_1(s)ds = \int_0^{h_\theta(Z_2)} Z_2(s)ds$. Taking derivatives with respect to θ yields:

$$Z_1(h_\theta(Z_1))h'_\theta(Z_1) = Z_2(h_\theta(Z_2))h'_\theta(Z_2) \Rightarrow$$
$$\theta h_\theta(Z_1)h'_\theta(Z_1) = \theta h_\theta(Z_2)h'_\theta(Z_2) \Rightarrow$$
$$h_\theta(Z_1)h'_\theta(Z_1) = h_\theta(Z_2)h'_\theta(Z_2)$$

Next we take the integral from $s = \theta$ to $s = +\infty$:

$$\int_\theta^{+\infty} h_s(Z_1)h'_s(Z_1)ds = \int_\theta^{+\infty} h_s(Z_2)h'_s(Z_2)ds$$

This leads to:

$$\left. \frac{h_s^2(Z_1)}{2} \right|_{s=\theta}^{s=+\infty} = \left. \frac{h_s^2(Z_2)}{2} \right|_{s=\theta}^{s=+\infty}$$

or, as

$$\lim_{s \to \infty} h_s^2(Z_1) = 0 = \lim_{s \to \infty} h_s^2(Z_2)$$

it follows that $h_\theta(Z_1) = h_\theta(Z_2)$, $\forall \theta$. From Theorem 1 we obtain: $Z_1 = Z_2$. $\square$

1.3 An Interesting Relation with the Inverse Function

The generalized h-index with parameter θ of a decreasing function $Z(x)$ is the abscissa of the intersection of the function $y = Z(x)$ with the line $y = \theta x$. Now, one may wonder if also the ordinate of this intersection has a meaning within the framework of parametrized measures, and if it leads to a CES. This will be investigated next.

Let $Z(x)$ be a positive, strictly decreasing, continuous function defined on an interval $[0, T]$ with $Z(T) = 0$. Then, clearly, this function is injective and hence has an inverse function, denoted as Z^{-1}.

Then $Z^{-1}(x)$ is also a strictly decreasing continuous function, now defined on the interval $[0, Z(0)]$. It can be visualized as the reflection of $Z(x)$ with respect to the first bisector. This is illustrated in Fig.1.1. In this figure we also show the line $y = \theta x$, the point $(h_\theta, Z(h_\theta))$ and the point $(Z(h_\theta), h_\theta)$ on the graph of Z^{-1}. Now we determine the slope of the line connecting the origin with the point $(Z(h_\theta), h_\theta)$. If $\theta = \tan(\phi)$, then the angle of the line connecting the origin with the point $(Z(h_\theta), h_\theta)$ is $\pi/2 - \phi$,

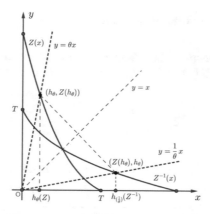

Fig. 1.1 Z its inverse function and related h_θ-indices.

and the corresponding slope is $\tan(\pi/2 - \phi) = \cot(\phi) = 1/\tan(\phi) = 1/\theta$. Hence $h_{1/\theta}(Z^{-1}) = Z(h_\theta)$. As this equality holds $\forall \theta \in \,]0, +\infty[$, we find:

$$h_\theta(Z^{-1}) = Z(h_{1/\theta}) = \frac{1}{\theta}h_{1/\theta}(Z)$$

We restate, and slightly generalize, this result as a theorem.

Theorem 7. If Z is a positive, strictly decreasing, continuous function, with inverse Z^{-1}, and if $dom(h_Z)$ denotes the domain of the function $\theta \rightarrow h_\theta(Z)$, then $h_\theta(Z^{-1}) = \frac{1}{\theta}h_{1/\theta}(Z), \forall \theta \in dom(h_Z)$.

Assume now that we have a set $\mathbf{Z}$ of functions which are all positive, strictly decreasing, continuous and such that $\forall Z \in \mathbf{Z}$ also Z^{-1} belongs to $\mathbf{Z}$, then, $\forall Z \in \mathbf{Z}$ we define $m_h^{-1}(Z)$ as the function $\theta \rightarrow h_\theta(Z^{-1}) = \frac{1}{\theta}h_{1/\theta}(Z), \forall \theta \in dom(\theta_Z)$. Now, $m_h^{-1}(Z_1) = m_h^{-1}(Z_2)$ implies $h_\theta(Z_1^{-1}) = h_\theta(Z_2^{-1}), \forall \theta$. Then we know that $Z_1^{-1} = Z_2^{-1}$ and hence $Z_1 = Z_2$. This shows that also the ordinate of the intersection of the function $y = Z(x)$ with the line $y = \theta x$ leads to a CES.

1.4 Examples of Cases that do not Lead to a CES

1.4.1 A Trivial Case

Let $f(x)$ be a given function and $\mathbf{Z}$ a set of functions, defined on $[0, +\infty[$, containing a least two elements. Define now, $\forall Z \in \mathbf{Z}$ and $\forall \theta \in [0, +\infty[: m_\theta(Z) = f(\theta)$. Then, clearly $m_\theta(Z_1) = m_\theta(Z_2) = f(\theta)$ does not imply that $Z_1 = Z_2$.

1.4.2 An Example Involving Quotients

Let $\mathbf{Z}$ be a set of functions including two different constant functions, namely $Z_1(x) = a > 0$ and $Z_2(x) = b > 0$, with same domain. Now, we define $\forall \theta$ and $\forall Z \in \mathbf{Z}$:

$$m_\theta(Z) = \frac{R_\theta(Z)}{h_\theta(Z)} = \frac{\sqrt{\int_0^{h_\theta(Z)} Z(s)ds}}{h_\theta(Z)} \tag{1.8}$$

We now determine $m_\theta(Z_j)$, $j = 1, 2$.

We know that $x = h_\theta(a)$ implies $a = \theta x$ and hence $h_\theta(a) = a/\theta$; similarly $h_\theta(b) = b/\theta$. Further:

$$R_\theta(a) = \sqrt{\int_0^{h_\theta(a)} a\,ds} = \sqrt{\int_0^{h_{a/\theta}} a\,ds} = \sqrt{\frac{a^2}{\theta}} = \frac{a}{\sqrt{\theta}}$$

and similarly:

$$R_\theta(b) = \frac{b}{\sqrt{\theta}}.$$

Hence:

$$m_\theta(Z_1) = \frac{a/\sqrt{\theta}}{a/\theta} = \sqrt{\theta} = m_\theta(Z_2)$$

while $Z_1 \neq Z_2$. This proves that this m is not a CES. In further investigations we will refer to this m as $m^{(1/2)}$, where the exponent ½ refers to the square root used in Eq.1.8. Replacing ½ by another power c, leads to a parameter set of measures $m^{(c)}$ studied further on.

Remark. For constant functions Z we know that $h_\theta(Z) = g_\theta(Z)$. Hence the previous example also shows that in general $\frac{R_\theta}{g_\theta}$ is not a CES.

Next we give a more subtle example which does not involve constant functions.

1.4.3 An Example which does not Make Use of Constant Functions

Let $\mathbf{Z}$ be a set of functions containing $Z_1(x) = ax^2$ and $Z_2(x) = bx^2$, defined on $\mathbf{R}+$, with $a \neq b$ (and both different from zero). If now $x_0 = h_\theta(Z_1)$, then $a(x_0)^2 = \theta x_0$ and hence $x_0 = \theta/a = h_\theta(Z_1)$. Similarly $h_\theta(Z_2) = \theta/b$.

Now, we have:

$$\int_0^{h_\theta(Z_1)} Z_1(s)ds = \int_0^{\theta/a} as^2 ds = \frac{a}{3}\left(\frac{\theta}{a}\right)^3 = \frac{1}{3a^2}\theta^3$$

(and a similar result for Z_2). Using the same m as in the example of previous section yields:

$$m_\theta(Z_1) = \frac{R_\theta(Z_1)}{h_\theta(Z_1)} = \frac{\sqrt{\int_0^{h_\theta(Z_1)} Z_1(s)\,ds}}{h_\theta(Z_1)} = \frac{\sqrt{\frac{1}{3a^2}\theta^3}}{\frac{\theta}{a}} = \frac{1}{\sqrt{3}}\sqrt{\theta}$$

As this result is independent of a, we also have $m_\theta(Z_2) = \frac{1}{\sqrt{3}}\sqrt{\theta}$. This shows that this m is not a CES on this set of functions $\mathbf{Z}$.

Note. It is easy to see that this example is also valid when the power function x^2 is replaced by another power function, even negative powers, as long as the integrals exist.

1.4.4 The Special Case of an Abstract Zipf Function

We reconsider the previous example for the case of an abstract Zipf function, by which we mean that we use an exponent strictly between -1 and 0. Concretely we consider the set of injective functions $\mathbf{Z}$ of the form $Z = ax^{-\gamma}$ with $0 < \gamma < 1$. Then we have the following interesting result.

Proposition 3. If $\mathbf{Z}$ is a set of injective functions and if, for each function $Z \in \mathbf{Z}$, the function $m_\theta(Z)$ is injective in θ, then it does not necessarily imply that m is injective, i.e. that m is a CES.

Proof. As announced, we will consider the set $\mathbf{Z}$ of injective functions of the form $Z = ax^{-\gamma}$ with $a > 0$ variable, and γ fixed, $0 < \gamma < 1$. We next consider $Z_1(x) = ax^{-\gamma}$ and $Z_2(x) = bx^{-\gamma}$, defined on $]0, +\infty[$ with $a \neq b$ (and both strictly positive). If now $x_0 = h_\theta(Z_1)$, then $a(x_0)^{-\gamma} = \theta x_0$ and hence $x_0 = \left(\frac{\theta}{a}\right)^{-1/(\gamma+1)} = h_\theta(Z_1)$. Similarly $h_\theta(Z_2) = \left(\frac{\theta}{b}\right)^{-1/(\gamma+1)}$. Now:

$$\int_0^{h_\theta(Z_1)} Z_1(s)\,ds = \int_0^{\left(\frac{\theta}{a}\right)^{-1/(\gamma+1)}} \frac{a}{s^\gamma}\,ds = \frac{a}{1-\gamma}\left(\frac{\theta}{a}\right)^{\frac{\gamma-1}{\gamma+1}} = \frac{\theta^{\frac{\gamma-1}{\gamma+1}}}{1-\gamma}a^{2/(\gamma+1)}$$

(and a similar result for Z_2 with the variable a replaced by b). Still using the same m as in Sect. 1.4.3 we have:

$$m_\theta(Z_1) = \frac{R_\theta(Z_1)}{h_\theta(Z_1)} = \frac{\sqrt{\int_1^{h_\theta(Z_1)} Z_1(s)\,ds}}{h_\theta(Z_1)} = \frac{\sqrt{\frac{\theta^{\left(\frac{\gamma-1}{\gamma+1}\right)}}{1-\gamma}a^{2/(\gamma+1)}}}{\left(\frac{\theta}{a}\right)^{-1/(\gamma+1)}}$$

$$= \frac{1}{\sqrt{1-\gamma}}\sqrt{\theta}\,\frac{a^{1/(\gamma+1)}}{a^{1/(\gamma+1)}} = \frac{1}{\sqrt{1-\gamma}}\sqrt{\theta}$$

As this expression is independent of the variable a, it shows that $\forall \theta$, $m_\theta(Z_1) = m_\theta(Z_2)$, although $Z_1 \neq Z_2$. Hence this m is not a CES. Moreover, $\forall Z \in \mathbf{Z}, m_\theta(Z)$ is injective in θ.

We admit that m, Z and $m_\theta(Z)$ are functions acting between different sets and hence there is no a priori reason why Z and $m_\theta(Z)$ being injective would imply that m is injective.

If, however, we take a set $\mathbf{Z}$ of functions of the form $ax^{-\beta}$, again with a as a variable and $\beta > 0$ fixed, but now defined on an interval $[c, T]$, with $c > 0$, then we do have a CES.

Proposition 4. If $\mathbf{Z}$ is a set of functions of the form ax^β, with a as a variable and $\beta > 0$ fixed, defined on an interval $[c, T]$, with $c > 0$, then m defined through

$$m_\theta(Z) = \frac{R_\theta(Z)}{h_\theta(Z)} = \frac{\sqrt{\int_c^{h_\theta(Z)} Z(s)ds}}{h_\theta(Z(s))}, \ Z \in \mathbf{Z} \text{ and } c \le h_\theta(Z), \text{ is a CES on this set } \mathbf{Z}.$$

Proof. Let $Z_1(x) = ax^{-\beta}$ and $Z_2(x) = bx^{-\beta}$, defined on $[c, +\infty[$ with $a \ne b$, $a, b > 0$. We already know that $\left(\frac{\theta}{a}\right)^{-1/(\beta+1)} = h_\theta(Z_1)$. And similarly $h_\theta(Z_2) = \left(\frac{\theta}{b}\right)^{-1/(\beta+1)}$. Assuming that these generalized h-values are larger than c, we have:

$$\int_c^{h_\theta(Z_1)} Z_1(s)ds = \int_c^{\left(\frac{\theta}{a}\right)^{-1/(\beta+1)}} \frac{a}{s^\beta} ds = \frac{a}{1-\beta} \left(\left(\frac{\theta}{a}\right)^{\frac{\beta-1}{\beta+1}} - c^{1-\beta} \right)$$

$$= \frac{1}{1-\beta} \left(\theta^{\left(\frac{\beta-1}{\beta+1}\right)} a^{2/(\beta+1)} - ac^{1-\beta} \right)$$

(and a similar result for Z_2 with the variable a replaced by b). Using a small variant of the m used in Sect. 1.4.3 (but with the same notation) we have:

$$m_\theta = \frac{R_\theta(Z_1)}{h_\theta(Z_1)} = \frac{\sqrt{\int_c^{h_\theta(Z_1)} Z_1(s)ds}}{h_\theta(Z_1)} = \frac{\sqrt{\frac{1}{1-\beta} \left(\theta^{\left(\frac{\beta-1}{\beta+1}\right)} a^{2/\beta+1} - ac^{1-\beta} \right)}}{\left(\frac{\theta}{a}\right)^{-1/(\beta+1)}}$$

and a similar expression for Z_2 with a replaced by b. If now $m(Z_1) = m(Z_2)$ then, after deleting common terms and factors, $a^{\left(\frac{\beta+3}{2(\beta+1)}\right)} = b^{\left(\frac{\beta+3}{2(\beta+1)}\right)}$, implying that $a = b$ and hence $Z_1 = Z_2$. $\qquad\square$

1.5 Using Derivatives

Let $\mathbf{Z}_d$ be a set of differentiable functions defined on $D = [b, +\infty]$ where the value in $+\infty$ is defined as a limit: $Z_0(+\infty) = lim_{\theta \to \infty} Z_0(\theta)$. Now we associate with every $Z \in \mathbf{Z}_d$ a function $m_d(Z)$ as follows:

$$\theta \in D, m_d(Z) : D \subset \ ]0, +\infty[\ \to \mathbf{R} : \theta \to Z'(\theta)$$

Theorem 8. If $\mathbf{Z_d}$ is the set of differentiable functions defined above and m_d is defined by differentiation as described above, then m_d is a CES unless the set $\mathbf{Z_d}$ contains two or more functions which are vertical translates.

Proof. Let, for Z_1 and $Z_2 \in \mathbf{Z_d}$, $m_d(Z_1) = m_d(Z_2)$, i.e. $\forall \theta \in D, Z_1'(\theta) = Z_2'(\theta)$. Then, the functions Z_1 and Z_2 differ by a fixed constant C such that: $\forall \theta \in D, Z_1(\theta) = Z_2(\theta) + C$. This means that if $C \neq 0$, then Z_1 and Z_2 are vertical translates and m_d is not a CES. Yet, if there exists $d \in D$ such that $Z_1(d) = Z_2(d)$, then $C = 0$ and the equality between Z_1 and Z_2 holds everywhere. In that case m_d is a CES. In particular, if Z_1 and Z_2 are defined in a neighborhood of $+\infty$ and $lim_{\theta \to \infty} Z_1(\theta) = lim_{\theta \to \infty} Z_2(\theta)$, then too m_d is a CES.

Remark. It is, of course, possible that some functions Z_1 and $Z_2 \in \mathbf{Z_d}$ intersect.

1.6 Another Example Related to Injectivity

Given $a, b > 0$. Let $\mathbf{Z_c}$ be a set of strictly decreasing continuous function defined on $[0, +\infty[$ such that $Z(b) = a$, i.e. all graphs of functions in $\mathbf{Z_c}$ have a common point. Hence every $Z \in \mathbf{Z_c}$ is an injective function from its domain to its range, passing through the point (b, a).

Now we define m_f through functions $m_f(Z)$ as follows:

$$m_f(Z): \; [0, +\infty[\; \to \mathbf{R} : \theta \to m_f(Z) = (Z(\theta) - a)^2$$

Theorem 9. The function m_f defined on the set $\mathbf{Z_c}$ of strictly decreasing continuous functions defined on $[0, +\infty[$ such that $Z(b) = a$, is a CES.

Proof. Let $m_f(Z_1) = m_f(Z_2)$, then $\forall \theta \in [0, +\infty[: (Z_1(\theta) - a)^2 = (Z_2(\theta) - a)^2$. Now, as Z_1 and Z_2 are strictly decreasing, we have that $Z_1(\theta) > a$ iff $Z_2(\theta) > a$ (case $\theta < b$) and similarly, $Z_1(\theta) < a$ iff $Z_2(\theta) < a$ (case $\theta > b$). Hence, we see that $\forall \theta \in [0, +\infty[\; : Z_1(\theta) - a = Z_2(\theta) - a$ and conclude that $Z_1 = Z_2$, implying that m_f is injective, i.e. is a CES.

Now, we note that the functions $m_f(Z)$ are not injective. For each Z, $m_f(Z)$ is first strictly decreasing, reaches the value zero in the point $\theta = b$, and then is strictly increasing. By continuity there exist points θ_1, θ_2, where $\theta_1 < b < \theta_2$ such that $(m_f(Z))(\theta 1) = (m_f(Z))(\theta)$, which shows that none of the functions $m_f(Z)$ is injective. $\qquad \square$

1.7 A Detailed Study of $m^{(c)}$

We first recall the definition of an homothetic image and prove an easy result in its connection.

Definition 2 A two-dimensional homothetic transformation with homothetic center in the origin is a mapping:

$$\mathbf{R}^2 \to \mathbf{R}^2 : (x_1, x_2) \to k(x_1, x_2), \; \text{with} \; k > 0$$

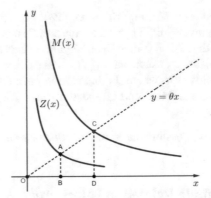

Fig. 1.2 Illustration of the relation between $h_\theta(Z)$ and $h_\theta(M)$, with M an homothetic image of Z.

If we apply this homothetic transformation on the graph of the function $Z(x)$, then the points $(x, Z(x))$ are mapped to the points $(kx, kZ(x))$. The resulting image is the graph of the function $M(t) = (t, kZ(t/k))$.

If now Z is a decreasing continuous function defined on a subset of $]0, +\infty[$, with generalized h-index $h_\theta(Z)$, then we investigate the relation between the generalized h-index of Z and that of its homothetic image M.

Lemma. If M is the homothetic image of Z with ratio k, then $h_\theta(M) = kh_\theta(Z)$, $\forall \theta \in \{Z(x)/x; x \in dom(Z)\}$.

Proof. Consider Fig.1.2. We see that $OB = h_\theta(Z)$ and $OD = h_\theta(M)$. As the triangles OAB and OCD are similar we have:

$$\frac{|OB|}{|OD|} = \frac{|OA|}{|OC|} \quad \text{or} \quad \frac{h_\theta(Z)}{h_\theta(M)} = \frac{1}{k}$$

This shows that $h_\theta(M) = kh_\theta(Z)$.   □

Let $\mathbf{Z}$ be a set of continuous, decreasing functions Z defined on a subset of $[u, +\infty[$ with $u > 0$, such that the integrals $\int_u^{h_\theta(Z)} Z(s)ds$ exist. Now we consider, $\forall c > 0$, the function $m^{(c)}$ on Z defined through:

$$m_\theta^{(c)}(Z) = \frac{\left(\int_u^{h_\theta(Z)} Z(s)ds\right)^c}{h_\theta(Z)}$$

$\forall Z \in \mathbf{Z}$ and θ in an appropriate subset of $]0, +\infty[$.

We will consider a set $\mathbf{Z_{hom}}$ such that if Z belongs to $\mathbf{Z_{hom}}$, then also its homothetic image with ratio k (fixed, $k \neq 1$), belongs to $\mathbf{Z_{hom}}$. Of course, if $k = 1$ this is trivial and of no importance further on.

Theorem 10. If $c = 1/2$, then $m^{(c)}$ is not a CES.

Proof. Let Z_2 be the homothetic image of Z_1 with ratio k, then $h_\theta(Z_2) = k h_\theta(Z_1)$. This is the denominator of $m_\theta^{(c)}(Z)$. Hence, we wonder if it is possible that $\left(\int_u^{h_\theta(Z_2)} Z_2(s)ds \right)^c = k \left(\int_u^{h_\theta(Z_1)} Z_1(s)ds \right)^c$ (the numerator of $m_\theta^{(c)}(Z)$). In that case $m_\theta^{(c)}(Z_1)$ would be equal to $m_\theta^{(c)}(Z_2)$ $\forall \theta > 0$, while obviously $Z_1 \neq Z_2$. Now, $\left(\int_u^{h_\theta(Z_2)} Z_2(s)ds \right)^c = k \left(\int_u^{h_\theta(Z_1)} Z_1(s)ds \right)^c$ implies:

$$\int_u^{h_\theta(Z_2)} Z_2(s)ds = k^{1/c} \left(\int_u^{h_\theta(Z_1)} Z_1(s)ds \right) \tag{1.9}$$

As, $\forall Z \in \mathbf{Z}_{\text{hom}}$, $\{ h_\theta(Z); \theta \in \,]0, +\infty[\, \} = dom(Z)$, we can rewrite Eq.1.9 as:

$$\int_u^{ky} Z_2(s)ds = k^{1/c} \left(\int_u^{y} Z_1(s)ds \right) \quad \text{for } y \in dom(Z)$$

Taking the derivative with respect to y yields:

$$k Z_2(ky) = k^{1/c} Z_1(y) \Rightarrow Z_2(ky) = k^{(1/c)-1} Z_1(y) \Rightarrow Z_2(x) = k^{(1/c)-1} Z_1(x/k)$$

Next, we calculate $h_\theta(Z_2)$ based on this equality. We now have:

$$x_0 = h_\theta(Z_2) \Leftrightarrow Z_2(x_0) = \theta x_0 \Leftrightarrow k^{(1/c)-1} Z_1(x_0/k) = \theta x_0 \Leftrightarrow$$

$$Z_1(x_0/k) = k^{1-(1/c)} \theta x_0 = k^{2-(1/c)} \theta \left(\frac{x_0}{k} \right) \Leftrightarrow$$

$$\frac{x_0}{k} = h_{k^{2-(1/c)}\theta}(Z_1) \Leftrightarrow x_0 = k h_{k^{2-(1/c)}\theta}(Z_1)$$

Yet, we know that $x_0 = k h_\theta(Z_1)$. From this we conclude that this equality can only hold if $c = 1/2$.

This shows that the approach we took to show that $m^{(c)}$ is not a CES only works for $c = 1/2$. □

We note though that, because we only considered one, rather simple, approach we have no information about $m(c)$ for other values of c.

Corollary 1. Given a function Z and its homothetic image with ratio k : H_Z^k, then we know that, $\forall \theta$ for which $h_\theta(Z)$ exists, $h_\theta(H_Z^k) = k h_\theta(Z)$. Then:

$$\left(\int_u^{h_\theta(H_Z^k)} H_Z^k(s)ds \right)^c = k \left(\int_u^{h_\theta(Z)} Z(s)ds \right)^c \quad \text{iff } c = 1/2.$$

Corollary 2. For two different constant functions $Z_1 = a$ and $Z_2 = b$ we have that: $m^{(1/2)}(Z_1) = m^{(1/2)}(Z_2)$. Indeed $Z_2 = (b/a)Z_1$. This is the example of Sect. 1.4.2.

Corollary 3. If $Z_1(x) = ax^\beta$ and $Z_2(x) = bx^\beta$, then $m^{(1/2)}(Z_1) = m^{(1/2)}(Z_2)$, (assuming that integrals exist) because also here $Z_2 = (b/a)Z_1$. This is essentially the example of Sect. 1.4.3.

Corollary 4. If $Z_1(x)$ is a circle with center in the origin and radius $C > 0$, restricted to the first quadrant and $Z_2(x)$ is another circle with center in the origin and radius $D > 0$, also restricted to the first quadrant, then $m^{(1/2)}(Z_1) = m^{(1/2)}(Z_2)$. This follows immediately from the fact that $Z_2(x)$ is the homothetic image of $Z_1(x)$ with ratio D/C. In this case a direct calculation would be non-trivial.

We will begin our derivation that $m^{(c)}$ is never a CES $(c \neq 0)$. For this we use the following approach. For the case $c = 1/2$ we tried to find a fixed value k such that $\forall \theta$ in an appropriate domain, $h_\theta(Z_2) = kh_\theta(Z_1)$, and $\left(\int_u^{h_\theta(Z_2)} Z_2(s)ds \right)^c = k \left(\int_u^{h_\theta(Z_1)} Z_1(s)ds \right)^c$. As it is in general not possible to find an example with constant k (as proved above), we now try a similar approach with k replaced by a function of $\theta : k(\theta)$.

As the steps are rather complicated, we begin with the case $c = 1$ and $Z_2(x) = \sqrt{x} \, Z_1(\sqrt{x})$.

Lemma. $h_\theta(Z_2) = (h_\theta(Z_1))^2$

Proof. We have:

$$x_0 = h_\theta(Z_2) \Leftrightarrow Z_2(x_0) = \theta x_0 \Leftrightarrow \sqrt{x_0} \, Z_1(\sqrt{x_0}) = \theta x_0 \Leftrightarrow Z_1(\sqrt{x_0}) = \theta \sqrt{x_0}$$

Hence, we derive:

$$\sqrt{x_0} = h_\theta(Z_1) \Leftrightarrow h_\theta(Z_2) = (h_\theta(Z_1))^2$$

$\square$

From this lemma we see that $k(\theta)$ is given by $h_\theta(Z_1)$.

Let $\mathbf{Z_u}$ be a set of continuous, decreasing functions Z defined on the subset of $[u, +\infty[$ with $u > 0$, such that the integrals $\int_u^{h_\theta(Z)} Z(s)ds$ exist. Now we consider the function $m^{(1)}$ on $\mathbf{Z_u}$, defined through:

$$m_\theta^{(1)}(Z) = \frac{\int_u^{h_\theta(Z)} Z(s)ds}{h_\theta(Z)} \quad \forall Z \in \mathbf{Z_u}$$

and θ in an appropriate subset of $]0, +\infty[$.

Theorem 11. The function $m^{(1)}$ is not a CES on $\mathbf{Z_u}$.

Proof. With $Z_1(x)$ and $Z_2(x) = \sqrt{x} \, Z_1(\sqrt{x}) \in \mathbf{Z_u}$, (such that also Z_2 is a continuous decreasing function) and inspired by the previous lemma we show that there exist a and b such that:

$$\int_a^{h_\theta(Z_2)} Z_2(s)ds = h_\theta(Z_1) \int_b^{h_\theta(Z_1)} Z_1(s)ds \tag{1.10}$$

The left-hand side of Eq.1.10 is equal to:

$$\int_a^{(h_\theta(Z_1))^2} \sqrt{s}\, Z_1(\sqrt{s})\, ds = \int_{\sqrt{a}}^{h_\theta(Z_1)} 2\, t^2\, Z_1(t)\, dt$$

Requiring that:

$$\int_{\sqrt{a}}^{(h_\theta(Z_1))} 2t^2\, Z_1(t)\, dt = h_\theta(Z_1) \int_b^{h_\theta(Z_1)} Z_1(s)\, ds$$

and using the fact that the domain of Z_1 is equal to the set of all $h_\theta(Z_1)$ we come to:

$$\int_{\sqrt{a}}^x 2t^2\, Z_1(t)\, dt = x \int_b^x Z_1(s)\, ds$$

Taking $x = b$ (right-hand side is then equal to zero) leads to the requirement $a = b^2$. Now we take the derivative with respect to x of both sides. This leads to:

$$2\, x^2\, Z_1(x) = \int_b^x Z_1(s)\, ds + xZ_1(x)$$

If we now set $I(x) = \int_b^x Z_1(s)\, ds$ (and note that $I(b) = 0$), then we have: $2\, x^2\, I'(x) = I(x) + xI'(x)$ and hence $I'(x) = \frac{I(x)}{x(2x-1)}$ or $I(x) = \frac{2x-1}{x}$, where we omit constants as we are only trying to find one example.

Since $I(x) = \int_b^x Z_1(s)\, ds = \frac{2x-1}{x}$, this leads to $Z_1(x) = \frac{1}{x^2}$ on the interval $[b, +\infty[$. As $I(b) = 0$, we see that $b = 1/2$.

This leads to $Z_2(x) = \sqrt{x}\, Z_1(\sqrt{x}) = \sqrt{x}\frac{1}{x} = \frac{1}{\sqrt{x}}$ on $[a, +\infty[$. Finally, as $a = b^2$, we have $a = 1/4$. We note that $Z_1(1/2) = 4$ and $Z_2(1/4) = 2$. The points $(1/2, 4)$ and $(1/4, 2)$ are both situated on the line $y = 8x$.

Thus, we conclude that if $Z_1(x) = \frac{1}{x^2}$ on the interval $[\frac{1}{2}, +\infty[$ and $Z_2(x) = \frac{1}{\sqrt{x}}$ on $[\frac{1}{4}, +\infty[$, then:

$$h_\theta(Z_2) = k(\theta)\, h_\theta(Z_1)$$
$$\int_{1/4}^{h_\theta(Z_2)} Z_2(s)\, ds = k(\theta) \int_{1/2}^{h_\theta(Z_1)} Z_1(s)\, ds$$

with $\theta \in\,]0, 8]$ and $k(\theta) = h_\theta(Z_1)$, then $m_\theta^{(1)}(Z_1) = m_\theta^{(1)}(Z_2)$. As $Z_1(x) \neq Z_2(x)$ this shows that $m^{(1)}$ is not a CES on $\mathbf{Z_u}$.   □

We note that $m^{(1)}(Z_1)$ and $m^{(1)}(Z_2)$ have the same θ-domain, but $dom(Z_1) \neq dom(Z_2)$.

Remark. We did not yet calculate $h_\theta(Z_1)$ (and did not need its exact value). By calculating now $h_\theta(Z_1)$ and $h_\theta(Z_2)$ directly from the definition we will check that indeed, $h_\theta(Z_2) = (h_\theta(Z_1))^2$. For $Z_1(x)$ we have: $\frac{1}{x^2} = \theta x$ and hence $h_\theta(Z_1) = \frac{1}{\theta^{1/3}}$ and for $Z_2(x)$: $\frac{1}{\sqrt{x}} = \theta x$, leading to $h_\theta(Z_2) = \frac{1}{\theta^{2/3}} = (h_\theta(Z_1))^2$.

We are able to show that also for other values of c, $m^{(c)}$ is not a CES. This is done by using the functions $Z_2(x) = x^p Z_1(x^q)$, with $p + q = 1$ and replacing

$k(\theta) = h_\theta(Z_1)$ from the case $c = 1$, by $k(\theta) = (h_\theta(Z_1))^{1/q-1}$. The domains of Z_1 and Z_2 too must be adapted (as they were for the case $c = 1$). As the construction is quite long we do not include these examples here. The constructions can be obtained from the authors.

1.8 An Infinite Number of Non-CES Examples

In this section we provide a mathematically interesting construction, leading to countably infinite non-CES cases. Let $n > 0$ be a natural number and let $\forall n > 0, \mathbf{Z_n}$ be a set of real-valued functions Z with domain $[0, 2^{n-1}]$, such that Z^n, i.e. $Z \circ Z \circ \ldots Z$ (n times a composition) also belongs to $\mathbf{Z_n}$. We will provide an example of a non-CES case for each $n > 0$, i.e. each $\mathbf{Z_n}$. Define for all $\theta > 0$ and $Z \in \mathbf{Z_n} : m_\theta(Z) = Z^n(\theta)$.

Let $Z_1(x) = 0.5$ and let $Z_2(x) = \max(0.5, x/2)$ both defined on $[0, +\infty[$.

Lemma. $\forall n > 0 :$

$$Z_2^n = \begin{cases} 0.5 & \text{for } x \leq 2^{n-1}, \\ x/2^n & \text{for } x > 2^{n-1} \end{cases}$$

Proof. By induction. The result is clearly correct for $n = 1$. Assume now that Z_2^n equals 0.5 for $x \leq 2^{n-1}$, and equals $= x/2^n$ for $x > 2^{n-1}$. Then we check the formula for $n + 1$:

$$Z_2^{n+1}(x) = (Z_2(Z_2^n(x)) = \max(0.5, 0.5(Z_2^n(x))) =$$

$$= \begin{cases} \max(0.5, 0.25) = 0.5 & \text{for } x \leq 2^{n-1} \\ \max(0.5, \frac{x}{2^{n+1}}) = 0.5 & \text{for } 2^{n-1} < x \leq 2^n \\ \max(0.5, \frac{x}{2^{n+1}}) = \frac{x}{2^{n+1}} & \text{for } x > 2^n \end{cases}$$

This proves the lemma.   □

Now, we restrict Z_1 and Z_2 to $[0, 2^{n-1}]$ so that both belong to $\mathbf{Z_n}$. Then we see, by the lemma, that $\forall \theta > 0$ and $\forall n > 1 : m_\theta(Z_1) = m_\theta(Z_2)$ or $Z_1^n(\theta) = Z_2^n(\theta)$ on $[0, 2^{n-1}]$. As $Z_1 \neq Z_2$ on $[0, 2^{n-1}]$ this shows that we do not have a CES.

Note that if $n = 1$, then $Z_1(x)$ and $Z_2(x)$ defined on $[0, 1]$ are equal.

1.9 An Example of an Integral Transformation: The Laplace Transformation

Ending our constructions on a positive note we consider a CES of another type. In what follows we restrict ourselves to a special case of Laplace transforms, namely real-valued transforms.

Definition 3 Let $Z(x)$ be a real-valued function defined on $[0, +\infty[$. Then the function:

$$\mathcal{L}(Z) : S_Z \subset R \rightarrow R : s \rightarrow \int_0^{+\infty} e^{-sx} Z(x)\, dx$$

with $S_Z = \{s \in R : \int_0^{+\infty} e^{-sx}|Z(x)|dx < +\infty\}$, is called the Laplace transform of Z. In general s is a complex variable, but we restrict the Laplace transform to a real variable.

Let $Z(x)$ be a continuous function defined on $[0, +\infty[$ and assume that there exist real numbers $a_Z, M_Z \geq 0$, and $N_Z \geq 0$ such that:

$$\forall x > N_Z : |Z(x)| \leq M_Z . e^{a_Z x}$$

Then Z is said to be of *exponential order with constants* (a_Z, M_Z, N_Z). Now, if Z is of exponential order with constants (a_Z, M_Z, N_Z) it can be shown that $\mathcal{L}(Z)$ exists for $s > a_Z$, i.e., the set S_Z defined above is equal to $]a_Z, +\infty[$.

Such functions also have an inverse transform. Indeed, it can be shown that if $\mathcal{L}(Z_1) = \mathcal{L}(Z_2)$ on an interval $]s_0, +\infty[$, then $Z_1 = Z_2$ [2].

Using the Laplace transformation we can make the following CES.

Let $\mathbf{Z_L}$ be a set of continuous functions of exponential order and let m_L be defined as follows:

$$m_L(Z) : \ ]a_Z, +\infty[\ \rightarrow R : 0 \rightarrow \int_0^{+\infty} e^{-\theta x} Z(x)\, dx$$

If now $m_L(Z_1) = m_L(Z_2)$ then, by the existence of the inverse Laplace transform, $Z_1 = Z_2$. This shows that m_L is a CES on $\mathbf{Z_L}$. It is possible to construct other CES-systems based on other integral transformations, such as the Fourier transformation, but we will not go into this here.

1.10 Mathematics in the Information Sciences

Encouraged by the comments of the reviewers of this article we take some time to comment on the role of mathematics in science and the information sciences in particular.

In the context of applications to economics [1] recalled that mathematics is a set of techniques for the exposition and discovery of relationships among quantities. In more general terms, the National Research Council of the USA wrote [13]:

> One aspect of the mathematical sciences involves unearthing and understanding deep relationships among abstract structures. Another aspect involves capturing certain features of the world by abstract structures through the process of modeling, performing formal reasoning on the abstract structures or using them as a framework for computation, and then reconnecting back to make predictions

about the world. ... The mathematical sciences are always innovating, and adapting existing knowledge to new contexts is a never-ending process.

Need we remind the reader of Wigner's belief of the unreasonable effectiveness of mathematics, bordering on the mysterious [19]? The answer to this 'mystery' is of course that mathematicians were motivated by the internal, aesthetic considerations of their, at the time, completely theoretic investigations, which later, often much later, turned out to be useful to industry and society at large. Indeed, mathematics provides a powerful tool for global understanding and communication. Using it, scientists can make sense of the world and solve complex and real problems. Mathematics is moreover a kind of language: a tool for stating one's thoughts more precisely and in shorthand notation, with powerful means for checking the internal consistency of statements. Maybe it is not a coincidence that Eugene Garfield, one of the founders of bibliometrics, got his Ph.D. in structural linguistics.

Concerning the information sciences we note that already in the 1930s, Otlet deplored that contrary to developments in other fields, librarianship did not use mathematical techniques [15]. To counter this poor state of affairs he proposed the term bibliometrics and provided – elementary – examples of the insights mathematics could bring for his profession.

We further recall that in the nineties we compared informetrics with econometrics and expressed the hope that the role of mathematics in the information sciences would be comparable to the role of mathematics in economics [16]. Indeed, although nowadays most economists would think favorable of mathematics and its applications this was certainly not the case a hundred years ago [9].

1.11 Conclusion

We showed that under certain circumstances it is possible to save all information contained in a production function. This was done by introducing measures m_θ. Such parametrized systems of evaluation measures were derived from the h-index, the g-index, and other informetric measures. This results in a new class of general systems referred to as complete evaluation systems. Two different functions in a CES can never have the same value for every value of the parameter. Examples are given of systems that are CES and other systems that are not.

This contribution can certainly be placed on the mathematical side of the informetric spectrum, as was e.g. our recent study of minimal impact arrays [8], but having its origins, via the h-index, in the applied side of the field.

Acknowledgements The authors thank Guido Herweyers for drawing the figures in this contribution. We thank the editors of this book to allow the inclusion of an article that stands on its own as a theoretical contribution, but which, when it comes to future applications, is just a promise (as do most mathematics-inspired articles).

References

1. Boulding, K.: Samuelson's foundations: The role of mathematics in economics. Journal of Political Economy **56**(3), 187–199 (1948)
2. Doetsch, G.: Handbuch der Laplace-Transformation. Birkhäuser, Basel (1956)
3. van Eck, N., Waltman, L.: Generalizing the h- and g-indices. Scientometrics **2**(4), 263–271 (2008)
4. Egghe, L.: Theory and practise of the g-index. Scientometrics **69**(1), 131–152 (2006)
5. Egghe, L.: A theory of pointwise defined impact measures. Journal of Informetrics (2021)
6. Egghe, L.: An improvement of the h-index: The g-index. ISSI Newsletter **2**(1), 8–9 (206)
7. Egghe, L., Rousseau, R.: Solution by step functions of a minimum problem in $L^2[0, T]$, using generalized h- and g-indices. Journal of Informetrics **13**(3), 785–792 (2019)
8. Egghe, L., Rousseau, R.: Minimal impact one-dimensional arrays. Mathematics **8**(5), 811 (2020)
9. Franklin, J.: Mathematical methods of economics. American Mathematical Monthly **90**(4), 229–244 (1983)
10. Hirsch, J.E.: An index to quantify an individual's scientific research output. Proceedings of the National Academy of Sciences **102**(46), 16569–16572 (2005)
11. Jin, B.H., Liang, L.M., Rousseau, R., Egghe, L.: The R- and AR-indices: Complementing the h-index. Chinese Science Bulletin **52**(6), 855 863 (2007)
12. Narin, F.: Evaluative Bibliometrics: The Use of Publication and Citation Analysis in the Evaluation of Scientific Activity. Computer Horizons, Cherry Hill, NJ (1976)
13. National Research Council (2013): The mathematical sciences in 2025 (2015)
14. Ossowska, M., Ossowski, S.: Nauka o nauce. Nauka Polska **20**, 1–12 (1935)
15. Otlet, P.: Traité de Documentation, le Livre sur le Livre. D. Van Keerberghen & Sons, Brussels (1934)
16. Rousseau, R.: Similarities between informetrics and econometrics. Scientometrics **30**(2–3), 385–387 (1994)
17. Rousseau, R., Egghe, L., Guns, R.: Becoming Metric-Wise. A Bibliometric Guide for Researchers. Chandos (Elsevier), Kidlington (2018)
18. Waltman, L.: A review of the literature on citation impact indicators. Journal of Informetrics **10**(2), 365–391 (2016)
19. Wigner, E.: The unreasonable effectiveness of mathematics in the natural sciences. Communications of Pure & Applied Mathematics **13**(1), 1–14 (1960)

Chapter 2
Extrinsic Factors Affecting Citation Frequencies of Research Articles

Natsuo Onodera and Fuyuki Yoshikane

Abstract In this chapter, we discuss what factors, and to what extent, exert influence on citation rates of research articles, mainly addressing to the influence of those factors that are not directly related to the quality or content of articles (extrinsic factors). First, extrinsic factors that may influence the citation rate are classified, and those which are focused on in this chapter and the measures for each of these factors are selected. Next, the main studies having investigated the influence of the selected factors on the citation frequency of articles are comprehensively reviewed. Then, the research done by the authors is described that used the selected measures for those extrinsic factors. In this research we aimed to examine whether there are some general trends across subject fields regarding the factors that affect the citation frequency of articles. Finally, the results obtained by our research are discussed together with those by other studies.

2.1 Introduction

The application of citation data to research evaluation has attracted a great deal of attention in every country in the world. The concept involves obtaining a quantitative measure of the importance (in terms of scientific impact) of an article using its citation rate. Moed in [45] discussed in detail the application of citation data to research evaluation in terms of its frameworks, analytical methodology, information obtained from analysis, and attention in use.

Natsuo Onodera (correspondence author)
University of Tsukuba, 1-2, Kasuga, Tsukuba, Ibaraki 305-8550, Japan
e-mail: nt.onodera@y5.dion.ne.jp

Fuyuki Yoshikane
Faculty of Library, Information and Media Science, the University of Tsukuba, 1-2, Kasuga, Tsukuba, Ibaraki 305-8550, Japan
e-mail: fuyuki@slis.tsukuba.ac.jp

However, there are many criticisms regarding the use of citation data for research evaluation based on the following reasons [38, 41, 42, 43, 44]:

- A low citation rate does not necessarily mean a low quality of research.
- Articles may be utilized in other ways besides for citations. For example, technological articles may be used for design or manufacturing, but many of them are seldom cited in normal articles.
- There are various reasons or motivations for citations, depending on the citers and situations. While some citations may be essential, others may be "perfunctory" or "negative". It is questionable what a simple summation of such citations means.
- The coverage of the main information sources of the databases used for citation analysis, such as the Web of Science Core Collection (called "WoS" hereafter) and Scopus might bias in terms of subject fields, language, and countries.

Although these criticisms include truth to a considerable extent, it is undeniable that the citation rate gives a relatively appropriate measures of an aspect of importance of research (the degree of impact or utilization of articles).

It should be noted, of course, that research must be evaluated from various aspects, and citation rates provide valuable data as one of these aspects. The measures based on citations are not objective indicators of evaluation themselves but complementary information for subjective peer review [29].

In addition, sufficient attention should be paid to the following points on the use of citation data.

- Citation data contain a significant number of errors and ambiguous expressions (e.g., homonyms of author names) that make accurate citation counting difficult.
- Not only the citation counts of articles but also various impact indicators derived from them show extremely skewed distributions [57].
- The citation rate of an article is affected by many factors that are not directly related to its quality or content ("extrinsic factors"[1]).

The theme of this chapter is related to the above third point. It is well known that the citation rate of an article is influenced by the subject field, type of article (original article, short report, review, etc.), and language. Therefore, it is almost meaningless to simply compare articles of different types that belong to different fields based on their raw (not normalized) citation counts. In addition, many studies have been conducted about various "extrinsic" factors that may influence the citation rate of an article.

In the following sections, we discuss what extrinsic factors, and to what extent, exert influence on articles' citation rates, considering the following items:

1. Potential influencing factors and their operationalization.
2. The influence of extrinsic factors on the citation frequency of articles.

---

[1] The essential factors that influence the citation rate of an article are related to its quality and content. However, various other factors are known to affect the citation rate, such as the subject field, the language used, and the number of authors of the article. The former and the latter are called "intrinsic factors" and "extrinsic factors", respectively [12, 50]

In the rest of this chapter, Section 2.2 discusses the above item (1). Section 2.3 presents a review of the main studies on item (2), whereas Section 2.4 describes our studies on item (2). The main parts of this chapter originate in the study by Onodera and Yoshikane [47]. We reused this article with permission from the publishers of the journals in which the article was published, making some additions and modifications. In particular, we added the descriptions on the many studies published after the publication of that article and the discussion of the relation between these added studies and ours.

2.2 Intrinsic and Extrinsic Factors

As mentioned in the footnote of Section 2.1, within many potential factors affecting the citation rate of an article, those related to its quality and content are called "intrinsic factors", and those not directly related to its quality and content (e.g., the number or reputation of authors, the number of cited references, etc.) are referred to as "extrinsic factors" [12, 50]. Supposing the citation rate of an article to be one of the important measures of its scientific influence, it is thought that principal influencing factors are intrinsic ones. However, various extrinsic factors are also known to possibly affect the citation rate.

Here, we classify these factors as follows.

1. Intrinsic factors:

    (a) Quality or importance of an article. This aspect involves academic value, practicality or applicability, and logicality and exactness of description. These matters are items evaluated by peer review.
    (b) Theme or topic dealt with by the article.
    (c) Methodology used (e.g., theoretical/experimental/computational).
    (d) Degree of significance of the results.
    (e) Interdisciplinarity or internationality of the study.

2. Extrinsic factors:

    (a) Subject field to which the article belongs.
    (b) Type of article (original paper/review/short note/others).
    (c) Language describing the article.
    (d) Attributes concerning author collaborations. This aspect regards the number of authors of the article, the number of institutions that the authors are affiliated with, and the number of countries where the institutions are located.
    (e) Attributes concerning cited references of the article. This aspect relates to the number of references, the recency of references, and the average citation impact of references.
    (f) Attributes concerning the visibility of an article. This aspect concerns the length of the article, the number of figures and tables, and the characteristics of terms used in the title or abstract.

(g) Attributes concerning authors. This aspect involves reputation, achievements in the past (number of articles and citations received), status, institution or country affiliated with, gender, and age.

(h) Attributes concerning the journal in which the article was published. This aspect involves the number of articles published, the citation impact, and the year of release.

It is actually difficult to discriminate distinctly between intrinsic and extrinsic factors because some extrinsic factors involve partly intrinsic features, such as attributes concerning an author's collaboration and those concerning references of the article. However, they are treated as extrinsic factors here, as in many existing studies.

Intrinsic factors lie outside of the scope of this chapter as it aims at revealing the extent to which various extrinsic factors affect the citation rate of articles.

Extrinsic factors examined in this chapter
In the next, the aforementioned signs (a)-(h) are used below for the case of extrinsic factors, which are dealt with as follows.

First, subject field to which an article belongs (a), type of the article (b), and the language used in the article (c) are not discussed in the literature review in Section 2.3 because it is already well known that the citation rate of an article is greatly affected by these factors.

In the research described in Section 2.4 (research conducted by the authors), to clearly understand the differences between the subject fields, we adopt an approach in which analysis is individually conducted for several different subject fields (a) in an attempt to identify some tendencies common in those fields. Concerning the factors (b) and (c), we focus on original articles written in English as representatives of journal articles. The influences of other extrinsic factors (d)-(h) on the citation rate are the principal target of literature review in Section 2.3 and the research by the authors in Section 2.4.

Selection of the measures representing the individual factors
Here, we determine the measures selected for each of the extrinsic factors (operationalization of factors) examined in the research described in Section 2.4. In principle, we selected at least two measures for each of the factors (d)-(h) because sometimes different measures for the same factor result in different influence patterns on the citation rate.

(1) The measures for the factor concerning the author's collaboration
First, we chose the number of authors, which is investigated in most related studies, and added the two measures, the number of institutions the authors are affiliated with, and the number of countries where the institutions are located.

(2) The measures for the factor concerning cited references of the article
We selected the number of references, which is investigated in most related studies, and the recency of references, which is not so used. Price index (i.e., the percentage of references with publication dates within five years of the publication year of the article in question) was used as the specific measure of the recency of references.

(3) The measures for the factor concerning the visibility of an article
Three measures were selected in addition to article length, which is often used in related studies: the number of figures, the number of tables, and the number of numbered equations in the article.

(4) The measures for the factor concerning authors
The following three measures, all of which concern the first authors' past achievements, were chosen:

- the author's productivity, measured by the number of articles published by the first author until the publication year of the article;
- the author's impact, measured by the number of citations received by the articles mentioned above until the publication year of the article; and
- the author's active years, measured by elapsed years from the publication of the first article of the first author to the publication of the article in question.

These three measures represent the authors' "cumulative achievement", which depends on the length of the authors' career. In addition, measures representing "efficient achievement" obtained by normalizing the measures of cumulative achievement were also considered. As for this kind of measures, the number of articles published per annum and the median of the number of citations received per annum by each published article were selected in this research.

(5) The measures for the factor concerning the journal in which the article was published.
We decided to use not the journal impact indicators, including the journal impact factor (JIF), but a set of dummy variables representing the journals that published the articles in question. The reason behind this decision is that the articles examined were sampled from only four journals for each subject field, limiting the value of the journal measure to four alternatives. Therefore, a set of dummy variables was considered to be rather preferable to a specific impact indicator because the former could involve various aspects of journals in the estimated values of the dummy variables.

2.3 Literature Review

Recently, Tahamtan et al. comprehensively reviewed this topic [64]. They referred to about 200 articles, classifying 28 factors into three categories; "paper-related factors", "journal-related factors" and "author(s)-related factors". In this section we develop an orthogonal presentation of the literature.

2.3.1 Citation Analyses Considering Various Potentially Influencing Factors Together

The multiple regression analysis is the most commonly used approach to separate the effects of individual factors (explanatory variables) and to identify the factors that are significant with respect to the citation rate. In this subsection, we outline several studies using this method.

Studies on the main determinants of the citation rates of articles

Peters and van Raan investigated the extent to which various factors influence the number of citations of articles in the field of chemical engineering [52]. They selected 18 internationally reputed scientists from the field and counted the number of citations received within five years after publication by each of the articles ($n = 226$) published by those scientists between 1980 and 1982. A multiple regression analysis using 14 factors as explanatory variables showed that the highly significant explanatory variables were as follows (in decreasing order of partial correlation coefficients): (a) the scientist's rank according to the number of articles published between 1980 and 1982, (b) the number of references, (c) language, (d) the reputation of the publishing journal, (e) influence weight (Narin's indicator of journal influence), and (f) the Price index. Interestingly, the scientist's rank, which had extremely high explanatory power, did not show a significant relationship with the number of citations when the correlation between these two variables was simply calculated, which indicates the importance of an integrated analysis considering various factors that may influence the citation rates of articles.

Didegah and Thelwall investigated the determinants of citation rates using over 50,000 articles published from 2007 to 2009 in the fields of nanoscience and nanotechnology [18]. They selected eight factors as the independent variables, two of which had not been considered until that time-internationality of the publishing journal and of the references. The JIF and the mean citations of referenced publications were the most strongly influential factors, and the number of references, the internationality of references, and the number of institutions of affiliation were also significant predictors for all article sets.

Van Wesel et al. investigated the relationships between the citation count and various article attributes, using articles from 1996 to 2005 in journals with high JIFs in three subject fields (sociology, general and internal medicine, and applied physics) [69]. The length and readability of the abstract, the length of the full text, and the number of references were identified as the factors that positively influence the citation rate of articles.

What is dominant among the features of author, journal, and article?

Some scholars have divided potential factors that influence the citation rates into the factors of author, journal, and article itself to determine which group is the most dominant. Stewart tried to predict the citation count received by each of 139 articles published in 1968 in the field of geoscience, and concluded that the article features were more important than the author features [62]. The article variables that contributed to a high number of citations included the number of references, the article length, and the time from acceptance to publication. Significant author variables included average citations per article published in the past by the author(s) and the proportion of authors with a university affiliation.

Walters predicted the citation counts of 428 articles published in 12 prime psychology journals in 2003, with nine explanatory variables including author, article, and journal characteristics [65]. The results revealed significant positive effects of the average citations of the first author's past publications, the first author's nationality, multi-authorship, and the journal impact. From these results, Walters suggested that the author characteristics might be more powerful for citation prediction than the journal and article characteristics.

Haslam et al. analyzed the citation counts of 308 articles published in three major journals of social-personality psychology in 1996 [27]. Thirty factors were classified into four groups (i.e., characteristics of the author, institution, article organization, and research approach). The main factors that increased citations were: (a) number of past publications of the first author, (b) the existence of a co-author with higher productivity than the first author, (c) a high level of journal prestige, (d) more pages, (e) more references, and (f) the recency of references.

Peng and Zhu used 18,580 social science articles about Internet studies [50]. They carried out two-stage multiple regression analyses, using article and author characteristics as explanatory variables at the initial stage and adding journal characteristics at the second one. The results indicated a stronger effect of the journal characteristics, especially the JIF. Significant predictors, however, included some article characteristics such as article length, number of authors, the proportion of highly-cited publications in references, and the active years of the first author.

Yu et al. developed a regression model extracting effective predictors of the citation impact from many variables, including article, author, journal, and early citation features [73]. Using 5-year citation data for 1,025 articles published in 2007 in 20 journals in the field of library and information science, they finally extracted six variables as significant factors from 24 variables entered in the model. The number of citations received in the first two years was the highest predictor, followed by the number of references and the five-year JIF.

Xie et al. conducted similar research [71]. They identified 66 potential factors that may influence the citations of articles, including article, author, reference, and citation features. Using 566 articles in the field of library and information science, they finally selected six variables as the most important factors predicting the citation count. The originality of this study is to show in earlier time the predictive power of the number of downloads, which is obtained in a short period after publication, for the long-term citation count.

"Signals" bringing quick attention to an article

Van Dalen and Henkens examined which factors influence the citation impact of articles in the field of demography, focusing on the roles of author and journal reputation as "signals" that brought quick attention to an article [14, 15]. They counted citations received by each of the 1,371 articles in this field with citation windows of 5 and 10 years after publication. The variables regarding journal reputation, such as the JIF, the journal circulation number, and the average number of citations obtained by the editorial board members had an extensive influence on the citation rate, while the influence of the variable regarding author reputation, which was measured by the accumulated number of citations obtained by the author with the highest accumulated citations, was significant but less influential. In other variables, the number of pages showed highly significant, and the author's nationality and the number of authors showed moderately significant relations to the citation rate.

Hanssen and Jorgensen also focused on the author's experience [26]. They analyzed the effect of various article attributes on the citation count until 2012 of articles (n = 779) published between 2000 and 2004 in the transportation field. The articles whose author(s) had published more previous articles tended to receive higher numbers of future citations, while the number of prior citations was not significant. Among the control variables, the number of references exhibited a positive relationship to the citations.

Studies addressing to measures of the quality or content of articles

Although this chapter mainly focuses on the extrinsic factors, the several studies addressing to the intrinsic factors are briefly discussed below.

To investigate whether the peer-review system of refereed journals fulfills its objective of selecting superior work, Bornmann and Daniel compared the citation rates between 878 accepted articles and 959 articles that were initially rejected but later accepted by another journal, both of which were submitted to *Angewandte Chemie International Edition (ACIE)* in 2000 [9]. The results of a negative binomial regression analysis controlling the possible effect of various influencing factors revealed that the accepted articles had an advantage over the rejected articles by 40%–50% in the average number of citations.

Bornmann and Leydesdorff used about 10,000 articles evaluated by reviewers of F1000Prime (a post-publication peer-review site) as the sample, to compare the predictive power between the F1000 scores (three levels) and some bibliometric variables [10]. It was revealed that the influence of the JIF of the publishing journal exceeds that of the peer's judgment for both the long and short citation windows.

It is difficult to represent the quality of an article by a quantitative measure that does not rely on self-evaluation or peer review. A study by Chen is notable in this regard [12]. Chen proposed a model predicting the potential of an article in terms of the degree to which it alters the intellectual structure of state-of-the-art (a "boundary-spanning" ability), and, to measure this ability, introduced three metrics quantifying the change in the existing intellectual network structure. Using these three "intrinsic" attributes and three traditional "extrinsic" attributes (the numbers of authors, references, and pages) as the explanatory variables, he predicted the citation rates of articles in several document sets in different fields. The results revealed that the cluster linkage, one of the intrinsic variables, was a much stronger predictor than the three extrinsic variables and that another intrinsic variable (the centrality divergence) might also have a boundary-spanning ability.

Yan et al. [72] conducted a study based on an idea similar to that of Chen. They represented an article's novelty according to two dimensions: new combinations and new components. The new combinations are defined as new pairs of knowledge elements in a related research area, and the new components as new knowledge elements that have never appeared in a related research area. They presented a methodology quantifying these two concepts using keywords assigned to articles published in the wind energy field between 2002 and 2015. It was revealed that the citation count of the articles increases with the two measures of novelty until a threshold value and begins to decrease after reaching the threshold.

Other studies

Lokker et al. tested the predictability of the citation counts of clinical articles from data obtained within three weeks of publication [39]. The significant predictors of citations included the average relevant scores by raters and the selection by EBM (evidence-based medicine) synoptic journals with regard to article quality; the number of databases indexing the journal and the proportion of articles recorded in EBM synoptic journals concerning journal quality; and the number of authors, the number of references, and others with regard to bibliographic attributes.

Under the hypothesis that open access (OA) articles have a higher citation impact than non-open access (NOA) articles because of biases toward OA (e.g., self-selection by the authors), Davis et al. conducted a randomized controlled trial (RCT) that compared the citation rates between randomly assigned OA and NOA articles [17]. The result revealed no evidence that OA articles received higher rates of citations than NOA articles. Among the control variables considered, the number of authors, the inclusion of author(s) from the United States, the JIF, and the number of references were significant, while article length was not significant.

Intending to argue against the accepted view that internationally co-authored articles have a higher citation rate compared to domestic ones, He applied a negative binomial regression to articles by biomedical researchers in New Zealand using several co-authorship variables as the explanatory variables [28]. The results revealed that adding one author from the same institute increased the citation count of an article comparable to adding one foreign (i.e., from outside New Zealand) author. Among the control variables used, more cited references boosted the citation count.

Fu and Aliferis predicted the long-term citation counts (10 years after publication) of 3,788 articles about internal medicine published between 1991-1994 [22]. Within the variables selected with a supervised learning model from many content-based features and bibliometric features, they further identified significant variables using a logistic regression analysis. Among the bibliometric features, only the JIF and the accumulated number of citations obtained by the last author were significant. The effective content-based features varied greatly according to the threshold value of the citation count in the logistic regression.

2.3.2 Analyses of the Individual Factors Potentially Influencing Citations

This subsection summarizes the findings that have been reported about the main factors that potentially influence citations including those from the integrated analyses mentioned in Section 2.3.1.

Does collaboration boost the citation rate of articles?

Many researchers have investigated the relationship between the citation rate and the number of authors of an article. These studies, at least partially, rely on the hypothesis that adding authors to an article leads to more citations because of the authors' different scientific influences. This idea was expanded to examine the relationship between the number of collaborative institutions or countries in articles and the citation rates.

Authors in [5, 20, 23, 61] focused on the relationship between the citation rates and the number of authors, their institutions, and/or countries. Their results suggest that, in many cases, these features have a positive effect on the citation rate.

Persson et al. explored the relationship between the number of authors and the citation rate received by the articles published in 1980 and 2000 [51]. For both years, they clearly noted a rise of average citations with increase of authors. On the other hand, the average number of citations for articles with the same number of authors increased by eight in the three-year citation window after publication from 1980 to 2000, which implies that the spread of research collaboration is not the only cause of the growth of citations during this period.

Larivière et al. analyzed the relationship between collaboration and the citation impact using the number of authors, the number of institutions, and the number of countries derived from articles published in the long period from 1900 to 2011, showing that an increase in these collaboration indicators leads to an increase in the citation impact through the whole period examined [35]. However, the impact of the same number of collaborators gradually diminished each year.

In addition, many studies have reported a positive correlation between the number of authors and the citation rate [1, 3, 8, 10, 12, 17, 19, 36, 39, 50, 55, 58, 73]. However, some studies using multiple regression analyses with numerous explanatory variables demonstrated that the ability of the number of authors to predict the citation impact is weak [9, 14, 15, 52, 65, 71] or insignificant [18, 22, 26, 62, 69, 72]. Based on their observation that there is no significant relationship between the number of authors and the citation rate within article sets in which only the articles by highly-cited authors were extracted, Levitt and Thelwall suggested that the apparent positive correlation seen in a mixed article may reflect a positive correlation between the average number of co-authors and the average citation count of individual authors [37].

Some have suggested that articles with international collaboration are more highly cited than those with only local or domestic collaboration [7, 26, 31, 32, 49, 51, 54, 58, 61]. However, some studies indicate that an article co-authored internationally cannot simply be said to acquire higher citations, because the effect of the number of authors or the countries to which the co-author belongs should be considered. From this viewpoint, He assigned three collaboration variables (i.e., numbers of foreign, domestic, and local co-authors) to 1,860 articles published by 65 biomedical scientists at a university in New Zealand, and indicated, through a negative binomial regression analysis, that the effect of adding one local co-author on the citation impact was comparable to that of adding one international co-author [28]. (Domestic collaboration was not significantly associated with the citation impact.)

Puuska et al., from an analysis of Finnish articles between 1990 and 2008 divided into subject fields, found that, although an article co-authored internationally tends to acquire higher citations than domestic co-authored ones, the difference is not significant if considering the effect of the number of authors [53]. Sud and Thelwall, from the three-year citation data of biochemistry articles after their publication, showed that, although an increase in co-authors certainly raises the citation impact, the effect of international co-authorship is different depending on the country of co-authors, that is, positive in the case of the United States and the UK, while negative in the case of other most countries [63].

Do articles determine the journal impact, or is the reverse true?
As mentioned in Section 2.3.1, van Dalen and Henkens showed that journal reputation measures such as the JIF, the average number of citations obtained by the editorial board members, and the circulation numbers had a strong positive influence on the citation impact [14, 15]. Some other studies described in Section 2.3.1 took the JIF (or other impact indicators) of the journal publishing an article as one of the most important factors behind increasing the citation rate of the article [17, 18, 22, 50, 52, 73]. However, the explanatory power of the JIF was not as strong according to [65, 71].

Authors in [3, 8, 11, 59, 68, 72] also reported an association between the citation rate of articles and the JIF of the journals in which they were published. Moreover, Larivière and Gingras employed a unique method of comparing 4,532 pairs of "duplicate" articles with the same title, the same first author, and the same number of references published in two different journals [34]. They reported that articles published in a higher-impact journal obtained, on average, twice as many citations as their counterpart published in a lower-impact journal. The obvious difference in the citation count between identical articles is strongly suggestive of the halo effect of journal prestige on the scientific impact of articles.

As described in Section 2.3.1, Bornmann and Leydesdorff found that the JIF of the journal in which an article is published shows a higher correlation with both short- and long-term citations than the peer-judged score given to the article, and also than other bibliometric variables added to the regression analysis [10].

Is there a halo effect of authors, institutions, or countries?
There have been many discussions about the halo effect on scientific impact, suggesting that articles written by authors with high achievement levels or authors affiliated with famous institutions attract more citations than those written by other authors. The measures of the reputation or achievement of authors include indicators based on the number of publications in the past, the citation count received in past publications, active years, and the present status. Peters at al. [52] and Haslam and van Raan [27] determined past publications to be an effective predictor of citations, but Fu and Aliferis [22] and Yu et al. [73] did not find it to be significant. As the indicators of past citations, aggregated citations [22, 73], average citations per article [62, 65], and the h-index [28, 67, 68] were significant predictors in all cases. Some reports claimed that articles by senior authors tended to receive more citations [50, 59], whereas others contradicted the claim [28, 62].

Danell investigated whether the citation rate of an article in the future can be predicted from the number of previous publications and the previous citation rate of the authors, using two article sets of limited subject areas (episodic memory and Bose–Einstein condensation) [16]. Using quantile regression models, he found that the previous citation rate was a significant predictor at most quantiles of the dependent variables (future citation rate) and was more significant at higher quantile values, while the previous publication number was not significant at most quantiles, except in some near the median.

Hanssen and Jorgensen revealed that an article whose author(s) published more previous articles tends to receive higher citations [26]. However, this relation is not linear but tends to saturate beyond a threshold value. Conversely, the effect of the number of prior citations is not significant, different from those of many other studies.

Based on the citation data of about 37,000 articles published in the journals of the American Physical Society between 1980 and 1989, Wang et al. proposed a methodology predicting the possibility that an article will become a Top 1%, Top 5%, or Top 25% article defined by ESI [66]. They used a logistic regression analysis including four predicting variables: (a) the impact (prior citations) of the first author, (b) the impact of the co-author with the highest impact, (c) the total impact of all co-authors, and (d) the relevance of the article to the authors' past articles. The important factors are (b) and (d) for a short citation window (three years after publication) and (c) and (d) for a long one (10 years after publication).

Regarding bias toward particular countries in citation rates, it has been suggested that articles by authors in a few highly productive countries tend to acquire higher numbers of citations because authors tend to favorably cite articles by other authors from their own country.

Using several sets of articles in the field of ecology, Leimu and Koricheva discovered that the annual citation rate was positively associated with authors from English-speaking nations compared with those from non-English-speaking nations and with US authors compared with European authors [36]. Cronin and Shaw showed that, in the field of library and information science, the proportion of uncited articles was lower in the case of a first author from the United States, the UK, or Canada than from other countries [13]. Other reports have also demonstrated an association of articles by authors from the United States or western/northern Europe with higher citation rates [5, 14, 15, 17, 50, 58, 65]. Nevertheless, [27, 39, 52] reported that the affiliation country of authors was not an important factor for predicting the citation rate.

Pasterkamp et al. examined the relationship of the affiliation countries of corresponding authors between articles published in 1996 in six cardiovascular journals and references cited by those articles [48]. They found that articles were cited by authors from the same country as much as 32% more frequently than expected, even excluding self-citations. Schubert and Glänzel also reported evidence of the tendency toward same-country citations [56]. However, some re-searchers denied this tendency based on a modified method for calculating the same-country (or same-language) citation rate [6, 45].

From a viewpoint different from the halo effects of authors or countries, Frandsen and Nicolaisen examined if an article cited earlier maintains its superiority in future citations [21]. They made nearly 70,000 article pairs, in each of which both articles were published in the same year and cited by the same review of the Cochran Reviews, but one had been cited by other articles cited by the review and another had been not. They found that the article cited earlier obtained three times as many citations as the other one, but the difference diminished with time.

Other potential factors that might influence the citation rate

As shown in Section 2.3.1, several studies that used multiple regression analyses considering various factors included the number of references as one of the explanatory variables and found it to be a significant predictor of citations [17, 18, 26, 27, 28, 39, 52, 62, 69, 73]. Many other studies have demonstrated that articles with a greater number of references tend to be cited more often [7, 12, 36, 55, 60].

The recency of the references was included as one of the latent factors in the multiple regression models in a few studies. Stewart [62] and Peters and van Raan [52] took the proportion of references within three and five years (Price index), respectively, and both found these variables to be a moderate predictor . Haslam et al. demonstrated that the newer the mean year of the references, the more citations the article obtained [27].

Peng and Zhu [50], Didegah and Thelwall [18] and Bornmann and Leidesdorff [10] asserted that articles with references that have a higher citation impact tend to acquire more citations. Larivière and Gingras analyzed the effects of interdisciplinarity on the citation impact of articles published 2000 in 14 subject areas, defining the indicator of the interdisciplinarity of an article as the percentage of its cited references in other subject areas [33]. In most subject areas, the citation rate became low at both extremes of high and low inter-disciplinarity. Didegah and Thelwall showed that the internationality of references was a significant predictor of the citation rates in the fields of nanoscience and nanotechnologyc [18].

Ahlgren at al. analyzed the effect of four reference features (i.e., the number of references, the share of references covered by WoS, the mean age of references, and the mean citation rate of references) on the field-normalized citation rate of articles published in 2009 and found that the all four features correlated significantly with the citation rate [2].

Some studies indicated a positive association between the number of citations and article length (number of pages) [10, 14, 15, 27, 36, 50, 52, 62, 69], while others showed no significant correlation [17, 55, 59, 60, 65].

As the relationship between article length and citations is controversial, Xie et al. performed a random-effects meta-analysis, in which they synthesized 24 effect sizes in 18 studies [70]. This data set includes a total of more than 1.5 million articles. Finally, they detected a moderate, positive correlation between article length and citations ($r = 0.31$), and they noted that the correlation can be moderated by citation windows and the perceived quality of journals to a significant extent.

Snizek et al. examined the relationship between the citation rate and the number of figures and tables, as well as the number of words in, and the readability of, author abstracts, but found no associations with any of these factors [60].

2.4 The Authors' Research on the Citation Rate of Articles

In this section, the research conducted by the authors (called "our research" or "this research") are described, mainly based on our study [47].

2.4.1 Objective of this Research

It is well known that citation behaviors, and therefore the distribution of the citation count of articles, differ greatly among subject fields. The objective of our research is to examine whether there are general trends across subject fields regarding the extrinsic factors that affect the number of citations of articles and the extent to which each factor influences the citation rate.

As described in Section 2.3, many studies exist concerning the factors that affect the citation rate of articles. However, there is still no consensus about this topic because at least one of the following three cases applies to most existing studies:

(a) analyzing a sample in which articles from different scientific fields are mixed,
(b) focusing on a single factor (or multiple factors as mutually independent) without considering interactions among different factors, or
(c) restricting the subject or source of the sample articles to a specific area or a specific affiliated country although considering multiple factors in an integrated way.

We carried out a systematic negative binomial regression analysis applying to several selected subject fields and aimed to find any common tendencies across fields.

2.4.2 Data and Methods Used

Data sources
We divided more than 200 Subject Categories used in WoS into 11 broad areas and identified six broad areas containing more journals, including physics, chemistry, engineering, biology, basic medicine, and clinical medicine. From each of the six, we selected one Subject Category regarded as a representative of the broad area, as follows:

- Physics: Condensed-matter physics (CondMat)
- Chemistry: Inorganic and nuclear chemistry (Inorg)
- Engineering: Electric and electronic engineering (Elec)
- Biology: Biochemistry and molecular biology (Biochem)
- Basic medicine: Physiology (Physiol)
- Clinical medicine: Gastroenterology and hepatology (Gastro)

These Subject Categories are called "fields", and the abbreviations shown in parentheses will be used hereinafter.

We selected four journals from each field (see Table 2.1), taking the followings into account:

(a) journals classified only in the relevant categories (nearly 50% of the journals included in WoS are classified in two or more categories)

Table 2.1 Selected subject fields and journals (modified from [47]).
(a) The journal titles at the time of 2000, although some were changed after that.

Subject field	Journal title[a]	Publishing country	Normal articles	Articles in sample
Condensed Matter Physics (CondMat)	Physical Review B	USA	4738	55
	Journal of Physics Condensed Matter	GBR	813	56
	European Physical Journal B	DEU	538	60
	Physica B	NLD	148	59
Inorganic & Nuclear Chemistry (Inorg)	Inorganic Chemistry	USA	931	53
	Journal of the Chemical Society - Dalton Transactions	GBR	682	54
	Inorganica Chimica Acta	CHE	546	60
	Transition Metal Chemistry	NLD	139	60
Electric & Electronic Engineering (Elec)	IEEE Transactions on Microwave Theory & Techniques	USA	295	59
	IEEE Transactions on Circuits & Systems I - Fundamental Theories & Applications	USA	218	60
	Signal Processing	NLD	178	59
	IEE Proceedings - Circuits, Devices & Systems	GBR	52	51
Biochemistry & Molecular Biology (Biochem)	Journal of Biological Chemistry	USA	5504	60
	Journal of Molecular Biology	USA	875	60
	European Journal of Biochemistry	GBR	788	60
	Journal of Biochemistry (Tokyo)	JPN	275	60
Physiology (Physiol)	Journal of General Physiology	USA	110	60
	Journal of Physiology - London	GBR	472	58
	Pflugers Archive European Journal of Physiology	DEU	238	58
	Japanese Journal of Physiology	JPN	72	60
Gastro-enterology (Gastro)	Gastroenterology	USA	259	59
	Gut	GBR	277	56
	American Journal of Gastroenterology	USA	430	58
	Journal of Gastroenterology	JPN	124	60

(b) journals using English only
(c) journals published in different countries included in each field
(d) journals with relatively high and low-impact factors included (to ensure a representative sample of all articles in the field)

All of the normal articles (assigned the document type "article" in WoS) published in 2000 in the 24 journals chosen above were extracted for analysis, excluding those simultaneously classified as "proceedings papers", those shorter than three pages, and those without an author's name. The numbers of the articles from each journal are also shown in Table 2.1.

Articles published in 2000 were selected as a sample. The citation frequencies received by the sample articles were measured in October 2006 and December 2011 using WoS. The length of the citing windows is 6–7 years and 11–12 years, respectively.

Multiple regression analysis

A negative binomial multiple regression (NBMR) analysis was performed for each of the six subject fields shown in this subsection.

The NBMR analysis has been demonstrated to successfully work for predicting citations in several studies [9, 10, 12, 14, 15, 17, 18, 26, 28, 65, 72] because the citation frequency as a response variable is a non-negative integer, its distribution is remarkably skewed, and the variance is usually larger than the mean. A linear multiple regression (LMR) model with a logarithm of citation frequency (in many cases, $\log[C + 1]$) as the response variable has also been frequently utilized [5, 20, 27, 62, 69]. However, we adopted the NBMR model because, after trying both, it provided us with results much better than those of the LMR model.

Sample articles. As the sample in the NBMR analysis, for each of the six fields, 230–240 articles (60 or somewhat fewer from each of the four journals shown in Table 2.1) were sampled randomly.

The response variables. The citation frequencies in the citing windows of 6–7 years and 11–12 years (hereafter called $C6$ and $C11$, respectively) obtained by the method described in this subsection served as the response variables. These citation frequencies include self-citations[2], since the possibility that the inclusion of self-citations biases the results of a macroscopic analysis under a long citation window is not high [4, 24, 25].

The explanatory variables. The measures selected in Section 2.2 function as the explanatory variables. Below we give them variable names and describe the procedure to obtain the data for these variables.

- Authors' collaborative degree
  (a) *Authors*: Number of authors of the article
  (b) *Institutions (Insts)*: Number of institutions the authors are affiliated with
  (c) *Countries*: Number of countries where the institutions are located.

---

[2] Our analysis was made for both cases, including and excluding self-citations, but a significant difference affecting the results was not observed.

- Cited references
  (d) *References (Ref)*: Number of references cited in the article
  (e) *Price*: Price index (percentage of the references whose publication year is within five years before the publication year of the article)
- Article's visibility
  (f) *Figures*: Number of figures in the article
  (g) *Tables*: Number of tables in the article
  (h) *Equations (Eqs)*: Number of numbered equations in the article
  (i) *Length*: Number of normalized pages of the article
- Authors' past achievements
  (j) *Published articles (Publ)*: Number of articles published by the first author of the article up to the year 2000
  (k) *Cited*: Number of citations received by the published articles (*Publ*) up to the year 2000
  (l) *Age*: Active years (elapsed years from the year of the first article publication to the year 2000) of the first author
  (m) *Rate of publication (RatePubl)*: Number of articles published per annum by the first author during his/her active years (= *Publ/Age*)
  (n) *Median of the number of citations (MedCites)*: Median of the number of citations received per annum by each published article.
  The attributes (j), (k), and (l) are collectively called "cumulative achievement indicators", and the attributes (m) and (n) are called "efficient achievement indicators".
- Publishing journal
  (o) *Jnl-1*; *Jnl-2*; *Jnl-3*: Dummy variables representing the journals publishing the articles

The values of these attributes for the sample articles were acquired according to the procedures discussed below. Data on *Authors, Insts, Countries*, and *Refs* were obtained from the downloaded WoS records. The values of *Price* were obtained from the reference list in the WoS CR field of the article by counting references with publication years between 1996 and 2000 (i.e., within five years before the publication of the article). *Figures, Tables*, and *Eqs* were directly counted from the original documents. *Length* was defined as the number of converted pages under a normalization of 6,400 characters per page, and the value for an article was determined by measuring, from sampled pages, the average characters per page of the journal publishing the article.

The data on authors' past achievements (*Publ, Age*, and *RatePubl*) were collected from the WoS search with the first author names of the sample articles from 1970 to 2000. To eliminate a large number of articles by homonym authors, we developed a model for author disambiguation based on the similarity of each retrieved article to its originating article and extracted the retrieved articles to be discriminated as "true" [46]. Data on the articles that cited "true" retrieved articles until 2000 were purchased from Thomson Reuters to calculate the values of *Cited* and *MedCites*. Fractional counting was used for calculating the values of *Publ, Cited, RatePubl*, and *MedCites* because the fit to the NBMR was somewhat better with fractional counting than with full counting.

We introduced dummy variables representing the journals in which the sample articles were published. In each subject field, taking the journal having the lowest average citation frequency as the "baseline", the three dummy variables — *Jnl-1*, *Jnl-2*, and *Jnl-3* — were assigned to the other three journals.

The predicting function of NBMR. In the NBMR analysis, the value of the response variable y_i for a case i is supposed to be subject to negative binomial distribution, as follows — here, $\Gamma(\cdot)$ is a gamma function:

$$Pr(y_i = k) = \frac{\Gamma(k + \theta)}{\Gamma(\theta)\Gamma(k + 1)} \left(\frac{\theta}{\mu_i + \theta} \right)^{\theta} \left(\frac{\mu_i}{\mu_i + \theta} \right)^{k} \qquad (2.1)$$

The expected value (μ_i) of y_i is estimated from the following regression equation:

$$\ln(\mu_i) = \beta_0 + \beta_1 X_{i1} + \beta_2 X_{i2} + \ldots + \beta_p X_{ip} \qquad (2.2)$$

Estimated values of the partial regression coefficients $\beta_0, \beta_1, \ldots, \beta_p$ and the parameter θ are given on the basis of the input data $\{X_{i1}, X_{i2}, \ldots, X_{ip}; y_i\}$. The value of θ is supposed to be independent of i.

The response variable y_i in Eq. 2.1 is $C6$ or $C11$. The explanatory variables $X_{i1}, X_{i2}, \ldots, X_{ip}$ in Eq. 2.2 are the attributes (a)–(o), introduced in this subsection.

The Advanced Regression Model of SPSS/PASW Version 18 was used to perform the NBMR analysis. Variable selection was not chosen in regression, and variables showing a significant relation to $C6$ or $C11$ were identified from the regression results.

2.4.3 Results

Correlations between the explanatory variables

Table 2.2 shows the combinations of explanatory variables of which the Spearman's rank correlation coefficient is significant ($p < 0.05$) in more than half the fields. The three cumulative achievement variables (*Publ*, *Cited*, and *Age*) have the strongest correlation with each other; thus, their ρ-values are greater than 0.7 for most cases, while the correlation between each of these variables and each of the two efficient achievement variables (*RatePubl* and *MedCites*) is not as strong, except for that between *Publ* and *RatePubl*. Within the groups (i) *Authors*, *Insts*, and *Countries* and (ii) *Refs*, *Figures* and *Length*, moderately strong correlations ($\rho = 0.5\ldots0.7$) were observed in many cases, but a stronger correlation was not found.

Considering the problem of multicollinearity, we decided not to simultaneously include explanatory variables whose ρ-values are greater than 0.7 in most fields in a regression model. Subject to this decision, we designed the three regression models;

Table 2.2 The explanatory variable pairs of which Spearman's rank correlation coefficient is significant in many fields (reproduced from [47])
(** 1% significant, * 5% significant, (a) in Gastro field, $Eqs = 0$ for all sample articles.

Variable pair	CondMat	Inorg	Elec	Biochem	Physiol	Gastro
(Authors, Insts)	0.555**	0.550**	0.441**	0.474**	0.436**	0.169**
(Authors, Countries)	0.376**	0.324**	0.243**	0.274**	0.281**	0.134*
(Insts, Countries)	0.647**	0.627**	0.613**	0.498**	0.541**	0.487**
(Refs, Figures)	0.113	0.412**	0.196**	0.342**	0.497**	0.287**
(Refs, Length)	0.548**	0.634**	0.478**	0.614**	0.711**	0.705**
(Figures, Length)	0.494**	0.693**	0.486**	0.660**	0.763**	0.598**
(Eqs, Length)	0.421**	0.099	0.352**	0.160*	0.376**	- a
(Price, MedCites)	0.273**	-0.169*	0.122	0.196**	0.069	0.146*
(Publ, Cited)	0.821**	0.862**	0.731**	0.811**	0.818**	0.845**
(Publ, Age)	0.731**	0.779**	0.816**	0.762**	0.749**	0.638**
(Publ, RatePubl)	0.687**	0.785**	0.835**	0.706**	0.706**	0.743**
(Publ, MedCites)	0.308**	0.321**	0.126	0.328**	0.238**	0.167*
(Cited, Age)	0.747**	0.841**	0.817**	0.825**	0.756**	0.735**
(Cited, RatePubl)	0.442**	0.553**	0.470**	0.495**	0.467**	0.473**
(Cited, MedCites)	0.622**	0.625**	0.538**	0.691**	0.570**	0.473**
(Age, RatePubl)	0.180**	0.335**	0.544**	0.334**	0.259**	.073
(Age, MedCites)	0.310**	0.461**	0.383**	0.435**	0.283**	0.262**

A, B, and C. These models differ mutually only in terms of the variables about author's achievements included, as follows:

- Model A: Publ, MedCites
- Model B: Cited, RatePubl
- Model C: Age, RatePubl, MedCites

Comparison of the regression models
Several measures of goodness of fit of the NBMR model exist (see [40]). From those measures, the Akaike information criterion (AIC) and the adjusted pseudo coefficient of determination (pseudo Rc2) are taken and their values for the response variable $C6$ are compared between the three models in Table 2.3.

Table 2.3 Goodness of fit measures for three regression models (response variable: $C6$) (reproduced from [47]).

Field	Model A	Model B	Model C		Field	Model A	Model B	Model C
CondMat	1511.3	1511.2	1515.6		CondMat	0.261	0.261	0.251
Inorg	1466.4	1475.9	1470.9		Inorg	0.332	0.304	0.322
Elec	1214.0	1211.4	1211.9		Elec	0.276	0.285	0.287
Bio-chem	1811.4	1814.9	1814.1		Bio-chem	0.441	0.433	0.438
Physiol	1613.9	1613.2	1612.1		Physiol	0.531	0.533	0.537
Gastro	1814.7	1811.0	1813.4		Gastro	0.487	0.495	0.492

AICs (Akaike information criteria)   Adjusted pseudo R-squared measures

As seen in this table, a large difference does not exist among the models in every field, but Model C appears slightly better than the others because of the higher stability across the fields.

Of the authors' achievement measures, the three cumulative achievement measures (*Publ, Cited* and *Age*) were found not to influence citations in most cases. In contrast, the two efficient achievement measures (*RatePubl* and *MedCites*) have a positive influence in some fields. Therefore, model C, including both these variables, is preferable to the other two.

The significant predictors of citation rates: the results of Model C. Since Model C appears to be more appropriate compared to Models A and B from the results mentioned above, we describe the results using Model C.

Table 2.4 shows the results of the NBMR analysis for $C12$ as the response variable. (The result for $C6$ is almost similar.) In this table, the x-standardized regression coefficients of the explanatory variable j, which are the partial regression coefficients β_js multiplied by the standard deviations of the variable s_js, are reported so that the relative strength of influence on the response variable can be compared among the explanatory variables. The estimated values of the parameter θ are also displayed in this table.

From Table 2.4, *Price* is revealed to be the most important influencing factor on citations in every fields. Next, *Refs* is a strong, or moderately strong, predictor in half the fields. For *Authors, Figures, RatePubl* and *MedCites*, there are some fields in which they have a positive significant influence on the response variable, with the

Table 2.4 The x-standardized regression coefficients ($s_j\beta'_j$s) of NBMR for $C12$ as the response variable (reproduced from [47])
(** 1% significant, * 5% significant, +10% significant).

	CondMat	Inorg	Elec	Biochem	Physiol	Gastro
n	230	227	229	240	236	233
θ	1.27	1.86	1.04	2.58	2.44	11.69
$s_j\beta_j$						
Authors	0.175*	0.052	−0.143	0.136**	−0.019	0.095
Insts	−0.169+	0.085	0.175	−0.006	0.113+	−0.098
Countries	−0.094	−0.092	−0.066	−0.021	−0.043	0.162*
Refs	0.213**	0.046	0.343**	0.123*	0.102	0.099
Price	0.313**	0.200**	0.225**	0.393**	0.314**	0.199**
Figures	−0.065	0.108	0.095	−0.034	0.139+	0.165+
Tables	0.022	−0.012	0.008	0.044	−0.039	0.362**
Eqs	0.000	−0.130*	−0.074	0.032	−0.079	−
Length	0.123	0.256*	0.137	0.085	0.148	−0.236*
Age	0.053	−0.020	0.041	0.024	−0.107*	−0.018
RatePubl	0.115+	−0.125*	0.164*	0.010	0.073	0.099
MedCites	−0.012	0.126*	0.066	0.081+	0.026	−0.025
Dummy1	0.248**	0.096	0.577**	0.125*	0.407**	0.335**
Dummy2	0.040	0.006	0.534**	0.27**	0.411**	0.814**
Dummy3	0.325**	−0.004	−0.002	0.315**	0.205**	0.648**

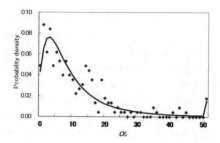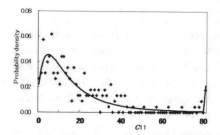

Fig. 2.1 Comparison of the probability distribution of citation counts predicted by NBMR (full line) with the observed distribution (points) in case of Inorg field. Left: Distribution of $C6$, Right: Distribution of $C11$ (reproduced from [47]).

exception of *RatePubl*, which is a negative significant predictor in the "Inorg" field. The estimated citation frequencies are considerably affected by the journal in which the articles were published in five fields, except the "Inorg" field.

Accuracy of prediction of the citation frequencies. The frequency distributions of $C6$ and $C11$ predicted from the NBMR analysis were obtained by the following procedure:

1. calculating the probability distribution of the citation frequency $Pr(C6i = k)$ or $Pr(C11i = k)$ for each article i based on Eq. 2.1 and
2. summing up the probability distributions of all articles.

The predicted frequency distributions obtained are compared to the observed distributions in Fig. 2.1. This figure shows the case of the "Inorg" field, for which the fitness of the NBMR model is moderate in the six fields. Although the observed distributions fluctuate considerably, the predicted distributions fit their smoothed curves well.

2.4.4 Discussion

Important factors influencing the citations of articles
In the six fields, we were able to predict, with acceptable accuracy, the citation frequency of an article within 6 years ($C6$) and 11 years ($C11$) after its publication, with 3–5 significant predictors. Price index (*Price*) was found to be the strongest influencing factor on citations. A few studies have taken notice of this kind of attribute (e.g., the recency measure of the references) as an influencing factor on citations [27, 52, 62]. These studies found a moderate positive correlation between the recency of references and the citation count, but their results were based on a relatively small sample ($n <\approx 300$) taken from a single subject field (see Section 2.3.2). It is our noticeable finding that the Price index is very important in every field when considering factors influencing citation rates. The second important explanatory factor was the number of references (*Refs*), which has been reported to have a significant relation with citation rates in many existing studies (see Section 2.3.2).

Although many scholars have reported that articles with more co-authors tend to obtain higher citations, such claims have not been very strongly supported by several systematic multiple regression analyses (see Section 2.3.2). Also in our NBMR analysis, the number of authors (*Authors*) was shown to be a (moderately) significant predictor in only two of the six fields, suggesting that the factor might not affect citation rates very strongly. Bornmann and Daniel reported that the correlation between the number of authors and that of citations diminished as the citing window became longer [9]. This may apply to our case since the citing window we used was relatively long (6 or 11 years).

Is there a halo effect of authors?
Several studies have claimed that an article written by higher performance author(s) – more publications and/or higher citations – have the possibility of receiving higher citations after its publication (see Section 2.3.2). In this research, hardly any effect of the three cumulative achievement indicators (*Publ, Cited* and *Age*) was found. On the other hand, the two efficient achievement indicators (*RatePubl* and *MedCites*) showed a significant, but not remarkable, influence in some fields. Our finding that the efficient achievement indicators are better predictors of citations than the cumulative ones agrees with those of Danell [16] and Hönekopp and Khan [30]. However, it may be because we used data only on the first authors that the effect was not as apparent in our analysis as in that shown in [16, 30]. It is considered to be a limitation of this research because it might weaken the halo effect of authors on citations.

2.5 Concluding Remarks

Here we summarize briefly the studies introduced in this chapter, comparing the main results of several researchers with ours.

In Table 2.5, the citation predictabilities of the extrinsic factors in the studies considering various variables in an integrated way (see Section 2.3.1) are compared with those in our research. The followings can be said from this table.

1. Factors confirmed to have a significant positive relation with citation rate from the results of many studies: the number of references; the citation impact of the publishing journal
2. Factors supposed to have a significant positive relation with citation rate from the results of some studies: the recency of references (e.g., Price index); the citation impact of authors
3. Factors of which the relation with citation rate is not determined since the studies claiming a significant relation and those denying it are competitive: the number of authors; the number of affiliated institutions; the number of affiliated countries; the article length; the productivity of authors; other authors' attributes (e.g., the active years and the type of affiliated institution)
4. Factors supposed to not have a significant relation with citation rate: the number of figures; the number of tables

Table 2.5 Correlation strength of various factors with citation rate demonstrated by multiple regression analysis studies (reproduced from [47]) A: Strong or definite predictor, B: Weak predictor or predictive power dependent on the model, C: Not-significant or negative predictor.

Work	#Authors	#Institutions	#Countries	#References	Recency of refs	#Figures	#Tables	Article length	Author's productivity	Author's citedness	Other status of author	Journal impact
Bornmann & Daniel (2008)	B							C			A	
Bornmann & Leydesdorff (2015)	A							A				A
Chen (2012)	A			A				C				
Davis et al. (2008)	A			A				C				A
Didegah & Thelwall (2013)	C	A	C	A								A
Fu & Aliferis (2010)	C	C		A					C	A		A
Hanssen & Jorgensen (2015)	C		B						A	C	A	
Haslam et al. (2008)	C			A	A	C	C	A	A			A
He (2009)	A		A	A	A	C		C		B	C	
Lokker et al. (2008)	A			A	A			C		B		A
Peng & Zhu (2012)	A			A				A			A	A
Peters & van Raan (1994)	B			A	A			B	A			A
Stewart (1983)	C			A	B			A		A	C	
van Dalen & Henkens (2001)	B							A	A	A		A
van Wesel et al. (2014)	C			A				B		A		
Walters (2006)	B							C		A		B
Xie et al. (2019)	B			B	B	B	B	B	B	B	A	B
Yan et al. (2020)	C	A							C	A		A
Yu et al. (2014)	A			A	A			C	C	A	C	A
This research	B	C	C	A	A	C	C	C	B	B	C	A

Although the broadened collaborative relations, especially international collaborations, is said to boost up citations, the effect is not clear in the studies using multiple regression analyses taking confounding among various variables into consideration, as shown in Table 2.5.

The fitness of the NBMR models obtained in our research was not very high but fair; the value of pseudo R_c^2 was in the range [0.25..0.5]. One would expect these results since all the explanatory variables used here are extrinsic factors that have no direct relation to the quality or content of articles. The aim of our research was not to develop a model with high fitness, but to seek a model working as a baseline of the expected values of the response variable for a given article based on such extrinsic factors. This aim was accomplished to some extent since some trends were common across different fields.

One of the challenges into the future is an analysis of the deviations of the observed citation measures from this baseline for individual articles. What attributes of articles do the deviations relate to? Are the attributes intrinsic and connected with the quality or content of articles?

References

1. Abramo, G., D'Angelo, C.A.: The relationship between the number of authors of a publication, its citations and the impact factor of the publishing journal: Evidence from Italy. Journal of Informetrics **9**(4), 746–761 (2015)
2. Ahlgren, P., Colliander, C., Sjögårde, P.: Exploring the relation between referencing practices and citation impact: A large-scale study based on Web of Science data. Journal of the Association for Information Science & Technology **69**(5), 728–743 (2018)
3. Aksnes, D.W.: Characteristics of highly cited papers. Research Evaluation **12**(3), 159–170 (2003)
4. Aksnes, D.W.: A macro-study of self-citations. Scientometrics **56**(2), 235–246 (2003)
5. Basu, A., Lewison, G.: Going beyond journal classification for evaluation of research outputs. Aslib Proceedings **57**(3), 232–246 (2005)
6. Bookstein, A., Yitzhaki, M.: Own-language preference: A new measure of 'relative language self-citation'. Scientometrics **46**(2), 337–348 (1999)
7. Bordons, M., Aparicio, J., Costas, R.: Heterogeneity of collaboration and its relationship with research impact in a biomedical field. Scientometrics **96**(2), 443–466 (2013)
8. Bornmann, L., Daniel, H.-D.: Selecting scientific excellence through committee peer review? A citation analysis of publications previously published to approval or rejection of post-doctoral research fellowship applicants. Scientometrics **68**(3), 427–440 (2006)
9. Bornmann, L., Daniel, H.-D.: Selecting manuscripts for a high-impact journal through peer review: A citation analysis of Communications that were accepted by Angewandte Chemie International Edition, or rejected but published elsewhere. Journal of the American Society for Information Science & Technology **59**(11), 1841–1852 (2008)
10. Bornmann, L., Leydesdorff, L.: Does quality and content matter for citedness? A comparison with para-textual factors and over time. Journal of Informetrics **9**(3), 410–429 (2015)
11. Callaham, M., Wears, R.L., Weber, E.: Journal prestige, publication bias, and other characteristics associated with citation of published studies in peer-reviewed journals. Journal of the American Medical Association **287**(21), 2847–2850 (2002)
12. Chen, C.: Predictive effects of structural variation on citation counts. Journal of the American Society for Information Science & Technology **63**(3), 431–449 (2012)

13. Cronin, B., Shaw, D.: Citation, funding acknowledgment and author nationality relationships in four information science journals. Journal of Documentation **55**(4), 402–408 (1999)
14. van Dalen, H.P., Henkens, K.: What makes a scientific article influential? The case of demographers. Scientometrics **50**(3), 455–482 (2001)
15. van Dalen, H.P., Henkens, K.: Signals in science - On the importance of signaling in gaining attention in science. Scientometrics **64**(2), 209–233 (2005)
16. Danell, R.: Can the quality of scientific work be predicted using information on the author's track record? Journal of the American Society for Information Science & Technology **62**(1), 50–60 (2011)
17. Davis, P.M., Lewenstein, B.V., Simon, D.H., Booth, J.G., Connolly, M.J.L.: Open access publishing, article downloads, and citations: Randomised controlled trial. BMJ **337**(7665), a568 (2008)
18. Didegah, F., Thelwall, M.: Determinants of research citation impact in nanoscience and nanotechnology. Journal of the American Society for Information Science & Technology **64**(5), 1055–1064 (2013)
19. Fanelli, D.: Positive results receive more citations, but only in some disciplines. Scientometrics **94**(2), 701–709 (2013)
20. Figg, W.D., Dunn, L., Liewehr, D.J., Steinberg, S.M., Thurman, P.W., Barrett, J.C., Birkinshaw, J.: Scientific collaboration results in higher citation rates of published articles. Pharmacotherapy **26**(6), 759–767 (2006)
21. Frandsen, T., Nicolaisen, J.: Citation behavior: A large-scale test of the persuasion by name-dropping hypothesis. Journal of the Association for Information Science & Technology **68**(5), 1278–1284 (2017)
22. Fu, L., Aliferis, C.: Using content-based and bibliometric features for machine learning models to predict citation counts in the biomedical literature. Scientometrics **85**(1), 257–270 (2010)
23. Gazni, A., Didegah, F.: Investigating different types of research collaboration and citation impact: A case study of Harvard University's publications. Scientometrics **87**(2), 251–265 (2011)
24. Glänzel, W., Thijs, B.: Does co-authorship inflate the share of self-citations? Scientometrics **61**(3), 395–404 (2004)
25. Glänzel, W., Thijs, B., Schlemmer, B.: A bibliometric approach to the role of author self-citations in scientific communication. Scientometrics **59**(1), 63–77 (2004)
26. Hanssen, T.E.S., Jorgensen, F.: The value of experience in research. Journal of Informetrics **9**(1), 16–24 (2015)
27. Haslam, N., Ban, L., Kaufmann, L., Loughnan, S., Peters, K., Whelan, J., Wilson, S.: What makes an article influential? Predicting impact in social and personality psychology. Scientometrics **76**(1), 169–185 (2008)
28. He, Z.-L.: International collaboration does not have greater epistemic authority. Journal of the American Society for Information Science & Technology **60**(10), 2151–2164 (2009)
29. Hicks, D., Wouters, P., Waltman, L., de Rijcke, S., Rafols, I.: The Leiden Manifesto for research metrics. Nature **520**(7548), 429–431 (2015)
30. Hönekopp, J., Khan, J.: Future publication success in science is better predicted by traditional measures than by the h index. Scientometrics **90**(3), 843–853 (2012)
31. Ibanez, A., Bielza, C., Larranaga, P.: Relationship among research collaboration, number of documents and number of citations: a case study in Spanish computer science production in 2000–2009. Scientometrics **95**(2), 689–716 (2013)
32. Katz, J.S., Hicks, D.: How much is a collaboration worth? A calibrated bibliometric model. Scientometrics **40**(3), 541–554 (1997)
33. Larivière, V., Gingras, Y.: On the relationship between interdisciplinarity and scientific impact. Journal of the American Society for Information Science & Technology **61**(1), 126–131 (2010)
34. Larivière, V., Gingras, Y.: The impact factor's Matthew Effect: A natural experiment in bibliometrics. Journal of the American Society for Information Science & Technology **61**(2), 424–427 (2010)

35. Larivière, V., Gingras, Y., Sugimoto, C.R., Tsou, A.: Team size matters: Collaboration and scientific impact since 1900. Journal of the American Society for Information Science & Technology **66**(7), 1323–1332 (2015)
36. Leimu, R., Koricheva, J.: What determines the citation frequency of ecological papers? Trends in Ecology & Evolution **20**(1), 28–32 (2005)
37. Levitt, J.M., Thelwall, M.: Citation levels and collaboration within library and information science. Journal of the American Society for Information Science & Technology **60**(3), 434–442 (2009)
38. Lindsey, D.: Using citation counts as a measure of quality in science: Measuring what's measurable rather than what's valid. Scientometrics **15**(3–4), 189–203 (1989)
39. Lokker, C., McKibbon, K.A., McKinlay, R.J., Wilczynski, N.L., Haynes, R.B.: Prediction of citation counts for clinical articles at two years using data available within three weeks of publication: Retrospective cohort study. BMJ **336**(7645), 655–657 (2008)
40. Long, J.S.: Hypothesis testing and goodness of fit. In: Regression Models for Categorical and Limited Dependent Variables, chap. 4. SAGE Publications (1997)
41. MacRoberts, M.H., MacRoberts, B.R.: Testing the Ortega hypothesis: Facts and artifacts. Scientometrics **12**(5–6), 293–295 (1987)
42. MacRoberts, M.H., MacRoberts, B.R.: Problems of citation analysis: A critical review. Journal of the American Society for Information Science **40**(5), 342–349 (1989)
43. MacRoberts, M.H., MacRoberts, B.R.: Problems of citation analysis. Scientometrics **36**(3), 435–444 (1996)
44. MacRoberts, M.H., MacRoberts, B.R.: Problems of citation analysis: A study of uncited and seldom-cited influences. Journal of the American Society for Information Science & Technology **61**(1), 1–13 (2010)
45. Moed, H.F.: Citation analysis in research evaluation. Springer (2005)
46. Onodera, N., Iwasawa, M., Midorikawa, N., Yoshikane, F., Amano, K., Ootani, Y., Kodama, T., Kiyama, Y., Tsunoda, H., Yamazaki, S.: A method for eliminating articles by homonymous authors from the large number of articles retrieved by author search. Journal of the American Society for Information Science & Technology **62**(4), 677–690 (2011)
47. Onodera, N., Yoshikane, F.: Factors affecting citation rates of research articles. Journal of the American Society for Information Science & Technology **66**(4), 739–764 (2015)
48. Pasterkamp, G., Rotmans, J.I., de Kleijn, D.V.P., Borst, C.: Citation frequency: A biased measure of research impact significantly influenced by the geographical origin of research articles. Scientometrics **70**(1), 153–165 (2007)
49. Peclin, S., Juznic, P., Blagus, R., Sajko, M.C., Stare, J.: Effects of international collaboration and status of journal on impact of papers. Scientometrics **93**(3), 937–948 (2012)
50. Peng, T.-Q., Zhu, J.J.: Where you publish matters most: A multi-level analysis of factors affecting citations of internet studies. Journal of the American Society for Information Science & Technology **63**(9), 1789–1803 (2012)
51. Persson, O., Glänzel, W., Danell, R.: Inflationary bibliometric values: The role of scientific collaboration and the need for relative indicators in evaluative studies. Scientometrics **60**(3), 421–432 (2004)
52. Peters, H.P.F., van Raan, A.F.J.: On determinants of citation scores: A case study in chemical engineering. Journal of the American Society for Information Science & Technology **45**(1), 39–49 (1994)
53. Puuska, H.-M., Muhonen, R., Leino, Y.: International and domestic co-publishing and their citation impact in different disciplines. Scientometrics **98**(2), 823–839 (2014)
54. van Raan, A.F.J.: The influence of international collaboration on the impact of research results. Some simple mathematical considerations concerning the role of self-citations. Scientometrics **42**(2), 423–428 (1998)
55. Rigby, J.: Looking for the impact of peer review: Does count of funding acknowledgements really predict research impact? Scientometrics **94**(1), 57–73 (2013)
56. Schubert, A., Glänzel, W.: Cross-national preference in co-authorship, references and citations. Scientometrics **69**(2), 409–428 (2006)

57. Seglen, P.O.: The skewness of science. Journal of the American Society for Information Science **43**(9), 628–638 (1992)
58. Sin, S.-C. J.: International coauthorship and citation impact: A bibliometric study of six LIS journals, 1980–2008. Journal of the American Society for Information Science & Technology **62**(9), 1770–1783 (2011)
59. Slyder, J.B., Stein, B.R., Sams, B.S., Walker, D.M., Beale, B.J., Feldhaus, J.J., Copenheaver, C.A.: Citation pattern and lifespan: A comparison of discipline, institution, and individual. Scientometrics **89**(3), 955–966 (2011)
60. Snizek, W.E., Oehler, K., Mullins, N.C.: Textual and non-textual characteristics of scientific papers. neglected science indicators. Scientometrics **20**(1), 25–35 (1991)
61. Sooryamoorthy, R.: Do types of collaboration change citation? Collaboration and citation patterns of South African science publications. Scientometrics **81**(1), 177–193 (2009)
62. Stewart, J.A.: Achievement and ascriptive processes in the recognition of scientific articles. Social Forces **62**(1), 166–189 (1983)
63. Sud, P., Thelwall, M.: Not all international collaboration is beneficial: The Mendeley readership and citation impact of biochemical research collaboration. Journal of the Association for Information Science & Technology **67**(8), 1849–1857 (2016)
64. Tahamtan, I., Afshar, A.S., Ahamdzadeh, K.: Factors affecting number of citations: A comprehensive review of the literature. Scientometrics **107**(3), 1195–1225 (2016)
65. Walters, G.D.: Predicting subsequent citations to articles published in twelve crime-psychology journals: Author impact versus journal impact. Scientometrics **69**(3), 499–510 (2006)
66. Wang, F., Fan, Y., Zeng, A., Di, Z.: Can we predict ESI highly cited publications? Scientometrics **118**(1), 109–125 (2019)
67. Wang, M., Yu, G., An, S., Yu, D.: Discovery of factors influencing citation impact based on a soft fuzzy rough set model. Scientometrics **93**(3), 635–644 (2012)
68. Wang, M., Yu, G., Yu, D.: Mining typical features for highly cited papers. Scientometrics **87**(3), 695–706 (2011)
69. van Wesel, M., Wyatt, S., ten Haaf, J.: What a difference a colon makes: How superficial factors influence subsequent citation. Scientometrics **98**(3), 1601–1615 (2014)
70. Xie, J., Gong, K., Cheng, Y., Ke, Q.: The correlation between paper length and citations: A meta-analysis. Scientometrics **118**(3), 763–786 (2019)
71. Xie, J., Gong, K., Cheng, Y., Li, J., Ke, Q., Kang, H., Cheng, Y.: A probe into 66 factors which are possibly associated with the number of citations an article received. Scientometrics **119**(3), 1429–1454 (2019)
72. Yan, Y., Tian, S., Zhang, J.: The impact of a paper's new combinations and new components on its citation. Scientometrics **122**(2), 895–913 (2020)
73. Yu, T., Yu, G., Li, P.-Y., Wang, L.: Citation impact prediction for scientific papers using stepwise regression analysis. Scientometrics **101**(2), 1233–1252 (2014)

Chapter 3
Remarks on Dynamics of Research Production of Researchers and Research Organizations

Nikolay K. Vitanov and Zlatinka I. Dimitrova

Abstract We discuss the dynamics of research production of researchers, research organizations, and systems of research organizations. We obtain several relationships for the production of papers by these individuals and units. Then, by means of the obtained relationships we analyze several scenarios for the dynamics of research production in research organizations and in systems of research organizations. The conclusions obtained on the basis of the studied model are that the recruitment of new researchers, conditions for researching in a research organization, and the influence of the environment on researchers and research organizations can lead to increase or decrease of the research production of researchers and research organizations.

3.1 Introduction

Research on complex systems increased much in the last decades. Several examples of new areas of this research are connected to sociology, economics, forest fires,network flows, and regional development [2, 5, 6, 21, 25]. An important part of this research is connected to the study of evolution and characteristics of complex systems containing researchers and research organizations. Related examples concern nuclear physics or other branches of the natural sciences and there exist many studies on such systems and their comparison [1, 4, 11, 12, 17]. There are many reasons for research on these systems and some of them are connected to the increasing costs to ensure advance in the modern science and the wish of governments to spend money for research effectively and efficiently. One of the most interesting topics in the above studies is the dynamics connected to research production [18, 19, 27]. Evidence

Nikolay K. Vitanov
Institute of Mechanics, Bulgarian Academy of Sciences, 1113 Sofia, Bulgaria
e-mail: `vitanov@imbm.bas.bg`

Zlatinka I. Dimitrova
Institute of Mechanics, Bulgarian Academy of Sciences, 1113 Sofia, Bulgaria

of this interest are the numerous books devoted to scientometrics, bibliometrics, informetrics, webometrics, scientometric indicators and their applications [3, 8, 9, 10, 13, 14, 15, 20, 22, 23, 24, 26], and especially to citations and citation analysis [7, 16].

This chapter discusses the research production of researchers, research organizations and systems of research organizations. The goal of each research organization is to increase its research production if possible. But increase or decrease of research production depends on many factors and several of these factors will be discussed below on the basis of mathematical formulations. These considerations allow also the calculation of different scenarios for future dynamics of research systems from the point of view of their staff and research production. We discuss several examples such as a scenario, including decrease of research staff and research production of an organization as a consequence of a recruitment strategy and worsening of the economic conditions. Many other scenarios can also be discussed on the basis of the presented mathematical theory.

This chapter is organized as follows. In Sect. 3.2 we obtain several relationships for the dynamics of research production of a single researcher. In Sect. 3.3 this theory is extended to the case of a research organization, and then we discuss the research production of a system of research organizations. In Sect. 3.4 we apply the obtained relationships for analysis of several simple scenarios to gain more information about the discussed quantitative characteristics of the research. In Sect. 3.5 we discuss in detail one of these scenarios which is connected to a strategy of recruitment of researchers combined with worsening of the conditions for research in the course of the years. In Sect. 3.6 we present several relationships for the research production in a system of research organizations. Several concluding remarks are summarized in Sect. 3.7.

3.2 Dynamics of the Research Production of a Single Researcher

We consider research organizations consisting of researchers who publish research papers. Let us denote the year by the index i. In this year we can consider $j = 1, \ldots, J_i$ research organizations. These organizations have research staff consisting of researchers. We assign a number to any of the researchers and let k be the index which denotes the corresponding number of the researcher from the j-th research organization. Let $\Delta_{k,j,i}$ be the number of research papers produced by the researcher k from the research organization j in the year i. If the researcher is not active in the year i then $\Delta_{k,j,i} = 0$. In addition, if in the year i the researcher is active, but is not anymore in the research organization j then again $\Delta_{k,j,i} = 0$.

With respect of the numbering of the researchers in the research organization j we note the following. In the case of her/his first entry into an organization the researcher obtains a number and keeps this number even if he/she leaves the organization (then the corresponding research production is set to 0). If the researcher returns to the research organization, then he/she obtains another number, etc. One researcher can work for more than one research organizations at the same time. In this case the researcher obtains a number in each of research organizations he/she works for. This manner of numbering of researchers includes automatically in the theory below: (i) the cases of re-entering of research organization, and (ii) the work for more than one research organization in the same time.

The total number of papers produced by the researcher k from his/her first year up to the year I (inclusive) for the research organization j is:

$$\pi_{k,j} = \sum_{i=1}^{I} \Delta_{k,j,i} \tag{3.1}$$

The total number of papers produced by the researcher k up to the year i is:

$$\Pi_k = \sum_{j=1}^{J} \sum_{i=1}^{I} \Delta_{k,j,i}, \tag{3.2}$$

where J is the number of the research organizations from the set of research organizations where the researcher can work in his/her career.

$\Delta_{k,j,i}$ depends on time and on other factors such as retirement, funding, etc. We assume that the researcher is active between years i_{k_s} (start of research activity) and i_{k_e} (end of research activity) and then the total number of papers he produces for the research organization j is:

$$\pi_{k,j} = \sum_{i=i_{k_s}}^{I-I_{k_e}} \Delta_{k,j,i}. \tag{3.3}$$

There are changes in the research production of the researcher in the course of the years. Let the produced papers in the year i_{k_s} be $\Delta_{k,j,i_{k_s}}$. The changes in the research production will be accounted by the function $f_{k,j,i}$ where:

$$\Delta_{k,j,i} = \Delta_{k,j,i_{k_s}} + f_{k,j,i}; \quad i_{k_s} \leq i \leq i_{k_e}, \tag{3.4}$$

and $f_{k,j,i_{k_s}} = 0$. We will rewrite Eq. 3.4 in the following way:

$$\Delta_{k,j,i} = \Delta_{k,j,i_{k_s}} + \overline{\overline{f}}_{k,j,i} + g_{k,j,i}; \quad i_{k_s} \leq i \leq i_{k_e}, \tag{3.5}$$

where:

$$\overline{\overline{f}}_{k,j,i} = \frac{\sum_{l=1}^{i-i_{k_s}} f_{k,j,l}}{i - i_{k_s}}, \quad i = i_{k_s} + 1, \ldots, i_{k_e} \tag{3.6}$$

and $\overline{\overline{f}}_{k,j,i_{k_s}} = 0$, whereas $g_{k,j,i}$ is the deviation of research production from the quantity $\Delta_{k,j,i_{k_s}} + \overline{\overline{f}}_{k,j,i}$.

We note that the above relationships hold if the researcher changes the research organization (perhaps even several times). The researcher is described by three quantities with respect to the research production: starting production $\Delta_{k,j,i_{k_s}}$; average excess production $\overline{\overline{f}}_{kji}$ over the starting production and deviation $g_{k,j,i}$ from the sum of starting production and the average excess production. In several more words the starting production accounts for education and previous experience of the researchers. The average excess production over the starting production reflects the conditions for work in the research organization and the deviation from the sum of starting production and the average excess production accounts for the influence of the environment of the research organization on the work of the researcher.

In such a way the total number of papers produced in the career of the researcher k is:

$$\Pi_k = \sum_{j=1}^{J} \sum_{i=i_{k_s}}^{i_{k_e}} [\Delta_{k,j,i_{k_s}} + \overline{\overline{f}}_{kji} + g_{k,j,i}] = \sum_{j=1}^{J} [(\Delta_{k,j,i_{k_s}} + \overline{\overline{f}}_{k,j,i_{k_e}})(i_{k_e} - i_{k_s})]$$

(3.7)

Let us assume that the researcher spends his entire career in the same research organization. This means that he/she enters the organization at the year i_{k_s}. Because of this we can denote this organization as j_{k_s} which will mean the organization where the k-th researcher begins his/her career at the years i_{k_s} and remains in this organization till the end of his/her research activity.

Then $\Delta_{k,j,i_{k_s}} \rightarrow \Delta_{k,j_{k_s},i_{k_s}}$ and $\overline{\overline{f}}_{k,j,i} \rightarrow \overline{\overline{f}}_{k,j_{k_s},i} = \dfrac{\sum\limits_{l=1}^{i-i_{k_s}} f_{k,j_{k_s},l}}{i - i_{k_s}}$, for $i = i_0 + 1, \ldots, i_k$, and

$$\Pi_k = (\Delta_{k,j_{k_s},i_{k_s}} + \overline{\overline{f}}_{k,j_{k_s},i_{k_e}})(i_{k_e} - i_{k_s}).$$

(3.8)

We can construct also other quantities, connected to the research production of a researcher. For example, the distribution of the research production in the researcher's career of the researcher working only in one research organization we have:

$$P_{k,j_{k_s},i} = \frac{\Delta_{k,j_{k_s},i}}{\Pi_k} = \frac{\Delta_{k,j_{k_s},i_{k_s}} + \overline{\overline{f}}_{k,j_{k_s},i} + g_{k,j_{k_s},i}}{(\Delta_{k,j_{k_s},i_{k_s}} + \overline{\overline{f}}_{k,j_{k_s},i_{k_e}})(i_{k_e} - i_{k_s})}.$$

(3.9)

Quantities like Eqs. 3.9 and 3.17 below show that the theory discussed in this chapter can have more applications than the applications mentioned below in the text.

3.3 Production of a Research Organization and of a System of Research Organizations

Let us now consider a research organization j having K_j researchers. The total number of papers produced by the researchers working in the research organization j in the year i are:

$$\phi_{j,i} = \sum_{k=1}^{K_j} \Delta_{k,j,i}. \tag{3.10}$$

We can study the different effects connected to the dynamics of researchers and research production in a research organization. To do this we assume the following. Every researcher of the research organization j has a research production $\Delta_{k,j,i_{ks}}$ during the first year of work in the research organization j. Further we assume that the research production per year of the researcher can change in the course of the years due to the increased experience of the researcher and due to other factors such as better conditions for work and life for an example. This change of the research production is further assumed to consist of two components: a linear increase based on the ensemble average over the researchers of the organization of the production change in their second year of work in the organization (the term $\alpha_{k,j}$ below in the text) and correction factors which correct the linear increase of the production to the actual value of the research production of the researcher for the corresponding year.

The production of a research organization in the year i is given by Eq. 3.10. For $\Delta_{k,j,i}$ we can use relationship similar to Eq. 3.5 as follows:

$$\Delta_{k,j,i} = \Delta_{k,j,i_{ks}} + \overline{f}_{k,j,i} + g_{k,j,i}; \quad i_{ks} \le i \le i_{ke}. \tag{3.11}$$

In Eq. 3.11 the term $\overline{f}_{k,j,i}$ is:

$$\overline{f}_{k,j,i} = \alpha_{k,j}(i - i_{ks}) + h_{k,j,i}, \tag{3.12}$$

where:

$$\alpha_{k,j} = \frac{\sum\limits_{k=1}^{K_j} f_{k,j,i_{ks}+1}}{K_j}, \tag{3.13}$$

is the average over all researchers of the organization of $f_{k,j,i_{ks}+1}$, which is the quantity for the second year of each researcher in the organization (we remember that $f_{k,j,i_{ks}} = 0$), whereas $h(k, j, i)$ is the deviation from $\alpha_{k,j}(i - i_{ks})$ for the researcher k of the organization j for the year i. In several more words, we assume that the deviation $\overline{f}_{k,j,i}$ of research production of the researcher k from the organization j for the year i has two components: a linear increase over the years with slope $\alpha_{k,j}$ and additional nonlinear component $h_{k,j,i}$. Thus Eq. 3.10 becomes:

$$\phi_{j,i} = \sum_{k=1}^{K_j} \Delta_{k,j,i} = \sum_{k=1}^{K_j} [\Delta_{k,j,i_{ks}} + \alpha_{k,j}(i - i_{ks}) + q_{k,j,i}], \tag{3.14}$$

where $q_{k,j,i} = h_{k,j,i} + g_{k,j,i}$. Then, from the point of view of the organization, the research production has a component depending on the quality of researchers at their entry to the research organization $\sum\limits_{k=1}^{K_j} \Delta_{k,j,i_{ks}}$. Another component depending on the conditions for work in the organization $\sum_{k=1}^{K_j} \alpha_{k,j}(i - i_{k_s})$; and finally a component depending on other factors which can affect the research production of the researchers $- \sum_{k=1}^{K_j} q_{k,j,i}$. The research production in some interval of years for such an organization is:

$$T_j = \sum_{i=i_1}^{i_2} \sum_{k=1}^{K_j} \Delta_{k,j,i} = \sum_{i=i_1}^{i_2} \sum_{k=1}^{K_j} [\Delta_{k,j,i_{k_s}} + \alpha_{k,j}(i - i_{k_s}) + q_{k,j,i}] \qquad (3.15)$$

If we consider a system of research organizations then the total number of papers for the year i for the set of all research organizations considered is:

$$\Phi_i = \sum_{j=1}^{J} \sum_{k=1}^{K_j} \Delta_{k,j,i} = \sum_{j=1}^{J} \sum_{k=1}^{K_j} [\Delta_{k,j,i_{k_s}} + \alpha_{k,j}(i - i_{k_s}) + q_{k,j,i}] \qquad (3.16)$$

We can define also probability distributions of the research production. For example, the probability distribution of the papers produced by a system of research organizations is:

$$Z_i = \frac{\Phi_i}{\sum_{i=i_1}^{i_2} \Phi_i} = \frac{\sum_{j=1}^{J} \sum_{k=1}^{K_j} \Delta_{k,j,i}}{\sum_{i=i_1}^{i_2} \sum_{j=1}^{J} \sum_{k=1}^{K_j} \Delta_{k,j,i}}; \quad i_1 \le i \le i_2 \qquad (3.17)$$

3.4 Several Simple Scenarios

We analyze the research production of an organization on the basis of Eq. 3.14. For simplicity first we assume that each of the $k = 1, \ldots, K_j$ researchers of the organization begins his/her research career at the year i_{k_s} and ends the career in the year i_{k_e}. We shall consider several possible scenarios of increasing complexity.

3.4.1 Scenario 1

All researchers begin their career in the same year i_{k_s}, stay in the same organization and end their career in the year i_{k_e}; there is no change in the research production of them through the years - $\alpha_{k,j} = 0$ and there are no additional factors which affect the research production - $q_{k,j,i} = 0$.

   In this case from Eq. 3.15 we obtain:

$$T_j = \sum_{i=i_{ks}}^{i_{ke}} \sum_{k=1}^{K_j} \Delta_{k,j,i_{ks}} = K_j (i_{ke} - i_{ks}) \overline{\Delta}_{j,i_{ks}} \tag{3.18}$$

where $\overline{\Delta}_{j,i_{ks}} = \frac{1}{K_j} \sum_{k=1}^{K_j} \Delta_{k,j,i_{ks}}$ is the average production of the researchers in the organization for the first year of their research career. Thus, if: (i) we consider two organizations of the same size in terms of the number of researchers; (ii) for the same period; and in addition (iii) the influence of the environment of the research production of the researchers is the same, then the organization which manages to recruit more productive researchers will be the one with the larger research production.

3.4.2 Scenario 2

All researchers begin their career in the same year i_{ks}, stay at the same organization and end their career in the year i_{ke}; there are changes in the research production for each/all of them through the years - $\alpha_{k,j} \neq 0$ and there are no additional factors which affect the research production - $q_{k,j,i} = 0$.
In this case from Eq. 3.15 we obtain:

$$T_j = \sum_{i=i_{ks}}^{i_{ke}} \sum_{k=1}^{K_j} [\Delta_{k,j,i_{ks}} + \alpha_{k,j}(i - i_{ks})]$$

$$= K_j (i_{ke} - i_{ks}) \left[\overline{\Delta}_{j,i_{ks}} + \frac{1}{2} \overline{\alpha}_j (i_{ke} - i_{ks} + 1) \right] \tag{3.19}$$

where $\overline{\alpha}_j = \frac{1}{K_j} \sum_{k=1}^{K_j} \alpha_{kj}$ is the average (over the researchers of the organization) slope of the linear increase of the deviation of the production from the initial production of the researchers of the organization.

Eq. 3.19 shows the presence of another factor which can influence the production of research organization: the conditions for work in the organization which result in different values of $\overline{\alpha}_j$ for different organizations j. Now if we have two organizations: $j = 1$ and $j = 2$ of the same size and consider them in the same time period, the ratio of research productions of these organizations is:

$$\frac{T_1}{T_2} = \frac{\overline{\Delta}_{1,i_{ks}} + \frac{1}{2}\overline{\alpha}_1 (i_{ke} - i_{ks} + 1)}{\overline{\Delta}_{2,i_{ks}} + \frac{1}{2}\overline{\alpha}_2 (i_{ke} - i_{ks} + 1)}. \tag{3.20}$$

If the recruitment in the two organizations has the same result - $\overline{\Delta}_{1,i_{ks}} = \overline{\Delta}_{2,i_{ks}}$ then the organization which has better work conditions (larger $\overline{\alpha}$) will have a larger research production. Another two conclusions are that the better work conditions can compensate for not so good recruitment and organizations with good recruitment can compensate for comparatively not so good work conditions. The best variant is however to make good recruitment and to offer good working conditions for the researchers.

3.4.3 Scenario 3

All researchers begin their career at the same year i_{k_s}, stay at the same organization and end their career at the year i_{k_e}; there are changes in the research production for each of them through the years - $\alpha_{k,j} \neq 0$; there are additional factors which affect the research production - $q_{k,j,i} \neq 0$, and these factors do not change their influence through the years.

The factors accounted by $q_{k,j,i}$ are external (unrelated) to the organization which can influence the research production, e.g., the conditions for life in corresponding town or the country. In this case we can introduce the average influence of the additional factors as $\overline{q}_{ji} = \frac{1}{K_j} \sum\limits_{k=1}^{K_j} q_{k,j,i}$ and thus we obtain:

$$
\begin{aligned}
T_j &= \sum_{i=i_{k_s}}^{i_{k_e}} \sum_{k=1}^{K_j} [\Delta_{k,j,i_{k_s}} + \alpha_{k,j}(i - i_{k_s}) + q_{k,j,i}] \\
&= K_j(i_{k_e} - i_{k_s}) \left[\overline{\Delta}_{j,i_{k_s}} + \frac{1}{2}\overline{\alpha}_j(i_{k_e} - i_{k_s} + 1) + \overline{q}_j \right]
\end{aligned}
\tag{3.21}
$$

The presence of the external factors introduces additional parameter which can influence the research production - $\overline{q}_j$. Thus, we can expect that if we have two organizations of the same size, and consider them for the same interval of time and if the results from recruitment and the conditions for the work are the same then the organization which provides better additional conditions for researchers will have a larger research production (e.g., is positioned in more developed country). Thus, the best recruitment, better work conditions and the good conditions for life lead to an increase in research production of the research organization.

3.5 More Complicated Scenario of Dynamics of Research Production of a Research Organization

3.5.1 General Theoretical Remarks

Let us now consider a scenario connected to result of recruitment efforts which change in the course of the years. In this case the averages with respect to the number of researchers of the organization will depend on time. We start from Eq. 3.10 but we are going to consider the more complicated situation when the researchers start and end their career in this organization in different years. This means that the parameters i_{k_s} and i_{k_e} will be different for the different researchers. The number of researchers in the j-th organization changes in the course of the time and because of this K depends not only on j but also on i: $K_{j,i}$. The number of papers produced by the researchers of the j-th organization in the year i is:

$$\phi_{j,i} = \sum_{k=1}^{K_{j,i}} \Delta_{k,j,i} \qquad (3.22)$$

To mark the entry and the departure of the k-th person we shall use the Heaviside theta-functions: $\Theta_{i,i_{k_s}}^{(k,j)}$ for entry of the organization and $1 - \Theta_{i,i_{k_e}+1}^{(k,j)}$ for the leave of the organization. The upper indices of the Θ-function mean that this function applies to the k-th person from the j-th research organization. Then we can write for the k-th researcher:

$$\Delta_{k,j,i} = \Theta_{i,i_{k_s}}^{(k,j)}(1 - \Theta_{i,i_{k_e}+1}^{(k,j)})[\Delta_{k,j,i_{k_s}} + \overline{f}_{k,j,i} + g_{k,j,i}]; \quad i_{k_s} \le i \le i_{k_e} \qquad (3.23)$$

We use again the relationships Eqs. 3.12 and 3.14 and obtain:

$$\Delta_{k,j,i} = \Theta_{i,i_{k_s}}^{(k,j)}(1 - \Theta_{i,i_{k_e}+1}^{(k,j)})[\Delta_{k,j,i_{k_s}} + \alpha_{k,j}(i - i_{k_s}) + q_{k,j,i}] \qquad (3.24)$$

Then the research productions of the researchers from the organization j in the year i is:

$$\phi_{j,i} = \sum_{k=1}^{K_{j,i}} \Theta_{i,i_{k_s}}^{(k,j)}(1 - \Theta_{i,i_{k_e}+1}^{(k,j)})[\Delta_{k,j,i_{k_s}} + \alpha_{k,j}(i - i_{k_s}) + q_{k,j,i}] \qquad (3.25)$$

We shall consider three classes of researchers and according to this we separate this research production into three parts. The classes of researchers are as follows

1. Researchers who start work in the research organization at the year i.
2. Researchers who leave the research organization at the year $i - 1$.
3. All other researchers from the research organization.

We consider the year i. There are researchers who are starting to work in the research organization in this year, and we assume for simplicity that this number is not equal to 0. As in the previous year the number of researchers in the organization was $K_{j,i-1}$, the new researchers are the researchers with numbers $K_{j,i-1} + 1, \ldots, K_{j,i}$. The number of papers of these new researchers is:

$$\phi_{j,i}^g = \sum_{k=K_{j,i-1}+1}^{K_{j,i}} \Delta_{k,j,i}. \qquad (3.26)$$

Superscript g means gain (gain of papers due to recruitment).

Second class is the class of researchers who leave the research organization in the year $i - 1$. For these researchers the year $i - 1$ is the year i_{k_e} and the corresponding research production is not anymore available for the research organization in the year $i_{k_e} + 1$. The number of papers of these researchers in the i-th year is:

$$\phi^l_{j,i} = \sum_{k=1}^{K_{j,i}} \Theta^{(k,j)}_{i,i_{k_s}} (1 - \Theta^{(k,j)}_{i,i_{k_e}+1}) \delta_{i-1,i_{k,j}} [\Delta_{k,j,i_{k_s}} + \alpha_{k,j}(i - i_{k_s}) + q_{k,j,i}], \quad (3.27)$$

where $\delta_{i-1,i_{k,j}}$ is the Kronecker delta-symbol for the k-th researcher of the j-th organization. $i_{k,j}$ is the year in which the k-th researcher leaves the j-th organization. The superscript l above means loss (loss of research production because of leaving of researchers for different reasons). Note that the sum here is up to $K_{j,i}$ as some newcomers can leave the organization in the same year.

The researchers who are not in the classes of newcomers or who do not leave, continue to produce papers for the research organization. This production can increase or decrease. The production of these researchers is $\phi^r_{j,i}$ (superscript r means to remain in the organization)

$$\phi^r_{j,i} = \sum_{k=1}^{K_{j,i-1}} \Theta^{(k,j)}_{i,i_{k_s}} (1 - \Theta^{(k,j)}_{i,i_{k_e}+1}) \prod_{m=1}^{i-1} (1 - \delta_{m,i_{k,j}}) [\Delta_{k,j,i_{k_s}} + \alpha_{k,j}(i - i_{k_s}) + q_{k,j,i}]$$
$$(3.28)$$

The change in the production of these researchers for the i-th year in comparison to the previous $(i - 1)$-th year is $\phi^r_{j,i} - \phi^r_{j,i-1}$.

The change $C_{j,i}$ in the research production of the j-th research organization in the i-th year because of leaving and entrance of new researchers as well as because of the increased experience of the research staff is:

$$C_{j,i} = \phi^g_{j,i} - \phi^l_{j,i} + \phi^r_{j,i} - \phi^r_{j,i-1} = \sum_{k=K_{j,i-1}+1}^{K_{j,i}} \Delta_{k,j,i} -$$

$$\sum_{k=1}^{K_{j,i}} \Theta^{(k,j)}_{i,i_{k_s}} (1 - \Theta^{(k,j)}_{i,i_{k_e}+1}) \delta_{i-1,i_{k,j}} [\Delta_{k,j,i_{k_s}} + \alpha_{k,j}(i - i_{k_s}) + q_{k,j,i}] +$$

$$\sum_{k=1}^{K_{j,i-1}} \Theta^{(k,j)}_{i,i_{k_s}} (1 - \Theta^{(k,j)}_{i,i_{k_e}+1}) \prod_{m=1}^{i-1} (1 - \delta_{m,i_{k,j}}) [\Delta_{k,j,i_{k_s}} + \alpha_{k,j}(i - i_{k_s}) +$$

$$q_{k,j,i}] - \sum_{k=1}^{K_{j,i-2}} \Theta^{(k,j)}_{i-1,i_{k_s}} (1 - \Theta^{(k,j)}_{i-1,i_{k_e}+1}) \prod_{m=1}^{i-1} (1 - \delta_{m-1,i_{k,j}}) [\Delta_{k,j,i_{k_s}} +$$

$$\alpha_{k,j}(i - 1 - i_{k_s}) + q_{k,j,i-1}]$$
$$(3.29)$$

If the management of the research organization wants to increase the research production with respect to the previous year, then $C_{j,i}$ has to be positive. This can be reached by successful recruitment (first term in Eq. 3.29) and by increased research production of the available researchers who will not leave (difference between the third and the fourth term there). Finally the leaving of researchers contributes to decrease of research production.

On the basis of Eq. 3.29 we are going to discuss several possible reasons for a large decrease of research production in some research organizations. First, we consider the first term in Eq. 3.29. This term is connected to the recruitment and its contribution to the research production can be increased in two ways. The first way is the quantitative one: the management has the possibility and resources to recruit more researchers (there is additional money for additional research staff). The second way is the qualitative: selection of good researchers for the research organization. The combined effect of these two cases can be large. But the combination of these two ways is possible in the most cases when the times are good from the point of view of money for research. If the times are not good, then the things can worsen. The number of recruited new researchers can be very small and even zero and the quality of these researchers can be low due to lack of motivation (because of low salaries for an example).

The second factor for the increase of the research production is the increase of research production in the staff who is available and is not leaving. The analysis of the corresponding terms in Eq. 3.29 shows that this is possible if the recruitment in previous years was good (sufficient new researchers arrived in the organization and their quality with respect to the quantity of produced papers is good). In addition, the research organization has to keep good working conditions for the researchers and the external influences on the work of researchers must be positive. However, all this can happen together in good times from the point of view of: (i) economic development of corresponding country, and (ii) the finances for the research work. If the times are not good, then the good recruitment from previous periods may have still positive influence, but the worsening of the conditions for work and the negative external influences can lead to decrease of research production of the available staff and this decrease can continue through the years.

Finally the leaving of researchers due to different reasons (starting another job, retirement, death, etc.) has a negative influence on the increase of research production. Usually the leave of personnel is balanced by recruitment and increase of production of the remaining staff. There are situations, however in which a large drop of research production is possible. One such situation will be discussed below.

3.5.2 An Example for Application of the General Theory

Numerous scenarios can be studied by means of the general theory presented in the previous sections of the present chapter. Below we are going to discuss in more detail just one of the possible scenarios. The brief description of this scenario is as follows. We consider a country where economic growth happens at the beginning of the time interval of interest for our scenario. As one of the consequences of the economic growth a new research organization is created and large amount of researchers are hired in the first 5 years of the time interval. The economic growth continues for 5 more years and in this time the hiring of new researchers is still large. From the year 11 on the economic situation in the country worsens and the number of recruited researchers per year drops almost two times for the next 30 years. The quality of the hired staff is different in the first 10 years and in the following 30 years. We assume that the staff hired in the first 10 years is more productive than the staff hired in the following 30 years and this is a consequence of the relative non-attractiveness of the profession of researchers in the times of economic difficulties. For simplicity, we will assume that the lower quality of the newcomers is the only effect of the worsened economic and social conditions and there are no other effects on the research production. We shall also assume for simplicity too, that the external environment does not influence the research production of the researchers from the research organization. Thus $q_{k,j,i} = 0$ and Eq. 3.29 becomes:

$$C_{j,i} = \phi_{j,i}^g - \phi_{j,i}^l + \phi_{j,i}^r - \phi_{j,i-1}^r = \sum_{k=K_{j,i-1}+1}^{K_{j,i}} \Delta_{k,j,i} -$$

$$\sum_{k=1}^{K_{j,i}} \Theta_{i,i_{k_s}}^{(k,j)} (1 - \Theta_{i,i_{k_e}+1}^{(k,j)}) \delta_{i-1,i_{k,j}} [\Delta_{k,j,i_{k_s}} + \alpha_{k,j}(i - i_{k_s})] +$$

$$\sum_{k=1}^{K_{j,i-1}} \Theta_{i,i_{k_s}}^{(k,j)} (1 - \Theta_{i,i_{k_e}+1}^{(k,j)}) \prod_{m=1}^{i-1} (1 - \delta_{m,i_{k,j}}) [\Delta_{k,j,i_{k_s}} + \alpha_{k,j}(i - i_{k_s})] -$$

$$\sum_{k=1}^{K_{j,i-2}} \Theta_{i-1,i_{k_s}}^{(k,j)} (1 - \Theta_{i-1,i_{k_e}+1}^{(k,j)}) \prod_{m=1}^{i-1} (1 - \delta_{m-1,i_{k,j}}) [\Delta_{k,j,i_{k_s}} + \alpha_{k,j}(i - 1 - i_{k_s})]$$

(3.30)

For simplicity, we shall consider the case where the only reason to leave research organization is retirement and the retirement age is the same for all researchers. In addition, again for simplicity we shall assume that the age of entry into the organization is the same for all researchers, in other words, they spend the same number of years, working in the research organization. We assume also that the first year (the entry year) research production is the same for all researchers who enter the organization in the corresponding year. We consider the length of the research career to be 40 years, and we consider two levels of entry research production per year: Δ_0 for the first 10 more productive years and $\Delta_1 < \Delta_0$ for the less productive 30 years that follow. In addition, we assume the following dynamics of the recruitment: 20 researchers on each of the first 3 years, then 5 researchers from the years from 4-th till 10-th inclusive and then 3 researchers per year for the next not so good decades. Let us consider the effects of the terms in Eq. 3.30. The first term is:

$$C_{j,i}^{(1)} = \sum_{k=K_{j,i-1}+1}^{K_{j,i}} \Delta_{k,j,i} \tag{3.31}$$

The values of the term $C_{j,i}^{(1)}$ are as follows. In the first year 20 researchers are recruited and the value of the term is $20\Delta_0$ and these researchers have to work up to the year 41 when they retire. In the second year of the organization additional 20 researchers are recruited, the value of the term is again $20\Delta_0$ and these researchers will work up to the year 42. In the third year, additional 20 researchers are recruited, the value of the term is again $20\Delta_0$ and these researchers will work up to the year 43 in which they will retire. From the years 4 to 10 5 researchers per year are recruited and the value of the term drops to $5\Delta_0$ and the recruited researchers will be active till the years 44,45,...,51 respectively. Then the situation worsens, and 3 researchers are recruited per year. The value of the term drops further to $3\Delta_1$ and remains at this level in the next decades.

Next we consider the researchers who leave the organization. The corresponding term in Eq. 3.30 is:

$$C_{j,i}^{(2)} = \sum_{k=1}^{K_{j,i}} \Theta_{i,i_{k_s}}^{(k,j)} (1 - \Theta_{i,i_{k_e}+1}^{(k,j)}) \delta_{i-1,i_{k,j}} [\Delta_{k,j,i_{k_s}} + \alpha_{k,j}(i - i_{k_s})] \tag{3.32}$$

Since we only consider researchers leaving due to retirement, this term is 0 up to the year 40. The first losses can be felt at the year 41 when the researchers who retired in the year 40 will not contribute anymore to the research production of the organization. Let us assume for simplicity that the coefficient $\alpha_{k,j}$ is the same for the all researchers in the organization and has constant values $\alpha_{k,j} = \alpha_0$ through the first 10 years and then $\alpha_{k,j} = \alpha_1$ for the remaining years. In addition, we assume $\alpha_1 < \alpha_0$. We note that by definition the only exception is the year i_{k_s} for which $\alpha_{k,j} = 0$. Thus, the researchers who retired at the year 40 have 39 years of increase of research production and at the year 40 their research production is $\Delta_0 + 39\alpha$. This is the loss of production for the organization from the fact that in the year 41 we have one researcher retired. But such researchers are 20. Then, the total loss for the organization is $20\Delta_0 + 780\alpha$. The same is the loss for the research organization at years 42 and 43 and in the year 44 we have 5 retirements and each of them is connected to loss of production equal to $\Delta_0 + 39\alpha$. The total loss of production for the year 44 is $5\Delta_0 + 195\alpha$. This is the loss of production for the year 45, 46 up to the year 50. The loss of production in the year 51 is due to the retired individuals in the year 50. These are the 3 individuals who started their careers at the year 11 and at the year 50 their research production is $\Delta_1 + 39\alpha_1$ per person. The total loss of production for the year 51 then is $3\Delta_1 + 117\alpha_1$ and this continues year by year.

From the above, we see that due to the recruitment strategy, there will be 3 critical years of the research organization and these are the years 41, 42, and 43 where the loss of production is large because of the retirement of the large number of researchers who have been hired at the years 1,2, and 3.

The loss of research production because of the retirements can be compensated by the increase of the research production of the staff who is not retired. This is described by the last two terms in Eq. 3.30. The situation with this increase is as follows. In the year 1 there is no increase of this kind. In the year 2 the increase is $20\alpha_0$ because of the staff hired previous year. In the year 3 the increase is $40\alpha_0$ and in the year 4 the increase is $60\alpha_0$. At the year 5 the increase is $65\alpha_0$ because of the new staff hired in the year 4. This increase by $5\alpha_0$ continues to the year 11 inclusive where the increase is $95\alpha_0$. The increase continues but becomes slower. In the year 12 the increase is $95\alpha_0 + 3\alpha_1$, at the year 13 the increase is $95\alpha_0 + 6\alpha_1$, etc., up to the year 40 when the increase is $95\Delta_0 + 87\alpha_1$. From the year 41 on the retirements begin and the increase of the production of the staff drops. It is $75\alpha_0 + 90\alpha_1$ for the year 41, $55\alpha_0 + 93\alpha_1$ for the year 42, $35\alpha_0 + 96\alpha_1$ for the year 43, $30\alpha_0 + 99\alpha_1$ for the year 44, and then the ease continues up to $10\alpha_0 + 111\Delta_1$ for the year 48 and $5\alpha_0 + 114\alpha_1$ for the year 49 and $117\alpha_1$ for the year 50 and for the next years.

On the basis of all above, we can distinguish three time periods of behavior of the research production of our example organization: period of increase and reaching a maximum of research production; period of decrease and reaching the equilibrium state; and period of stable production at this equilibrium state. In more detail the dynamics of the changes in the research production of the organization are shown in Table 3.1.

We observe the following. After the beginning of the research in the research organization and the first hiring of research staff there is an increase of the research production as the number of researchers in the organization increases. For example, in the year 6 of functioning of the research organization, it has 75 researchers and the research production is $\phi_6 = 75\Delta_0 + 255\alpha_0$. In the year 10 the research organization has 95 researchers and the research production increased to $\phi_{10} = 95\Delta_0 + 585\alpha_0$ papers per year. The research production continues to grow because of the hiring of new staff and because of the increasing experience of the available research staff. For example, in the year 12 the number of researchers in the research organization is 101 and the number of papers produced in this year is $\phi_{12} = 95\Delta_0 + 6\Delta_1 + 775\alpha_0 + 3\alpha_1$. The growth continues further and in the year 39 of functioning of the research organization the number of researchers is 182 and the research production in this year is $\phi_{39} = 95\Delta_0 + 87\Delta_1 + 3340\alpha_0 + 1218\alpha_1$. The maximum of the research production of the examined organization occurs in the year 40 when the number of researchers is 185 and the research production is $\phi_{40} = 95\Delta_0 + 90\Delta_1 + 3435\alpha_0 + 1305\alpha_1$. From the year 41 one begins to feel the consequences of the chosen recruitment strategy. The researchers who have been hired in the first year of functioning of the research organization, retire and this leads to a large drop of the research production. The organization has 168 researchers and the research production drops to $\phi_{41} = 75\Delta_0 + 93\Delta_1 + 2730\alpha_0 + 1395\alpha_1$. The large drop because of the retirement continues for the next two years. In the year 42 the number of researchers in the research organization is 151 and the production drops to $\phi_{42} = 55\Delta_0 + 96\Delta_1 + 2005\alpha_0 + 1488\alpha_1$. In the year 43 the number of researchers is 134 and the research production drops further to $\phi_{43} = 35\Delta_0 + 99\Delta_1 + 1260\alpha_0 + 1584\alpha_1$. The drop continues for the next 7 years. In the years 44 the researchers in the organization are 132 and the research

Table 3.1 Yearly change in research production of a research organization by the research production of newly hired research staff (newcomers), retired staff, and remaining staff. The last column contains the total change in the research production of the research organization for the corresponding year.

year	Change of the amount of the research production because of the newcomers	Change of the amount of the research production because of the retired	Change of the amount of the research production because of the remaining staff	Total change of the amount of the research production for the corresponding year
1	$20\Delta_0$	0	0	$20\Delta_0$
2	$20\Delta_0$	0	$20\alpha_0$	$20\Delta_0 + 20\alpha_0$
3	$20\Delta_0$	0	$40\alpha_0$	$20\Delta_0 + 40\alpha_0$
4	$5\Delta_0$	0	$60\alpha_0$	$5\Delta_0 + 60\alpha_0$
5	$5\Delta_0$	0	$65\alpha_0$	$5\Delta_0 + 65\alpha_0$
6	$5\Delta_0$	0	$70\alpha_0$	$5\Delta_0 + 70\alpha_0$
...	...	...	...	...
10	$5\Delta_0$	0	$90\alpha_0$	$5\Delta_0 + 90\alpha_0$
11	$3\Delta_1$	0	$95\alpha_0$	$3\Delta_1 + 95\alpha_0$
12	$3\Delta_1$	0	$95\alpha_0 + 3\alpha_1$	$3\Delta_1 + 95\alpha_0 + 3\alpha_1$
13	$3\Delta_1$	0	$95\alpha_0 + 6\alpha_1$	$3\Delta_1 + 95\alpha_0 + 6\alpha_1$
...	...	...	...	...
39	$3\Delta_1$	0	$95\alpha_0 + 84\alpha_1$	$3\Delta_1 + 95\alpha_0 + 84\alpha_1$
40	$3\Delta_1$	0	$95\alpha_0 + 87\alpha_1$	$3\Delta_1 + 95\alpha_0 + 87\alpha_1$
41	$3\Delta_1$	$-20\Delta_0 - 780\alpha_0$	$75\alpha_0 + 90\alpha_1$	$-20\Delta_0 + 3\Delta_1 - 705\alpha_0 + 90\alpha_1$
42	$3\Delta_1$	$-20\Delta_0 - 780\alpha_0$	$55\alpha_0 + 93\alpha_1$	$-20\Delta_0 + 3\Delta_1 - 725\alpha_0 + 93\alpha_1$
43	$3\Delta_1$	$-20\Delta_0 - 780\alpha_0$	$35\alpha_0 + 96\alpha_1$	$-20\Delta_0 + 3\Delta_1 - 745\alpha_0 + 96\alpha_1$
44	$3\Delta_1$	$-5\Delta_0 - 195\alpha_0$	$30\alpha_0 + 99\alpha_1$	$-5\Delta_0 + 3\Delta_1 - 165\alpha_0 + 99\alpha_1$
45	$3\Delta_1$	$-5\Delta_0 - 195\alpha_0$	$25\alpha_0 + 102\alpha_1$	$-5\Delta_0 + 3\Delta_1 - 170\alpha_0 + 102\alpha_1$
...	...	...	...	...
49	$3\Delta_1$	$-5\Delta_0 - 195\alpha_0$	$5\alpha_0 + 114\alpha_1$	$-5\Delta_0 + 3\Delta_1 - 190\alpha_0 + 114\alpha_1$
50	$3\Delta_1$	$-5\Delta_0 - 195\alpha_0$	$117\alpha_1$	$-5\Delta_0 + 3\Delta_1 - 195\alpha_0 + 117\alpha_1$
51	$3\Delta_1$	$-3\Delta_1 - 117\alpha_1$	$117\alpha_1$	0
52	$3\Delta_1$	$-3\Delta_1 - 117\alpha_1$	$117\alpha_1$	0
...	...	...	...	...
89	$3\Delta_1$	$-3\Delta_1 - 117\alpha_1$	$117\alpha_1$	0
90	$3\Delta_1$	$-3\Delta_1 - 117\alpha_1$	$117\alpha_1$	0
91	$3\Delta_1$	$-3\Delta_1 - 117\alpha_1$	$117\alpha_1$	0
...	...	...	...	...

production drops further to $\phi_{44} = 30\Delta_0 + 102\Delta_1 + 1095\alpha_0 + 1683\alpha_1$. In the year 49 there are 122 researchers in the organization and the research production is $\phi_{49} = 5\Delta_0 + 117\Delta_1 + 195\alpha_0 + 2219\alpha_1$. The last year of the drop of the production is the year 50 when the number of researchers in the organization becomes 120 and the research production is $\phi_{50} = 129\Delta_1 + 2336\alpha_1$. This level of publication activity remains constant in the following years and the research staff of the research organization remains also constant at level of 120 researchers.

Let us now examine the previous example by assigning values to the parameters. We assume that the newcomers to the organization produce in the first 10 years on average 1 paper per year, i.e., $\Delta_0 = 1$. The parameter has a lower value for the recruited researchers in the following years, and we assume that for these researchers $\Delta_1 = 0.8$. Let us assume that $\alpha_0 = 0.1$ which means that at the end of their research career the researchers recruited in the first 10 years produce 5 papers per year on average. For the other researchers $\alpha_1 = 0.08$ which means that at the end of their research career these researchers produce 4 papers per year. In this case in the year 6 of functioning of the organization the production is $\phi_6 = 100.5$ for 75 researchers, which means 1.34 papers per researcher. At the year 10 the research production is $\phi_{10} = 153.5$ for 95 researchers which means about 1.615 papers per researcher. From the year 11 the quality of the recruited researchers drops as the crisis times decreases the interest of young people to the profession of researcher. During this year the production of research organization is $\phi_{11} = 165.4$ and the number of researchers is 98. The production per researcher is 1.687. As we see the production per researcher continues to grow and this is because of the increased experience of the researchers who work already in the research organization. The growth continues till the year 40 when the research production of the organization is $\phi_{40} = 614.9$ and the number of researchers is 182. This means that the research production per researcher is about 3.378 papers per year. The retirements at the year 41 lead to drop in the total production and in the production per researcher. In this year the research production is $\phi_{41} = 534$ and the number of researchers drops to 168. This corresponds to research production of about 3.178 papers per researcher. The drop in the research production continues in the year 42 in which the number of researchers decreases further to 151. The research production is about $\phi_{42} = 451.4$ which means about 2.989 papers per researcher. The last year of the large drop in the research production was year 43 when the number of researchers is 134 and the research production is about $\phi_{43} = 367$ papers which means about 2.738 papers per researcher on average. The drop continues in the following 7 years. For example, in the year 44 the number of researchers in the research organization is 132 and the research production is about $\phi_{44} = 355.8$ papers or 2.695 papers per researcher. The drop continues till the year 50 when there are 120 researchers in the research organization and the research production is about $\phi_{50} = 282.9$ papers or about 2.357 papers per researcher. These values of the research production and for research production per researcher remain constant in the following years.

In summary, if we compare the best year for the research organization - the year 40 and the year 50 we observe a drop by about 34 % of the research staff, drop by about 54% of the total research production, and a drop by about 30% of the research production per researcher. Thus, certain kinds of recruitment strategies combined with unfavorable conditions can lead to a drop of the number of researchers in a research organization and this can influence the capability of the organization to work alone on projects which complexity is higher than a certain level (there are not enough researchers). In addition, this can lead to drop in research production and this drop can be with respect to total number of publications and with respect to number of publications per researcher.

3.6 Remarks on Research Production of a System of Research Organizations

The theory for one research organization can be easily extended to the case of system of J research organizations. The number of papers produced by researchers in a system of J research organizations for the year i is:

$$\phi_i = \sum_{j=1}^{J} \sum_{k=1}^{K_{j,i}} \Delta_{k,j,i} \tag{3.33}$$

which can be further written as:

$$\phi_i = \sum_{j=1}^{J} \sum_{k=1}^{K_{j,i}} \Theta_{i,i_{k_s}}^{(k,j)} (1 - \Theta_{i,i_{k_e}+1}^{(k,j)}) [\Delta_{k,j,i_{k_s}} + \alpha_{k,j}(i - i_{k_s}) + q_{k,j,i}] \tag{3.34}$$

We will separate this research production into three parts again. We consider the year i. We have researchers who enter the research system in this year. As previous year the number of researchers in each organization of the research system was $K_{j,i-1}$ the new researchers are the researchers with numbers $K_{j,i-1} + 1, \ldots, K_{j,i}$. The number of papers of these new researchers is:

$$\phi_i^g = \sum_{j=1}^{J} \sum_{k=K_{j,i-1}+1}^{K_{j,i}} \Delta_{k,j,i} \tag{3.35}$$

Superscript g means gain (gain of papers due to recruitment or researchers for the research system).

Second class is the class of researchers who left the system of research organizations in the year $i - 1$. For these researchers year $i - 1$ is the year i_{k_e} and the corresponding research production is not anymore available for the research organization in the year $i_{k_e} + 1$. The number of papers of these researchers in the i_k-th year is:

$$\phi_{j,i}^l = \sum_{j=1}^{J} \sum_{k=1}^{K_{j,i}} \Theta_{i,i_{k_s}}^{(k,j)} (1 - \Theta_{i,i_{k_e}+1}^{(k,j)}) \delta_{i-1,i_{k,j}} [\Delta_{k,j,i_{k_s}} + \alpha_{k,j}(i - i_{k_s}) + q_{k,j,i}] \tag{3.36}$$

where $\delta_{i-1,i_{k,j}}^{(k,j)}$ is the Kronecker delta-symbol for the k-th researcher of the j-th organization. The superscript l above means loss (loss of research production because of leaving of researchers for different reasons). Note that the sum here is up to $K_{j,i}$ as some of the newcomers can leave the organization in the same year.

The researchers who are not in the classes of newcomers or who do not leave, continue to produce papers for the system of research organizations. This production can increase or decrease. The production of these researchers is ϕ_i^r (superscript r means to remain in the organization)

$$\phi_i^r = \sum_{j=1}^{J} \sum_{k=1}^{K_{j,i-1}} \Theta_{i,i_{ks}}^{(k,j)}(1 - \Theta_{i,i_{ke}+1}^{(k,j)}) \prod_{m=1}^{i-1}(1 - \delta_{i-1,i_{k,j}})[\Delta_{k,j,i_{ks}} + \alpha_{k,j}(i - i_{ks}) + q_{k,j,i}]$$

(3.37)

The change in the production of these researchers for the i-th year in comparison to the previous $(i-1)$-th year is $\phi_i^r - \phi_{i-1}^r$.

The change C_i in the research production of the system of research organizations in the i-th year because of leaving and entrance of new researchers as well as because of the increased experience of the research staff is:

$$C_i = \phi_i^g - \phi_i^l + \phi_i^r - \phi_{i-1}^r = \sum_{j=1}^{J} \left\{ \sum_{k=K_{j,i-1}+1}^{K_{j,i}} \Delta_{k,j,i} - \right.$$

$$\sum_{k=1}^{K_{j,i}} \Theta_{i,i_{ks}}^{(k,j)}(1 - \Theta_{i,i_{ke}+1}^{(k,j)})\delta_{i-1,i_{k,j}}[\Delta_{k,j,i_{ks}} + \alpha_{k,j}(i - i_{ks})] +$$

$$\sum_{k=1}^{K_{j,i-1}} \Theta_{i,i_{ks}}^{(k,j)}(1 - \Theta_{i,i_{ke}+1}^{(k,j)}) \prod_{m=1}^{i-1}(1 - \delta_{m,i_{k,j}})[\Delta_{k,j,i_{ks}} + \alpha_{k,j}(i - i_{ks})] -$$

$$\sum_{k=1}^{K_{j,i-2}} \Theta_{i-1,i_{ks}}^{(k,j)}(1 - \Theta_{i-1,i_{ke}+1}^{(k,j)}) \prod_{m=1}^{i-1}(1 - \delta_{m-1,i_{k,j}})[\Delta_{k,j,i_{ks}} + \alpha_{k,j}(i -$$

$$\left. 1 - i_{ks})] \right\}$$

(3.38)

In analogy with the case of 1 research organization on the basis of Eq. 3.38 we can study the dynamics of research production, and we can calculate the results for different recruitment strategies and for cohorts of researchers with different abilities with respect to the production of research papers.

3.7 Concluding Remarks

In this chapter we presented several relationships that model the research production of singe researcher, research organization, and system of research organizations. In the case of single researcher, we construct the corresponding quantities in such a way that we are able to quantify the quality of the newcomers in the research organization with respect to the quantity of produced research production. In addition, we consider a characteristic quantity about the conditions for research in the research organization and another quantity about the influence of external (for organization) factors on the number of papers produced by a researcher from the research organization. Similar quantities can be defined for the researchers from the point of view of the research organization and some of these quantities are slightly different from the quantities defined in the case of single researcher.

On the basis of the presented theory we discuss several simple scenarios to supply more information connected to the defined quantities for the research production of the researchers. The presented theory allows us to discuss more complicated scenarios and one example of such a scenario is given in which the consequences of a typical recruitment strategy are discussed and the consequences of the combination of this recruitment strategy and the worsening of the social environment for research are presented.

Finally, we note that the theory presented in this chapter can be used for analysis and evaluation of large number of scenarios for dynamics of research production of research organizations. This chapter is devoted to the quantity of research production and the presented theory can help for planning of increase of the quantity of this production. The theory can account also for the quality of the research production. Such analysis deserves separate attention and it will be discussed elsewhere.

References

1. Andres, A.: Measuring Academic Research. How to Undertake a Bibliometric Study. Chandos, Oxford (2009)
2. Boccara, N.: Modeling Complex Systems. Springer, New York (2010)
3. Braun, T., Dujdodo, E., Schubert, A.. Literature of Analytical Chemistry: A Scientometric Evaluation. CRC Press, Boca Raton, FL (1987)
4. Braun, T., Glänzel, W., Schubert, A.: Citation Analysis in Research Evaluation. World Scientific, London (1985)
5. Castellani, B., Hafferty, F.: Sociology and Complexity Science. Springer, Berlin (2009)
6. Chian, A.C.L.: Complex Systems Approach to Economic Dynamics. Springer, Berlin (2007)
7. Cronin, B.: The Citation Process. The Role and Significance of Citations in Scientific Communication. Taylor Graham, London (1984)
8. Cronin, B., Sugimoto, C.R.: Beyond Bibliometrics: Harnessing Multidimensional Indicators of Scholarly Impact. MIT Press, Cambridge, MA (2014)
9. Ding, Y., Rousseau, R., Wolfram, D. (eds.): Measuring Scholarly Impact. Springer, Cham (2014)
10. Egghe, L.: Power Laws in the Information Production Process: Lotkaian Informetrics. Elsevier, Amsterdam (2005)
11. Egghe, L., Rousseau, R.: Introduction to Informetrics: Quantitative Methods in Library, Documentation, and Information Science. Elsevier, Amsterdam (1980)
12. Fisher, K.: Changing Landscapes of Nuclear Physics: A Scientometric Study. Springer, Berlin (1993)
13. Haitun, S.D.: Scientometrics: State and Perspectives (in Russian). Nauka, Moscow (1983)
14. Ingwersen, P.: Scientometric Indicators and Webometrics and the Polyrepresentation Principle in Information Retrieval. ESS Publications, New Delhi Bangalore, India (2012)
15. Leydesdorff, L.: The Challenge of Scientometrics: The Development, Measurement, and Self-organization of Scientific Communications. DSWO Press, Leiden (1995)
16. Moed, H.: Citation Analysis in Research Evaluation. Springer, Netherlands (2005)
17. Moed, H.F., Glänzel, W., Schmoch, U. (eds.): Handbook of Quantitative Science and Technology Research. Springer, Netherlands (2005)
18. Scharnhorst, A., Börner, K., van den Besselaar, P. (eds.): Models for Science Dynamics. Springer, Berlin (2012)

19. Small, H.: Bibliometrics of Basic Research. National Technical Information Service (1990)
20. Thelwall, M.: Introduction to Webometrics: Quantitative Web Research for the Social Sciences. Morgan & Claypool, San Rafael, CA (2009)
21. Treiber, M., Kesting, A.: Traffic Flow Dynamics: Data, Models, and Simulation. Springer, Berlin (2013)
22. Vinkler, P.: The evaluation of Research by Scientometric Indicators. Chandos, Oxford (2010)
23. Vitanov, N.K.: Science Dynamics and Research Production. Indicators, Indexes, Statistical Laws and Mathematical Models. Springer, Cham (2015)
24. Vitanov, N.K., Ausloos, M.: Models of Science Dynamics, chap. Knowledge Epidemics and Population Dynamics Models for Describing Idea Diffusion, pp. 69–125. Springer, Berlin (2012)
25. White, W., Engelen, G., Ulijee, I.: Modeling Cities and Regions as Complex Systems. The MIT Press, Cambridge, MA (2015)
26. Wolfram, D.: Applied Informatics for Information Retrieval Research. Libraries Unlimited, Westport, CT (2003)
27. Yablonskii, A.I.: Mathematical Methods in the Study of Science (in Russian). Nauka, Moscow (1986)

Chapter 4
Does Publicity in the Science Press Drive Citations? A Vindication of Peer Review

Manolis Antonoyiannakis

Abstract We study how publicity, in the form of highlighting in the science press, affects the citations of research papers. After a brief review of prior work, we analyze papers published in *Physical Review Letters* (PRL) that are highlighted across eight different platforms. Using multiple linear regression we identify how each platform contributes to citations. We also analyze how frequently the highlighted papers end up in the top 1% cited papers in their field. We find that the strongest predictors of medium-term citation impact—up to 7 years post-publication—are Viewpoints in *Physics*, followed by Research Highlights in *Nature*, Editors' Suggestions in PRL, and Research Highlights in *Nature Physics*. Our key conclusions are that (a) highlighting for importance identifies a citation advantage, which (b) is stratified according to the degree of vetting during peer review (internal and external to the journal). This implies that we can view highlighting platforms as predictors of citation accrual, with varying degrees of strength that mirror each platform's vetting level.

4.1 Introduction

Publishing is a selection process. Through a series of selections, or filtering, journal editors decide which papers merit external review, which papers to send back to referees for a second (or third) look, and finally, which papers to publish.

But increasingly over the past 20 years, the selection of papers does not stop at publication, and editors curate lists of their favorite accepted papers for their readers. Nor is such filtering limited to a journal's own papers, as some publishers (notably the *Science* and *Nature* series journals) curate lists of papers published

Manolis Antonoyiannakis[1,2]

[1] Department of Applied Physics & Applied Mathematics, Columbia University, New York, NY 10027, USA

[2] American Physical Society, Editorial Office, 1 Research Road, Ridge, NY 11961-2701, USA
e-mail: ma2529@columbia.edu

among other journals in their field. So, why do editors produce these selections? In view of an exponential increase in scientific publications [21] and the wide accessibility of papers through online repositories, the role of journals and editors is being redefined. By curating lists of accepted or recently published papers, editors add value to their journals, mitigate diminishing attention spans and retain a larger portion of their readers' attention. Indeed, these select sets of papers, or *highlights*, are deemed to be of higher quality, interest, or importance than the typical paper in their source journals. Some publishers invest further editorial resources to increase visibility of these select papers and accompany them with short summaries by editors, or longer commentaries by experts. Depending on publisher, journal, and type, highlights are called Editors' Choice, Research Highlights, Editors' Suggestions, Perspectives, News & Views, Viewpoints, etc. See Table 4.1. The community seems to pay attention. Indeed, researchers often promote (in their websites, resumés, cover letters, and progress reports) their highlighted papers, while funding agencies track their grantees' progress by monitoring coverage in various highlighting platforms.

Highlighting is a form of publicity in the science press, aimed at an expert audience of peers, from research scientists to science journalists. It is a different kind of publicity than University press releases, which are promotions by interested parties, or coverage of research in popular media (newspapers, magazines), whose readership is different from that of scholarly journals.

Table 4.1 Highlighting platforms, by publisher, type, and write-up coverage

Publisher	Highlight name	Type (intra- or inter-)	Write-up type
AAAS[a]	Editors' Choice	inter-highlight	summary
AAAS	Perspective	both	expert commentary
NPG[b]	Research Highlight	inter-highlight	summary
NPG	News & Views	both	summary
APS[c]	Editors' Suggestion	intra-highlight	none
APS	Viewpoint	intra-highlight	expert commentary
APS	Synopsis	intra-highlight	summary
IOP[d]	IOPselect	intra-highlight	none
OSA[e]	Spotlight on Optics	intra-highlight	none
ACS[f]	ACS Editors' Choice	intra-highlight	none
NAS[g]	Journal club	intra-highlight	summary
NAS	Commentary	intra-highlight	expert commentary
JPS[h]	Editors' Choice	intra-highlight	none

[a] American Association for the Advancement of Science
[b] Nature Publishing Group (Springer Nature)
[c] American Physical Society
[d] Institute of Physics
[e] Optical Society of America
[f] American Chemical Society
[g] National Academy of Sciences
[h] Japanese Physical Society

Given the efforts publishers undergo to select and promote their favorite papers, one would expect that highlighted papers stand out compared to other papers in their source journals. In particular, when highlighting is done for importance (however problematic it is to define "importance") we would expect it to translate into higher citation counts. Whether the highlighting process identifies or *causes* increased citations is a pertinent question. Here, we find both effects are at play but have different magnitudes: Papers that are highlighted randomly pick up a few additional citations, but nowhere as many as those that are deliberately highlighted by editors for importance. In contrast, when highlighting is done for an intrinsically interesting, cute or elegant paper, then citations may not be an appropriate measure of distinction. Citation metrics have been found to agree well with peer review at the aggregate level. For example, citation-based rankings of British universities correlate with the Research Excellence Framework (REF) power ranking at 0.97 [13], while another analysis reported correlations above 0.8 for a number of fields, including physics [22]. Therefore, it makes sense to explore whether the additional "round" of review on highlighting (beyond mere acceptance of a paper) correlates with citation counts of those select papers.

In this chapter, we address these issues quantitatively. As a case study, we analyze citation data for several publicity markers on papers published in the journal *Physical Review Letters* (PRL) of the American Physical Society (APS). Building on our previous work [4, 5, 6], we analyze here a much broader set of 8 highlighting markers, both internal and external to the APS. We analyze the citation behavior of various markers of publicity, from journal cover to highlighting in the source journal to highlighting in non-APS journals such as *Nature* or *Nature Physics*. Using multiple linear regression, we explore whether publicity in the science press is associated with higher citations, so we can use it as an indicator of future citation impact. We also explore how frequently the highlighted papers end up in the highly-cited-papers list of the Essential Science Indicators in Clarivate Analytics [9], i.e., the top 1% cited papers in their field, and whether we can use any of the highlighting markers as predictors of placement in this list.

We make a distinction of whether the paper is highlighted by a platform that belongs to its own publisher (its source journal, a sibling journal, or a blog) or to another publisher. In the former case, we speak of an *intra-highlight*, in the latter of an *inter-highlight*. Such a distinction is significant because when the editors select one of their own papers (intra-highlight) they have access to more information from the paper's peer review (referee reports, discussions with the editorial board, caveats that arose in the review process, etc.) to assist their decision. In contrast, when the editors select a paper in another journal (inter-highlight), they generally have no detailed knowledge of the paper's peer review other than its conclusion, namely, that the paper was accepted in its source journal and, for a published paper, that it was highlighted there. (Inter-highlighting editors have even less information if they select a paper from the arXiv for highlighting prior to acceptance or publication in any journal.) Why do we bother to distinguish between intra- and inter-highlights? Because even among papers in the same journal there is typically considerable variation of quality, importance, or interest. These attributes come up usually in the review process, enriching the editorial decision making process and aiding the editors in selecting their journal's highlights—an advantage that is lost to inter-highlighting platforms.

4.2 Previous Work

As early as 1991, Phillips et al. [18] reported a citation advantage for research articles in the *New England Journal of Medicine* that received press coverage in the *New York Times*. The effect lasted in each of the 10 years since publication, and was strongest in the first year. They also sought to determine whether publicity (coverage by the *Times*) itself increased citations, or merely earmarked outstanding articles that would have been well-cited anyway (*publicity* vs. *earmark* hypothesis). Due to a 12-week strike in 1978, the *Times* continued to print a limited edition but did not sell copies to the public. Scientific articles covered by the *Times* during the strike period were not cited more than the control group, thus giving support to the publicity hypothesis over the earmark hypothesis.

Kurtz et al. [15] analyzed data from the NASA Astrophysics Data System and from the arXiv e-print archive, in order to test three possible explanations for the citation advantage of papers that are freely available on the web. They found no causal link between open-access itself and citations. They did find a strong *early-access* effect, whereby papers pick up citations because they are available sooner; and also a strong *self-selection* bias effect, whereby authors tend to promote their most important and, thus, the most citable articles, by posting them on the arXiv.

Dietrich [10, 11] analyzed another kind of publicity for research articles, namely, placement in the top slots of the daily astrophysics (astro-ph) listing for e-prints published on the arXiv:astro-ph server. He found that e-prints appearing at or near the top of the astro-ph mailings receive significantly more citations than those further down the list. He identifies two causes for this citation boost: the *visibility* effect as more people see the top few papers in the list, and the *self-promotion effect* by authors who tend to promote their most important works by carefully timing their submission to the server in order to land in the top slots. (The visibility effect is essentially the publicity hypothesis of Phillips et al., while the self-promotion effect is essentially the self-selection bias effect of Kurtz et al.) Dietrich was able to separate the two effects, due to a serendipitous situation whereby some articles were accidentally placed in the top positions of listings. He concluded that increased visibility contributes to higher citation counts, but not as much as self-promotion by authors.

Ginsparg and Haque [12] confirmed and extended Dietrich's results, by analyzing downloads and citations for listings of papers in astrophysics (astro-ph) and two large subcommunities of theoretical high energy physics (hep-th and hep-ph) of the arXiv. They reported a strong correlation between article position in these initial announcements and later citation counts, due primarily to intentional *self-promotion* by authors. Articles that fortuitously appeared near the top also received more citations than had they appeared in a lower position, due to a *visibility bias*. The increase in citations due to the visibility bias was smaller than that due to the self-promotion bias.

By correlating download and citation data, the authors were also able to conclude that citations appear because of readership and lead to further readership.

Wainer, Eckmann, & Rocha [23] analyzed the citations of papers in prestigious Computer Science conferences that were selected on the basis of peer review as "best" for their year prior to presentation at the conference and publication. They found strong evidence that the selected papers are cited more than random papers at the conference, with a citation advantage that remained stable for at least two years following publication. They also reported that a significant number of the "best" papers are among the top cited papers in the conference.

Previously, we reported [4] a citation advantage of papers highlighted in *Physical Review B* (PRB) as Editors' Suggestions. We also found that Editors' Suggestions were six times more likely than other PRB papers to be cited in the top 1% in the field of physics, in annual lists of highly cited papers curated by Clarivate Analytics [9].

The potential of editorial selections for identifying outstanding research has been noted by experts at the European Research Council, Europe's prestigious funding agency of frontier research. Mugabushaka, Sadat & Dantas Faria [16] identified "editorial highlights" as one type of "recognition channel" and created an open dataset [17] of editorial highlights for use in bibliometrics research and evaluative bibliometrics, including papers that received highlighting in *Science* as "Breakthroughs of the year" or *La Recherche* as "les 10 découvertes de l'année", as well as coverage in *Nature* as "research highlights" and in *Science* as "editor's choice". In followup work, Sadat and Mugabushaka [19] observe that editorial recommendations carry a precious expert judgement of scientific quality that is missing from the current practices of assessing research impact shortly after publication, like altmetrics. They proposed comparing these editorial recommendations with currently established measures of research impact. As a case in point, they report that 5% of the recommended articles of Nature's "Research Highlights" were associated with projects funded by the European Research Council.

4.3 Methods

We study papers published in *Physical Review Letters* (PRL), the flagship journal of the American Physical Society (APS). The highlighting platforms include intra-highlights in APS venues (PRL itself and the APS publication *Physics* [2]) and inter-highlights in other journals and publishers, such as *Nature Physics* and *Nature*. The intra-highlights are (i) journal cover in PRL (i.e., choice of paper in the weekly cover of the journal), (ii) *Editors' Suggestion* in PRL, (iii) *Focus* in *Physics*, (iv) *Synopsis* in *Physics*, and (v) *Viewpoint* in *Physics*. The inter-highlights are (i) *Research Highlight* and (ii) *News & Views*. See Table 4.1.

To assess the citation performance of highlighted papers, we use citation data from the Web of Science of Clarivate Analytics. We analyze citation data in three ways. We perform a multiple linear regression for the various highlighting platforms, using citation data from 1–10 years after publication, and compare regression coefficients.

We also display box plots for the citation distributions of each highlighting platform, and compare medians. Finally, we compare the overlap of papers in each platform with the top 1% of cited papers, classified as highly cited papers in the Essential Science Indicators of Clarivate Analytics [9].

4.3.1 Examined Highlighting Platforms

Since every highlighting platform is different, it is useful to briefly describe the platforms we analyzed in this study.

- *PRL cover image*: The journal cover is chosen by the editors for aesthetic reasons mainly (beautiful figures in the paper). Placement of a paper in the cover does not generally reflect selection on the basis of importance.
- *Editors' Suggestions*. First launched in 2007, these are generally chosen for "potential interest in the results presented and, importantly, on the success of the paper in communicating its message" [1]. While importance is not the only reason why a paper is selected, Editors' Suggestions fare well "among many measures of importance, [...] impact, and intrinsic interest" [8].
- *Viewpoints*: First launched in 2008, these are commentaries commissioned by the *Physics* editors and written by experts. They explain why a paper is important to the field [2].
- *Focus stories*: These are journalist-written news stories aiming to explain some of the latest research to the broadest possible audience [2], i.e., to the educated non-physicists. First launched in 1998, Focus merged with *Physics* in 2011, where it continues to produce the same style of articles as before.
- *Synopses*: First launched in 2008, these are short summaries of newsworthy results written by journalists and *Physics* staff [2].
- *Research Highlights*: These are short summaries of papers written by journal editors in the *Nature* journals.
- *News & Views*: These are commentaries by experts or (less often) by journal editors in the *Nature* journals.

Note that Viewpoints, Synopses, and Focus stories are mutually exclusive. A paper highlighted by a Viewpoint does not get considered also for Synopsis or Focus.

4.3.2 What is Importance?

Since a central aspect of this work is the citation analysis of manuscripts marked "important" by the editors, it is necessary to say a few words about what we mean by importance and how such assessments are made in practice. We could use the working definition by the late Jack Sandweiss, former Lead Editor and Chairman of the Editorial Board of Phys. Rev. Letters, who used to say that "an important result is one that researchers in the field should not miss, while those in related fields would be interested in" [20]. A more detailed description is given in the journal's website: "Important results are those that substantially advance a field, open a significant new area of research, or solve—or take a crucial step toward solving—a critical outstanding problem and thus facilitate notable progress in an existing field." [3]. Of course, an element of subjectivity is always present in any notion of importance, and a judgment call by the editor is implied in the words "substantially," "significant," and "crucial" in the above definition of importance. Nevertheless, the above descriptions of importance are *practically* useful to editors in at least *shortlisting* potentially important papers, because they help them develop the mindset and set the stage for what kind of papers to look for.

Once a paper is accepted in Phys. Rev. Letters, the editors may shortlist it for potential highlighting. Decisions for Editors' Suggestion are made at the journal level. For further highlighting in *Physics*, the papers nominated by journal editors are discussed in a committee of *Physics* editors, who decide which papers get a Synopsis and which papers warrant further consideration as Viewpoints, in which case external experts are consulted, both for their opinion on the paper and on their availability as Viewpoint authors. So, by the time a PRL paper has received a Viewpoint, it has normally gone through three additional layers of post-acceptance scrutiny: consideration as an Editors' Suggestion by the journal, consideration as a Viewpoint by the *Physics* editors, and final vetting by external experts.

4.3.3 Variation in Coverage

In Fig. 4.1 we show the annual frequency of each highlighting platform used in this study. Editors' Suggestions are most frequent (about 30 per month), followed by Synopses (10 per month) and Viewpoints (5 per month). On the other hand, inter-highlights are the least frequent platforms, with an annual median count of just over a dozen (Research Highlights in Nature and Nature Physics) or a half-dozen (News & Views in Nature Physics). See Table 4.2. Inter-highlights display also higher variation from year to year, reflecting shifting priorities or focus of *Nature* journal editors in covering the most newsworthy developments in physics, the entry of new journals in the physics market, etc. See Fig. 4.2.

Table 4.2 Median annual frequency of each highlighting platform for PRL papers published from 2010–2018

CVR	Focus	Viewpoint	LSUGG	Synopsis	RHNatPhys	NVNatPhys	RHNature
50	40	64	354	111	14	6	14

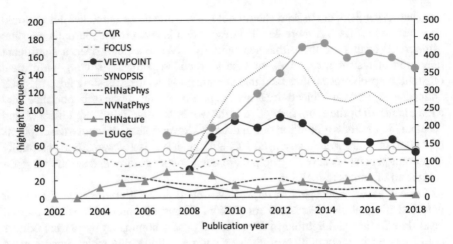

Fig. 4.1 Annual frequency of each highlighting platform. Editors' Suggestions (LSUGG) are plotted on the secondary (right) axis. Note: Editors' Suggestions started in 2007, Viewpoints and Synopses in 2008.

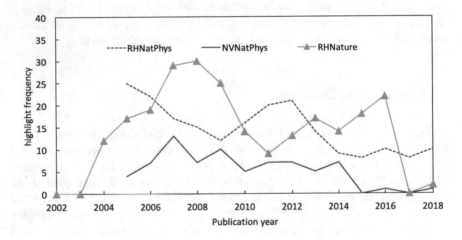

Fig. 4.2 Detail of Fig.4.1. Annual frequency of the inter-highlighting platforms analyzed here

4.4 Results and Discussion

We analyze the citations of highlighted papers using three tools: (i) multiple linear regression, (ii) boxplots of citation distributions, and (iii) lists of highly cited papers by Clarivate Analytics. Multiple linear regression is our central tool.

4.4.1 Multiple Linear Regression

We perform Multiple Linear Regression on the citation data of the papers highlighted in the seven different highlighting platforms described in Section 4.3.1. We collect citation data for each paper each year for a range (citation window) of 1–10 years after publication, in order to allow dynamic observation of the citation performance of highlights. Obviously, for recently published papers (< 10 years), not all citation windows can be counted.

As a first sample to analyze, we chose papers published from 2008–2012, and counted their cumulative citations from 1–7 years after publication. This amounts to 246 papers in the journal cover, 203 in Focus, 354 Viewpoints, 1232 Editors' Suggestions, 556 Synopses, 84 Research Highlights in *Nature Physics*, 36 News & Views in *Nature Physics*, and 91 Research Highlights in *Nature*. Note that Viewpoints and Synopses first started in 2008, and Editors' Suggestions in 2007. In Table 4.3 we display the regression coefficients for each highlighting type. We use a single star (*) for values with $P < 0.05$, two stars (**) for $P < 0.01$, and three stars (***) for $P < 0.001$. We find statistical significance for all data of Viewpoints, Suggestions, Synopses, and Research Highlights in *Nature*, and for 3 years of Covers. The data are also plotted in Fig. 4.3 for clarity.

The first thing to note from Table 4.3 and Fig. 4.3 is the strong association between a Viewpoint and high citations. In all citation windows, i.e., from 1 to 7 years after publication, the Viewpoint marker is clearly a stronger predictor of citations than any other marker of publicity. All else being equal, a Viewpoint marker predicts 10 additional citations to a PRL paper during the first year after publication, 44 citations within 5 years, and 58 citations within 7 years. This strong position of Viewpoints as the leading marker for citation accrual is found across all years and samples, so it is a very stable result.

Table 4.3 Multiple linear regression coefficients for the citation data of papers highlighted in the various platforms, over a range (cumulative) of 1–7 years after publication. The papers were published from 2008–2012. Shown in the 2nd column are the numbers of papers in each platform. CVR = PRL journal cover, LSUGG = PRL *Editors' Suggestion*, RHNatPhys = Research Highlight in *Nature Physics*, NVNatPhys = News & Views in *Nature Physics*, RHNature = Research Highlight in *Nature*. Statistical significance: * denotes $P < 0.05$, ** denotes $P < 0.01$, *** denotes $P < 0.001$. No star denotes $P > 0.05$.

Highlight	# papers	1Y	2Y	3Y	4Y	5Y	6Y	7Y
CVR	246	0.6	2.2	5.1*	6.3*	7.5*	8.5	9.7
FOCUS	203	-0.1	0.2	0.6	1.5	2.4	3.2	4.1
VIEWPOINT	354	10.3***	20.4***	29.0***	37.4***	44.3***	51.3***	58.0***
LSUGG	1232	3.4***	6.7***	10.0***	12.9***	15.7***	18.5***	20.9***
SYNOPSIS	556	1.7**	4.0***	5.5***	7.2***	8.6***	9.8**	10.7*
RHNatPhys	84	2.4*	5.8*	8.4*	11.4*	15.1*	19.0*	23.8*
NVNatPhys	36	2.9	6.3	8.7	10.7	13.0	14.7	15.8
RHNature	91	2.3*	6.6**	11.7***	17.6***	24.8***	32.5***	39.0***

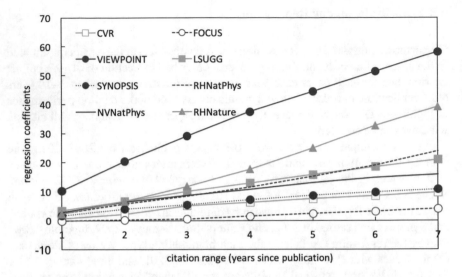

Fig. 4.3 Coefficients of multiple linear regression for each highlighting platform. Papers were published from 2008–2012. Citations were collected from 1–7 years after publication

The second strongest association with citations is found for Research Highlights in *Nature*. 5 years within publication, this marker predicts almost 25 additional citations, all else being the same. In the first 1–3 years since publication, this effect is not clearly pronounced compared to the other markers, but from the 4th year onward it can be clearly seen.

The next strongest effect is found for Research Highlights in *Nature Physics* and Suggestions, whose behavior is similar, so we group them together; and after that, for Synopses and Covers, which we also group.

Broadly speaking, we observe the following stratification of citation advantage for the different highlights:

$$\text{Viewpoint} > \text{RH } Nature > \left\{ \begin{array}{l} \text{Suggestion} \\ \text{RH } Nat. \ Physics \\ \text{[N\&V } Nat. \ Physics] \end{array} \right\} > \left\{ \begin{array}{l} \text{Synopsis} \\ \text{Cover} \\ \text{[Focus]} \end{array} \right\} \quad (4.1)$$

Here, we have grouped together highlights in terms of decreasing citation advantage. In the first group are Viewpoints, which are cited the most. In the second group are Research Highlights in *Nature*. In the third group we have Suggestions and Research Highlights in *Nature Physics*. And in the fourth group we have Synopses and Covers. For completeness, we included News & Views in *Nature Physics* and Focus papers, but put them in brackets, since their coefficients lack statistical significance.

Interestingly, the stratification in Eq. 4.1 follows also a hierarchical pattern of decreasing scrutiny, with regard to importance, during peer review. First, Viewpoints and Suggestions are intra-highlights, so the editors who select them have the full benefit of both internal and external to the journal review. Between these two, Viewpoints are clearly vetted more, since they are discussed not only among the PRL journal editors once the papers are accepted for publication, but also among a committee of Physics editors who then seek additional advice from external experts on whether such papers merit highlighting as Viewpoints. Synopses and Viewpoints are mutually exclusive, and since Viewpoints receive the highest level of scrutiny, it follows that Synopses do not meet the same standard in terms of importance during the vetting process—indeed, many intellectually curious, 'fun' or elegant papers are selected for Synopsis. Focus stories are aimed at an audience of educated non-physicists, and papers are often chosen based on educational value and intrinsic interest to non-specialists rather than on scientific merit, so, again, importance is not the key criterion for selection. We find that a Focus marker has no significant prediction on a paper's citations. (We discuss Covers further below.)

Research Highlights in *Nature* or *Nature Physics*, and News & Views in *Nature Physics* are all inter-highlights, which makes their selection more challenging to begin with, since the NPG editors do not have access to the peer review files from PRL to aid their decisions. These journals are also more constrained in terms of journal space since their highlights cover several journals, and, in the case of *Nature*, several fields. So it does not seem surprising that these highlighting platforms yield a smaller citation advantage than Viewpoints, for which it is reasonable to assume that papers receive the highest scrutiny among all markers presented here. As for the relative difference in citations between Research Highlights in *Nature* or *Nature Physics*, this may be due to the fact that *Nature* is more selective than *Nature Physics* since it covers all fields of science, and its highlighting slots may therefore be more vetted.

A Cover marker predicts a statistically significant but small citation boost. All else being equal, placement in the journal cover predicts no more than 1.5 additional citations per year (7.5 citations within 5 years of publication, almost 10 citations within 7 years). The small effect of cover placement may come as a surprise to some. For instance, the author has first-hand experience, from institutional visits throughout the world, of the pride many authors take in seeing their work appear in the cover of PRL. Indeed, authors often interpret placement in the cover as an editorial endorsement of their paper's higher importance among other papers in a journal issue. However, selection of a cover figure for PRL, as for several other scientific journals, is primarily done for aesthetics and does not typically imply an editorial endorsement of higher scientific merit. (Of course, if the editors have qualms about a paper's results or conclusions they are less likely to select it for the cover.) While the reasons behind cover selection seem to be largely lost to the community (according to our anecdotal experience), the prediction of the cover marker on citations actually falls in line with the editors' opinion: papers in the cover are cited only marginally more than other papers, all other things being the same. So, this type of *accidental or serendipitous* publicity brings a small citation advantage—in direct analogy to the effect of *visibility* bias reported by Dietrich [10, 11] and Ginsparg and Haque [12], for papers that accidentally end up in the top slot of the daily arXiv email listings. However, the citation advantage is clearly greater when publicity is deliberate and results from an endorsement of the paper's merit formed through peer review, as in a Viewpoint or any marker presented here other than Focus. Viewed from this lens, we can understand the *self-promotion* bias in Dietrich [10, 11] and Ginsparg and Haque [12] for papers whose authors deliberately and carefully engineered their placement in the top slot or arXiv lists, as a form of *internal* review: Clearly, authors aim to push their better papers in the top slot of arXiv listings, and this self-selection reflects an *endorsement* of higher quality and results in a citation advantage.

So, publicity alone does help in terms of citations, but does not make a lot of difference unless it is supported by an endorsement, formed through peer review (internal and/or external), that the paper has above-average merit.

To recap, we observe that increased editorial scrutiny of a journal's papers in terms of importance is associated with higher citation counts. Our findings are consistent with the notion that citations follow or mirror peer review [4, 13, 22]. Our analysis takes this premise a step further, since it shows that we can allocate relative degrees of importance and correspondingly predict citation accrual for papers published in the same journal but publicized by different highlighting platforms.

4.4.2 Box Plots and Medians

In the previous subsection, we used multiple linear regression to study the association between the various highlighting platforms and the citations of the highlighted papers, in order to identify the different prediction of each marker. Another way to visualize our data is to compare the citation distributions of the highlighted sets of papers. One caveat here is that some papers are co-highlighted in more than one platform (say, a Suggestion, Viewpoint, and Cover) and are thus present in all the corresponding distributions. With this caveat in mind, we proceed.

Because citation distributions are usually highly skewed, their comparison is done by visual inspection of the corresponding boxplots. A boxplot is a graphical depiction of a 5-number summary of the distribution, which in the Altman convention [14] that we follow here includes the 5th percentile, the 25th percentile (or 1st quartile), the median, the 75th percentile (or third quartile), and the 95th percentile. Outlier data points that lie below the 5th and above the 95th percentiles are shown as dots.

For example, in Fig. 4.4 we show the boxplots for the 5-year citations of each highlighted group, for papers published from 2008-2014. Once again, Viewpoints are clearly cited the most, whether one compares medians, means (shown as crosses), or interquartile ranges (Q1–Q3). If we rank the remaining groups by medians, in order to avoid the distorting effect of outliers (highly cited papers) on the means [7], we have News & Views in *Nature Physics* and Research Highlights in *Nature* closely together, followed by Suggestions, Synopses and Research Highlights in *Nature Physics* that are also on par. Finally, we have Cover and Focus articles.

We checked for statistical significance in the difference between any two medians among these 8 distributions of highlights using the nonparametric Mann-Whitney U test. (With the added caveat that Viewpoints, Synopses, and Focus articles are

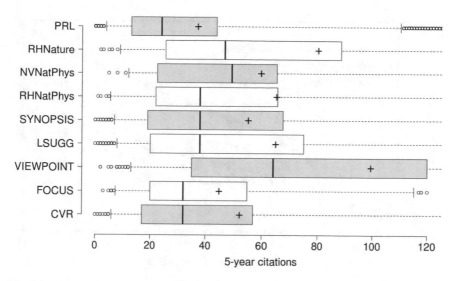

Fig. 4.4 Boxplots of citations of highlighted papers, plotted horizontally. Publication years = 2008–2014. Citations counted over a 5-year window since publication. The horizontal extent of each box spans the *interquartile range* IQR = Q1–Q3 (i.e., from the 25th–75th percentile). Medians are shown as line segments, means as crosses. Whiskers extend from the 5th to the 95h percentile (most 95th-percentile whiskers are not shown in this truncated plot). Outliers are shown as dots beyond the extent of whiskers. Alternate white/gray shading is used simply to increase legibility.

mutually exclusive and hence not quite independent.) Our null hypothesis is that the medians are equal and the one-sided alternative is that one median is greater than the other, at 5% significance level. For the 5-year citation distributions of Fig. 4.4, we could thus establish that, at 5% significance level:

- the citation median for an article with a Viewpoint is greater than for all other articles;
- the citation median for articles in Research Highlights in *Nature* is greater than for Suggestions, Synopses, Covers, or Focus articles; and
- the citation median for articles in Research Highlights in *Nature Physics*, News & Views in *Nature Physics*, Suggestions, and Synopses is greater than for Covers, or Focus articles.

The above ranking of citation medians essentially confirms the stratification of Eq. 4.1, which is reassuring. Any minor adjustments (e.g., Synopsis faring better than Covers) reflect small, gradual shifts and are thus inconsequential.

Fig. 4.4 utilizes just a part of our full dataset, i.e., papers published from 2008–2014 and cited over the next 5 years. To utilize the full dataset and reduce statistical uncertainties, we calculated the citation median for the citation range 1–10 years after publication for all combinations of publication years from 2008–2018. That is, to obtain the citation median for one year since publication, we took the median of citations from 2009, 2010, ..., 2019 to articles published respectively in 2008, 2009,

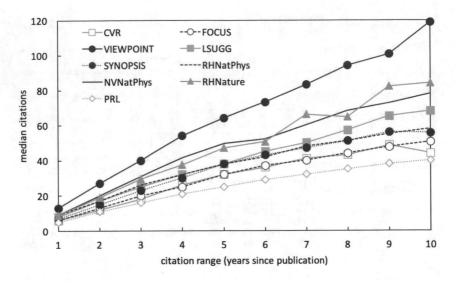

Fig. 4.5 Median citations for each highlighting platform and for the PRL journal. Papers were published from 2008–2018. Citations were collected from 1–10 years after publication.

..., 2018, etc. For two years since publication, we took the median of citations from 2009–2010, ..., 2018–2019 to articles published respectively in 2008, ..., 2017, etc. See Fig. 4.5.

We then processed these data a little further, as follows. For each year in the citation range and each highlighting marker, we calculated the relative increase of the citation median compared to the journal median, and took the average over all publication years. This is how we calculated the average enhancement shown in Fig. 4.6. For example, papers in the journal cover have a citation median that is on average 21% higher than the baseline (the whole journal) over the 10-year citation range.

Put together, the results from the figures presented here broadly confirm our findings from the previous subsection. Intra-highlights for importance (Viewpoints) bring the most citations (1st group), followed by inter-highlights for importance (News & Views in *Nature Physics*, Research Highlights in *Nature* and *Nature Physics*) and Editors' Suggestions (2nd group), followed by accidental publicity papers (covers) and Focus stories (3rd group).

Even though box plots do not correct for co-highlighting, as we mentioned before, the stratification of citations for Viewpoints, Suggestions, Synopses, and Focus, as well as Research Highlights in *Nature* and *Nature Physics*, broadly agrees with the regression analysis. This makes sense, for the reasons discussed in the previous subsection.

It is worth noting that the Suggestions collection contains many highly cited papers, even though its median (and mean) are considerably lower than for Viewpoints.

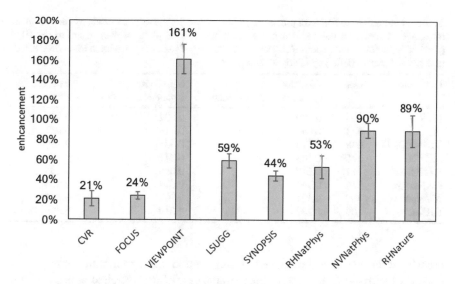

Fig. 4.6 Average value of enhancement of median citations for each highlighting platform compared to the PRL journal. Papers were published from 2008–2018. Citations were collected from 1–10 years after publication. Error bars show 99% confidence intervals.

4.4.3 Inclusion in Clarivate List of Highly Cited Papers

Here, we ask the question, *Can we use highlighting markers as predictors of a paper being highly cited?* This question takes us beyond the analysis of multiple linear regression, where we looked at how many citations the 'average' paper picked upon being highlighted. Placement of a paper in a *highly* cited list is identifying potential for extreme, not average, citation performance.

As a benchmark for highly cited papers, we use the *Highly Cited Papers (HCP)* [9] indicator of Clarivate Analytics. These papers are the top 1% cited in their subject per year. Lists of HCP papers going back 10 years are available for download from the Essential Science Indicators of Clarivate Analytics. We downloaded the list of HCP papers in physics published from 2010–2018, among which there are 1371 papers published in PRL. For each highlighting marker we calculate how often it "predicts" placement in the HCP list, i.e., how many highlighted papers are highly cited; this is the positive predictive value, or *precision*. We also calculate how often the HCP list contains marked papers, i.e., how many highly cited papers are highlighted; this is the true positive rate or *sensitivity*. The geometric mean of these two quantities (precision and sensitivity) is the *F1 score* [24], which is a measure of a marker's accuracy. See Table 4.4.

Evidently, 29% of Viewpoint papers are placed in the highly-cited-papers list of Clarivate Analytics. So, Viewpoints have the highest precision, followed by Research Highlights in *Nature* (23.9%), Research Highlights in *Nature Physics* (16.4%), and Editors' Suggestions (14.3%). The hierarchical pattern of Eq. 4.1 in terms of de-

Table 4.4 Summary statistics for precision (or positive predictive value), sensitivity (or true positive rate), and F1 score, for the performance of each highlighting platform with regard to identifying (predicting) highly cited papers. Publication years, 2010–2018. 1371 papers in PRL were listed as highly-cited-papers (HCP) by Clarivate Analytics.

Highlight	Count	Precision (positive predictive value)	Sensitivity (true positive rate)	F1 score
CVR	446	0.099	0.032	0.048
FOCUS	378	0.063	0.018	0.027
VIEWPOINT	638	0.290	0.135	0.184
LSUGG	3269	0.143	0.341	0.201
SYNOPSIS	1117	0.117	0.096	0.105
RHNatPhys	116	0.164	0.014	0.026
NVNatPhys	33	0.121	0.003	0.006
RHNature	109	0.239	0.019	0.035

creasing scrutiny is thus reproduced. Again, we find that highlighting for importance correlates with citations even for the extreme case of top-1% cited papers, as in the HCP lists.

With regard to sensitivity, the inter-highlighting platforms in *Nature* and *Nature Physics* score much lower than the intra-highlights, as expected (see below). For example, just under 0.3% of HCP papers in PRL are highlighted by a News & Views item in *Nature Physics*. For comparison, 3.2% of the PRL's HCP papers are PRL Covers, and 34.1% of PRL's HCP papers were marked by a Viewpoint. Why is this expected? Because inter-highlighting platforms such as Research Highlights and News & Views have many journals to cover and a limited number of slots. So, *Nature* or *Nature Physics* can only highlight a few PRL papers at any time. At the same time, the high sensitivity values of Viewpoints and especially Editors' Suggestions are remarkable: More than 1 in every 3 HCP papers in PRL is an Editors' Suggestion. This result echoes our earlier remark that the Suggestions list contains many highly cited papers.

From the precision and sensitivity values we obtain the F1 score (it is their geometric mean). The F1 score can be viewed as the overall accuracy of a highlighting marker's predictive power for placement in the HCP list. As shown in Table 4.4, Editors' Suggestions, Viewpoints, and Synopses dominate this score, with values 0.2, 0.18, and 0.1, respectively.

4.5 Conclusions

We have presented an analysis on the citation advantage of papers from *Physical Review Letters* (PRL) that are highlighted in various platforms, both within and beyond the journal. Among the various highlighting platforms we analyzed, we found that Viewpoints in *Physics* are the strongest predictors of citation accrual, followed by

Research Highlights in *Nature*, Editors' Suggestions in PRL, and Research Highlights in *Nature Physics*.

When a PRL paper is highlighted by a Viewpoint in *Physics* or a Research Highlight in *Nature*, its chances of being listed among the top 1% cited papers in physics are about 1 in 4.

Our key conclusion is twofold. First, highlighting for importance identifies a citation advantage. When editors select papers to highlight with importance as their main criterion, they can identify papers that end up well cited. Accidental or serendipitous publicity (i.e., mere visibility), whereby a paper is publicized for other reasons than importance, e.g., in the journal cover, gives a clearly smaller citation advantage.

Second, the stratification of citations for highlighted papers follows the degree of vetting of papers for importance during peer review (internal and external to the journal). So, citation metrics mirror peer review even *within* a journal. This implies that we can view the various highlighting platforms (e.g., Viewpoints) as predictors of citation accrual, with varying degrees of strength that mirror each platform's vetting level.

Acknowledgements I am grateful to Hugues Chaté, Jerry I. Dadap, Paul Ginsparg, and Jessica Thomas for stimulating discussions, and to Richard Osgood Jr. and Irving P. Herman for hospitality. This work uses data, accessed through Columbia University, from the Web of Science and InCites Essential Science Indicators of Clarivate Analytics.

Competing interests
The author is an Associate Editor in *Physical Review B*, a Contributing Editor in *Physical Review Research*, and a Bibliostatistics Analyst at the American Physical Society. He is also an Editorial Board member of the Metrics Toolkit, a volunteer position. He was formerly an Associate Editor in *Physical Review Letters* and *Physical Review X*. The manuscript expresses the views of the author and not of any journals, societies, or institutions where he may serve.

References

1. Anonymous: Announcement: Editors' suggestions: Another experiment. Physical Review Letters **98**(1), 010001 (2007)
2. Anonymous: *Physics* is a free, online magazine from the American Physical Society (2021). https://physics.aps.org. Cited 1 May 2021
3. Anonymous: (Physical Review Letters Editorial Policies and Practices). https://journals.aps.org/prl/authors/editorial-policies-practices. Cited 1 May 2021
4. Antonoyiannakis, M.: Editorial: Highlighting impact and the impact of highlighting: PRB Editors' Suggestions. Physical Review B **92**, 210001 (2015)
5. Antonoyiannakis, M.: Highlighting impact: Do editors' selections identify influential papers? In: APS March Meeting Abstracts, p. X43.008 (2016). https://ui.adsabs.harvard.edu/abs/2016APS..MARX43008A/abstract. Cited 1 May 2021
6. Antonoyiannakis, M.: Editorial highlighting and highly cited papers. In: APS March Meeting Abstracts, p. H15.012 (2017). https://ui.adsabs.harvard.edu/abs/2017APS..MARH15012A/abstract. Cited 1 May 2021
7. Antonoyiannakis, M.: Impact factor volatility due to a single paper: A comprehensive analysis. Quantitative Science Studies **1**, 639–663 (2020)

8. Chaté, H.: Editorial: A decade of editors' suggestions. Physical Review Letters **118**, 030001 (2017)
9. Clarivate-Analytics: Highly cited papers indicator (2021). `https://bit.ly/3jdw2Fg`. Cited 1 May 2021
10. Dietrich, J.P.: Disentangling visibility and self-promotion bias in the arxiv:astro-ph positional citation effect. Publications of the Astronomical Society of the Pacific **120**, 801–804 (2008)
11. Dietrich, J.P.: The importance of being first: Position dependent citation rates on arxiv:astro-ph. Publications of the Astronomical Society of the Pacific **120**, 224–228 (2008)
12. Ginsparg, P., Haque, A.: Positional effects on citation and readership in arxiv. Journal of the American Society for Information Science and Technology **60**, 2203–2218 (2009)
13. Harzing, A.W.: Running the REF on a rainy Sunday afternoon: Do metrics match peer review? (2017). `https://bit.ly/35b02wz`. Cited 1 May 2021
14. Krzywinski, M., Altman, N.: Visualizing samples with box plots. Nature Methods **11**, 119–120 (2014)
15. Kurtz, M.J., Eichhorn, G., Accomazzi, A., Grant, C., Demleitner, M., Henneken, E., Murray, S.S.: The effect of use and access on citations. Information Processing and Management **41**, 1395–1402 (2005)
16. Mugabushaka, A.M., Sadat J. andDantas-Faria, J.C.: In search of outstanding research advances: Prototyping the creation of an open dataset of "editorial highlights (2020). `https://arxiv.org/abs/2011.07910`. Cited 1 May 2021
17. Mugabushaka, A.M., Sadat J. andDantas-Faria, J.C.: An open dataset of scholarly publications highlighted by journal editors (version 01) [data set] (2020). `http://doi.org/10.5281/zenodo.4275660`. Cited 1 May 2021
18. Philips, D.P., Kanter, E.J., Bednarczyk, B., Tastad, P.L.: Importance of the lay press in the transmission of medical knowledge to the scientific community. New England Journal of Medicine **325**, 1180–1183 (1991)
19. Sadat, J., Mugabushaka, A.M.: In search of outstanding research advances – exploring editorial recommendations. (2021). `https://refresh20.infrascience.isti.cnr.it/files/RefResh_2020_paper_2.pdf`. Cited 1 May 2021
20. Sandweiss, J.: Private communication
21. Sinatra, R., Deville, P., Szell, M., Wang, D., Barabási, A.L.: A century of physics. Nature Physics **11**, 791–796 (2015)
22. Traag, V.A., Waltman, L.: Systematic analysis of agreement between metrics and peer review in the UK REF. Palgrave Communications **5**(29) (2019)
23. Wainer, J., Eckmann, M., Rocha, A.: Peer-selected "best papers" – are they really that "good"? PLoS ONE **10**, 1–12 (2005)
24. Wikipedia: The F1 score is a measure of a test's accuracy. `https://en.wikipedia.org/wiki/F1_score`. Cited 1 May 2021

Chapter 5
Ranking Papers by Expected Short-Term Impact

Ilias Kanellos, Thanasis Vergoulis, and Dimitris Sacharidis

Abstract The number of published scientific papers has been constantly increasing in the past decades. As several papers can have low impact or questionable quality, identifying the most valuable papers is an important task. In particular, a key problem is being able to distinguish among papers based on their short-term impact, i.e., identify papers that are currently more popular and would be cited more frequently in the future than others. Consequently, a multitude of methods that attempt to solve this problem have been presented in the literature. In this chapter, we examine methods that aim at, or have been evaluated based on, ranking papers by their expected short-term impact. First, we formally define the related ranking problem. Then, we present a high-level overview of current approaches, before presenting them in detail. We conclude by discussing previous findings regarding the effectiveness of all studied methods, and give pointers to related lines of research.

5.1 Introduction

In recent decades the growth rate of the number of published scientific papers has been constantly increasing [7, 30]. Meanwhile, it has been shown that among the vast number of published works, many are of low impact, or can even be characterized of questionable quality [23]. As a result, identifying the most valuable research papers

Ilias Kanellos
Athena Research Center, Athens, Greece
e-mail: ilias.kanellos@athenarc.gr

Thanasis Vergoulis
Athena Research Center, Athens, Greece
e-mail: vergoulis@athenarc.gr

Dimitris Sacharidis
Université Libre de Bruxelles, Belgium,
e-mail: dimitris.sacharidis@ulb.be

for each scientific topic, a task that dominates the researchers' daily routine, has become extremely tedious and time consuming.

In this era, predicting the attention a paper will attract in the next few years, i.e., its expected *short-term impact*, which reflects its current popularity [26], can be very useful for many applications. Consequently, a multitude of methods that attempt to solve related problems have been presented in the literature. In particular, there are two main problems in this field: *impact prediction*, and *ranking by impact*. The former concerns the estimation for *a specific paper* of the future value of some impact indicator (e.g., the citation count), and is typically addressed as a regression (e.g., [1, 31, 32, 34, 49, 52]) or a classification (e.g., [40, 43, 45]) task. The latter is the focus of this chapter and concerns the ranking of *all papers* according to their expected impact.

At this point, it should be highlighted that the term "impact" can have several diverse definitions [6]. In this chapter, the *short-term impact* of papers is measured by node centrality metrics (most notably in-degree) in the citation network formed by all citations that are made in the near future (see Section 5.2.1). Computing centrality metrics in the current state of the citation network introduces a *bias* against recently published papers that have had much fewer opportunities for citation. Hence, the vast majority of impact ranking methods are based on *time-aware* citation network analysis, i.e., they incorporate the dimension of time in their computational model aiming to overcome the aforementioned *age bias*. In this chapter, we present methods that directly attempt to counteract age bias, but also other ranking methods that were evaluated (either when proposed or in subsequent studies) in terms of the ground truth ranking by short-term impact.

The rest of this chapter is organized as follows: in Section 5.2 we introduce the notation used, we discuss the relevant background, and formally define the problem of ranking by expected impact. In Section 5.3 we present a high level categorization of ranking methods aiming to solve this problem. In Section 5.4, we elaborate on the most important methods. In Section 5.5, we provide some discussion, and we conclude in Section 5.6.

5.2 Background and Problem Definition

This section introduces preliminary background and notation, and further proceeds to describe the short-term impact-based paper ranking problem examined in the following sections.

5.2.1 Background

Citation Network. A *citation network* is a graph that has papers as nodes, and citations as edges. For a paper node, an outgoing (incoming) edge represents a reference to (from) another paper. A citation network is an evolving graph. While node out-degrees remain constant, in-degrees increase over time when papers receive new references.

A citation network of N papers can be represented by the adjacency matrix A, where the entry $A_{i,j} = 1$ if paper j cites paper i, i.e., there exists an edge $j \rightarrow i$ in the network, and 0 otherwise. We denote as t_i the *publication time* of paper i; this corresponds to the time when node i and its outgoing edges appear in the network. We additionally define two further network matrices that describe the relations of papers to their authors and their publication venues. The authorship matrix M, encodes the authors that wrote each paper in the network matrix, with $M_{i,j} = 1$ when paper i is written by author j and $M_{i,j} = 0$ otherwise. Similarly the venue matrix V, encodes which venue published each paper, with $V_{i,j} = 1$ when paper i is published in publication venue j and $V_{i,j} = 0$ otherwise.

We further use the following notations with respect to ranking method scores: we use s_i to denote the *rank score* of paper i, as calculated by each method. Many methods employ additional author and venue rank scores, which we denote by a_i, and u_i, for author and venue i, respectively. For methods that employ multiple paper, author, and venue scores, we use bracketed superscripts to differentiate between them, e.g., $s^{\{X\}}$; superscripts without brackets denote exponentiation.

Finally, we introduce additional notation to facilitate the description of the various ranking methods. We define scores of author *lists*, denoting the score of the i-th list by l_i. Additionally, we define functions v(i), which returns the publication venue of paper i and $\ell(i)$, which returns the author list of paper i. A summary of notation conventions for the current chapter is presented in Table 5.1.

Table 5.1: Summary of notations used throughout this chapter.

Symbol	Meaning
	Network Matrices
$A_{i,j}$	Citation Matrix. $A_{i,j} = 1$ when paper j cites paper i, otherwise 0.
$M_{i,j}$	Author Matrix. $M_{i,j} = 1$ when author j wrote paper i, otherwise 0.
$V_{i,j}$	Venue Matrix. $V_{i,j} = 1$ when venue j publishes paper i, otherwise 0.
	Scores
s_i	Score of paper i.
a_i	Score of author i.
v_i	Score of venue i.
l_i	Score of author list i.
$s_i^{\{X\}}/a_i^{\{X\}}/v_i^{\{X\}}$	Score of paper/author/venue i, based on X.
	Functions
v(i)	Venue that publishes paper i, i.e., $V_{i,v(i)} = 1$.
$\ell(i)$	Author list that wrote paper i, i.e., $l_{\ell(i)} = \{k : M_{i,k} = 1\}$.
	Other Notations
t_i	Publication time of paper i.

In the following, we overview two commonly used node centrality metrics for citation networks, which are central to the various ranking methods.

Citation Count. The *citation count* of a paper i is the in-degree of its corresponding node, computed as $k_i = \sum_j A_{i,j}$. We also use k_i^{out} to denote the out-degree of paper i, i.e., the number of references i makes to other papers.

PageRank score. PageRank [8,39] was introduced to measure the importance of Web pages. In the context of citation networks, it simulates the behaviour of a "random researcher" who first reads a paper, and then either picks one of its references to read, or chooses any other paper in the network at random. The PageRank score of a paper i indicates the probability of a random researcher reading it, and satisfies:

$$s_i = \alpha \sum_j P_{i,j} s_j + (1 - \alpha) u_i \qquad (5.1)$$

where P is the network's transition matrix, $P_{i,j} = A_{i,j} / k_j^{out}$, and k_j^{out} is the out-degree of node j; $(1 - \alpha)$ is the *random jump probability*, controlling how often the researcher chooses to read a random paper; and u_i is the *landing probability* of reaching node i after a random jump. Note, that the random jump and landing probabilities concern different aspects of PageRank's model. While the random jump probability $(1-\alpha)$ models the likelihood of a random researcher's next step during her random walk (i.e., whether she chooses a paper from a reference list, or any one from the citation network), the landing probability u_i, in contrast, is derived from a probability distribution over all papers and models the likelihood of choosing a particular paper, when performing a random jump. Typically, each paper is given a uniform landing probability of $u_i = 1/N$. Note that in the case of "dangling nodes", i.e., nodes j with out-degree $k_j^{out} = 0$, the value of the transition matrix is undefined. To address this, the standard technique is to set $P_{i,j} = 1/N$, or to some other landing probability, whenever $k_j^{out} = 0$.

In the case of citation networks, researchers [12,36] usually set $\alpha = 0.5$, instead of $\alpha = 0.85$, which is typically used for the Web. The assumption is that a researcher moves by following references once, on average, before choosing a random paper to read. An in-depth analysis of PageRank and its mathematical properties can be found in [29].

5.2.2 Ranking Papers Based on their Expected Short-Term Impact

Using the aforementioned node centrality metrics to capture the impact of a paper can introduce biases, e.g., against recent papers, and may render important papers harder to distinguish [12,22,54]. This is due to the inherent characteristics of citation networks, the most prominent of which is the delay between a paper's publication and its first citation, also known as "citation lag" [5,14,19,42]. Thus, impact metrics should also account for the evolution of citation networks.

The expected *short-term impact* of a paper, which is the impact aspect we consider in this chapter, is based on node centrality metrics computed using future states of the citation network. Therefore, all methods that target this aspect, need to predict the ranking of papers according to an unknown future state. Popularity reflects the level of attention papers enjoy *at present*, e.g., as researchers study them and base their work on them. The short-term impact can only be quantified by the citations a paper receives in the *near future*, i.e., once those citing papers are published. Exactly how long in the future one should wait for citations depends on the typical duration of the research cycle (preparation, peer-reviewing, and publishing) specific to each scientific discipline.

For what follows, let $A(t)$ denote the snapshot of the adjacency matrix at time t, i.e., including only papers published until t and their citations. Further, let t_c denote current time. Assuming the selected future citation accumulation period is T, the short-term impact of a paper is its centrality metric computed on the adjacency matrix $A(t_c + T) - A(t_c)$. Note that this matrix contains a non-zero entry only for citations made during the $[t_c, t_c + T]$ time interval.

If the centrality used is citation count, the short-term impact essentially represents the number of citations a paper receives in the near future. On the other hand, when it is defined by PageRank, it portrays the significance endowed to a paper by citation chains that occur during that time interval. Based on the above, the following problem can be defined:

Problem 1 (Ranking by Expected Short-Term Impact) Given the state of the citation network at current time t_c, produce a ranking of papers that matches their ranking by short-term impact, i.e., their expected centrality on adjacency matrix $A(t_c + T) - A(t_c)$, where T is a parameter.

5.3 Classification of Ranking Methods

The classification of approaches that follows, is based on two main dimensions. The first concerns the type of time-awareness the various methods employ. Here, we distinguish among no time-awareness, where simple adaptations of PageRank are proposed (Section 5.3.1); time-aware modifications in the adjacency matrix (Section 5.3.2); and time-awareness in the landing probability of a paper (Section 5.3.3). The second dimension is on the use of side information, where we discern between the exploitation of paper metadata (Section 5.3.4), and the analysis over multiple networks, e.g., paper-paper, paper-author, venue-paper (Section 5.3.5). Moreover, we cover methods that aggregate the results of multiple approaches that fall into one or several of the aforementioned categories (Section 5.3.6), and some that do not fit our classification (Section 5.3.7). Table 5.2 presents a classification summary.

Table 5.2: Classification of Ranking Methods

Method	Basic PR variants	Network Matrix	Landing Probability	Metadata	Multiple Networks	Ensemble	Other
		Time Aware					
Non-Linear PageRank (NPR) [53]	✓						
SPR [57]	✓						
Weighted Citation (WC) [51]		✓		✓			
Retained Adjacency Matrix (RAM) [18]		✓					
Timed PageRank [54,55]		✓		✓			
Effective Contagion Matrix (ECM) [18]		✓					
NewRank (NR) [15]		✓	✓				
NTUWeightedPR [13]		✓	✓	✓			
EWPR [35]		✓		✓		✓	
SARank [37]		✓		✓		✓	
CiteRank (CR) [47]			✓				
FutureRank (FR) [41]			✓	✓	✓		
MR-Rank [56]		✓		✓	✓		
YetRank (YR) [22]			✓	✓			
Wang et al. [48]			✓	✓	✓		
COIRank. [2]			✓	✓	✓		
MutualRank [25]					✓		
bletchleypark [21]		✓		✓		✓	
Age-Rescaled PR [38]							✓
Age- & Field- Rescaled PR [44]							✓
AttRank [27]			✓				
ArtSim [11]			✓				

5.3.1 Basic PageRank Variants

Here we refer to variations of PageRank's transition matrix P, which utilize neither metadata nor time-based information. For example, in each iteration, Non-Linear PageRank [53] computes a paper's score by summing the scores of its in-neighbors raised to the power of θ, and then taking its θ root, for some $0 < \theta < 1$. The effect is that the contribution from other important works is boosted, while the effect of citations from less important works is suppressed. SPRank [57], on the other hand, incorporates a similarity score in the transition matrix P, which is calculated based on the overlap of common references between the citing and cited papers. In this way it simulates a focused researcher that tends to read similar papers to the one she currently reads.

5.3.2 Time-Aware Adjacency Matrix

The adjacency matrix A can include time quantities as weights on citation edges. There are three time quantities of interest, denoted as τ_{ij}, concerning a citation $j \rightarrow i$:

- *citation age*, or citing paper age, the elapsed time $t - t_j$ since the publication of the citing paper j,

- *citation gap*, the elapsed time $t_j - t_i$ from i's publication until its citation from j, and
- *cited paper age*, the elapsed time, $t - t_i$ since the publication of the cited paper i.

The prevalent way to infuse time-awareness into the adjacency matrix is to weigh each non-zero entry $A_{i,j}$ by an exponentially decaying function of τ_{ij}:

$$A'_{i,j} = \kappa e^{-\gamma \tau_{ij}} A_{i,j},$$

where $\gamma > 0$ is the *decay rate*, and κ represents additional factors and/or a normalisation term.

If τ_{ij} is set to the citation age, recent citations gain greater importance. If τ_{ij} is set to the citation gap, citations received shortly after a paper is published gain greater importance. If τ_{ij} is set to the cited paper age, citations to more recently published papers gain greater importance. While it is possible to weigh citations based on any combination of these time quantities, we have not seen such an approach.

The effect of a time-aware adjacency matrix on degree centrality (citation count) is immediate: $\sum_j A'_{i,j}$ denotes a *weighted* citation count. This approach is taken by Weighted Citation [51] and MR-Rank [56] using citation gap, and by Retained Adjacency Matrix [18] using citation age.

In PageRank-like methods, the importance of a citation depends not only on the importance the citing paper carries, but also on a citation's time quantity. Timed PageRank [54, 55], and NewRank [15] adopt this idea using exponentially decayed citation, or cited paper age, and thus compute the score of paper i with a formula of the form:

$$s_i = \alpha \sum_j \kappa e^{-\gamma \tau_{ij}} P_{i,j} s_j + (1 - \alpha) u_i \tag{5.2}$$

Effective Contagion Matrix [18], is another time-aware method using citation age. However it is not based on PageRank, but on Katz centrality [33], i.e., compared to Equation 5.2 it uses the adjacency matrix A and does not calculate random jump probabilities. Other time-aware weights have also been proposed. For example, [13] uses a weight based on the ratio of the cited paper's number of citations divided by its age. Further, [35] and its extension, SARank [37] use a citation age-based exponentially decaying weight, but only when the cited paper has reached its citation peak, i.e., the year when it receives the most citations.

5.3.3 Time-Aware Landing Probabilities

In PageRank and several PageRank-like methods, papers are assumed to all have an equal landing probability, but in several cases non-uniform probabilities are assigned. We denote as u_i the landing probability assigned to paper i. Past works assign landing probabilities that decay exponentially with the paper's age, i.e.:

$$u_i = \kappa e^{-\gamma(t-t_i)}$$

This implies that newer papers have higher visibility than old ones. Note the contrast between the time quantities described in Section 5.3.2, which refer to *edges*, and the single time quantity, paper age, that concerns *nodes*.

We discern two ways in which landing probabilities can affect the network process. The first is in the probabilities of visiting a node after a random jump, similar in spirit to topic-sensitive [20] and personalised [24] PageRank. CiteRank [47], FutureRank [41], YetRank [22], and NewRank [15] compute the score of paper i with a formula of the form:

$$s_i = \alpha \sum_j P_{i,j} s_j + (1 - \alpha)\kappa e^{-\gamma(t-t_i)}$$

To be precise, NewRank also employs a time-aware transition matrix as per Section 5.3.2, while the process in FutureRank involves an additional authorship matrix for the paper-author network, as discussed in Section 5.3.5.

The second way is more subtle and concerns dangling nodes. Recall that the standard approach is to create artificial edges to all other nodes assigning uniform transition probabilities. Instead, YetRank [22] assigns exponential decaying transition probabilities to dangling nodes as:

$$P_{i,j} = \kappa e^{-\gamma(t-t_i)}, \quad \forall i, j : k_j^{out} = 0$$

5.3.4 Paper Metadata

Ranking methods may utilize paper metadata, such as author and venue information. Scores based on these metadata can be derived either through simple statistics calculated on paper scores (e.g., average paper scores for authors or venues), or from well-established measures such as the Journal Impact Factor [17], or the Eigenfactor [4].

The majority of approaches in this category incorporates paper metadata in Page-Rank-like models, to modify citation, or transition matrices and/or landing probabilities. Weighted Citation [51] modifies citation matrix A using weights based on the citing paper's publication journal. Thus, the method gives higher importance to citations made by papers published in high rank venues. YetRank [22] modifies PageRank's transition matrix P and landing probabilities u. Particularly it uses journal impact factors to determine the likelihood of choosing a paper when starting a new random walk, or when moving from a dangling node to any other paper. This way, it simulates researchers that prefer choosing papers published in prestigious venues when beginning a random walk. NTUWeightedPR [13] modifies transition matrix P and landing probabilities u. It uses weights based on the cited paper's author, venue, and citation rate information, to simulate a "focused" researcher, who prefers following references to, or initiating a random walk from papers that are written by well-known authors, published in prestigious venues, and which receive many citations per year.

An alternative to the above approaches is presented in Timed PageRank [54, 55], which calculates the scores of recent papers, for which only limited citation information is currently available, solely based on metadata, while using a time-aware PageRank model for the rest. Particularly, scores for new papers are calculated based on averages (or similar statistics) of their authors' other paper scores, or based on average paper scores (or similar statistics) of other papers published in the same venue. A similar approach is followed by ArtSim [11], which relies on topic and author metadata. ArtSim recalculates the scores produced by any other ranking method, for all recently published papers. Hence, it can be applied on top of any other method to further improve its effectiveness.

5.3.5 Multiple Networks

Ranking methods may also employ iterative processes on multiple interconnected networks (e.g., author-paper, venue-paper networks) in addition to the basic citation network. We can broadly discern two approaches: the first approach is based on mutual reinforcement, an idea originating from HITS [28]. In this approach ranking methods perform calculations on bipartite graphs where nodes on either side of the graph mutually reinforce each other (e.g., paper scores are used to calculate author scores and vice versa), in addition to calculations on homogeneous networks (e.g. paper-paper, author-author, etc). In the second approach, a single graph spanning heterogeneous nodes is used for all calculations.

The first of the aforementioned approaches is followed by FutureRank [41], Wang et al. [48], and COIRank [2]. FutureRank combines PageRank on the citation graph with an author-paper score reinforcement calculation and an age-based factor. MR-Rank uses a bipartite paper-venue graph, along with PageRank calculations on paper and venue graphs to rank papers and venues using linear combinations of their scores. Wang et al. use paper-venue and paper-author bipartite networks, as well as time-based weights to rank papers, authors, and venues. COIRank extends this model by modifying paper citation edges when their authors have previously collaborated, or when they work at the same institution. The goal is to reduce the effect of artificially boosted citation counts.

The second approach to using multiple networks is used by MutualRank [25]. MutualRank uses an adjacency matrix comprised of 3 inter- and 6 intra-network individual adjacency matrices. The intra-networks are weighted, directed graphs of papers, authors, and venues, while the inter-networks consist of edges between the aforementioned graphs (i.e., edges between papers and authors, etc). MutualRank ranks all of the aforementioned nodes, based on an eigenvector calculation on this aggregated adjacency matrix.

5.3.6 Ensemble Methods

Ensemble methods implement multiple ranking methods, and combine their results to come up with a single score per paper. The majority of the 2016 WSDM Cup[1] methods fall in this category. The goal of the Cup was to rank papers based on their "query-independent importance" using information from multiple interconnected networks [46].

NTUEnsemble [10], combines scores from the metadata-based version of Page-Rank proposed in NTUWeightedPR [13], with the cup's winning solution [16], and a method based on Wang et al [48]. In EWPR [35] paper scores are a combination of time-weighted PageRank scores calculated on a paper graph, time-weighted PageRank venue scores, calculated on a venue graph, and author scores, calculated as averages of their authored papers' PageRank scores. SARank [37] extends EWPR by including an additional score for papers based on exponentially weighted citation counts, as well as additional scores for authors and papers based on averages of this aforementioned citation count-based score. Finally, in bletchleypark [21], paper scores result as a linear combination of citation counts, PageRank scores, paper age, author, and venue scores, where author and venue scores are based on aggregations derived from their respective papers.

5.3.7 Other Methods

We discern a handful of methods that do not fall into any of the aforementioned categories. Age-Rescaled PageRank [38] and Age- and Field-Rescaled PageRank [44] calculate simple PageRank scores and then rescale them. In the case of Age-Rescaled PageRank this is done based on the mean and standard deviation of the scores of n papers published before and after each paper in question. In the case of Age- and Field-Rescaled PageRank this rescaling is performed only based on papers published in the same field.

5.4 Ranking Methods

In the following we present the examined ranking methods in more detail. As discussed in Section 5.1, we examine methods which relate to the problem of impact-based ranking, where impact is defined as a centrality calculated on the citation network formed by citations made in the near-future. In particular the methods presented have at least one of the following characteristics:

- They are time-aware, i.e., they include the dimension of time in their computational model.

---

[1] http://www.wsdm-conference.org/2016/wsdm-cup.html

- They have been evaluated in their original work based on a ground truth ranking that uses centrality values of citation networks formed by near-future citations.
- They have been shown to perform well in ranking based on the expected paper impact by third party studies.

In the following we present the ranking methods of the literature based on their chronological order of publication.

5.4.1 Timed PageRank

Timed PageRank [54, 55] uses a PageRank variant that incorporates time-based weights, which depend on citation age (Section 5.3.3). It then calculates rank scores by applying an aging function on the scores of this time-weighted PageRank variant. The PageRank variant incorporates time-based weights w_i into PageRank as:

$$s_i^{\{T\}} = \alpha \sum_j w_j P_{i,j} s_j^{\{T\}} + (1 - \alpha) \tag{5.3}$$

where $\alpha \in [0, 1]$, as with simple PageRank. The citation weights w_i in Eq. 5.3, depend on *citation age* and are defined as follows:

$$w_i = d^{\frac{t_c - t_i}{12}} \tag{5.4}$$

where $d \in (0, 1]$ is called the *decay rate* and the time quantities $(t_c - t_i)$ are given in months. Incorporating these weights into PageRank's transition matrix reduces the effect of older citations on the score of a paper.

To calculate the final ranking scores, Timed PageRank, modifies the scores calculated by Eq. 5.3 based on whether a paper is considered as new. For papers that are not considered new, the scores are multiplied by one of two factors, which are called, depending on the method's implementation, the Aging [54] and Trend Factor [55]. For this set of papers, the final Timed PageRank scores of papers i are calculated as:

$$s_i = \begin{cases} f^{\{ageing\}}(i) s_i^{\{T\}} \\ f^{\{trend\}}(i) s_i^{\{T\}} \end{cases} \tag{5.5}$$

In Eq. 5.5, $f^{\{ageing\}}(i)$ returns a number in the range [0.5, 1], determined by a linearly decreasing function, based on paper i's age. On the other hand, $f^{\{trend\}}(i)$ is based on the so-called trend ratio, which is the quotient of the number of citations paper i received in the most recent quarter year, divided by the number of citations received in the preceding quarter, normalised to fit the range [0.5, 1]. Papers receiving less than a citation per month are assigned a minimum trend factor value by default.

In the case of papers considered as new, i.e., those that still haven't accumulated citations due to being published recently, the rank scores are calculated differently. In particular, the scores of papers i are calculated based on statistics derived from their authors or publication venues. Such statistics, for authors and venues can be calculated, for example, using the averages of the Timed PageRank values of all their papers.

5.4.2 CiteRank

CiteRank [47] is a time-aware, PageRank-based ranking method, with landing probabilities based on publication age (Section 5.3.3). Essentially, it modifies the behavior of random researchers so that they prefer to start their random walks from recently published papers. The CiteRank score of paper i is given as:

$$s_i = \rho_i + (1 - \alpha) \sum_j P_{i,j} \rho_j + (1 - \alpha)^2 \sum_j P_{i,j}^2 \rho_j + \cdots \qquad (5.6)$$

or equivalently [33]:

$$s_i = \alpha \sum_j P_{i,j} s_j + (1 - \alpha) \rho_i \qquad (5.7)$$

In Eq. 5.6-5.7, ρ is an exponentially decreasing function, based on publication age. It models how likely random researchers are to initially select a paper and biases them with a preference for recent ones. The value of ρ_i is calculated as $\rho_i = e^{-\frac{t_i}{\tau}}$, where τ is called the *Decay Time*, hence the value of ρ diminishes with paper i's age. This emulates the fact that researchers are most likely to read papers recently published in conferences they attended, or journals they read. Similarly to simple PageRank, with probability α, researchers follow references from papers they are reading, while with probability $1 - \alpha$ they start reading any paper in the citation network, with landing probabilities based on ρ.

5.4.3 FutureRank

FutureRank [41] scores papers based on calculations on the citation and the authorship network. The method is iterative and involves an author and a paper score calculation in each iteration, where author scores reinforce paper scores, and paper scores reinforce author scores (as inspired by HITS [28]). The FutureRank scores of authors and papers are calculated iteratively based on the following set of equations:

$$a_i = \sum_j M_{i,j}^{\mathsf{T}} s_j \qquad (5.8)$$

$$s_i = \alpha \sum_j P_{i,j} s_j + \beta \sum_j M_{i,j} a_j + \gamma \rho_i + (1 - \alpha - \beta - \gamma) \frac{1}{N} \qquad (5.9)$$

where M^{T} is the transpose of authorship matrix M. The function ρ_i is an exponentially decreasing weight, based on paper i's age, given as $e^{-c(t_c - t_i)}$, where parameter c is determined based on the empirical distribution of citations which papers receive each year after their publication. In the original work, the best fit to the empirical distribution was given for $\rho_i = e^{-0.62(t_c - t_i)}$. Parameters α, β, γ in Eq. 5.8 and 5.9 take values in $[0, 1]$. Based on this, Eq. 5.9 includes a uniform landing probability. However, in the original work, authors set $\alpha + \beta + \gamma = 1$, thus neutralizing the last factor of Eq. 5.9. Hence, FutureRank calculates author scores based on the scores of their authored papers, while it calculates paper scores by combining PageRank, author scores, and a time-based weight, based on publication age. By tuning parameters α, β, γ, it is possible to determine the contribution each of these components has in the paper score calculation.

5.4.4 Weighted Citation

Weighted Citation [51] is a variation of the Citation Count, where citations have a weight, based on the time interval between the publication and the citation of a paper (citation gap, see Section 5.3.2) and also based on the citing paper's publication venue. According to this method citations are viewed as academic votes and should not contribute equally to an article's score. In particular the status of an article is determined by: (a) the number of citations it receives, (b) the prestige of the venues of its citing papers, and (c) the time interval that passed until articles receive each of their citations. This last time interval is considered important, because articles that are cited soon after their publication are seen as: (a) breakthrough articles that produce important results, or (b) works of trusted authorities in a given scientific field. Therefore, shorter time intervals until a paper is cited are assumed to construe greater prestige. The method has been shown to perform well in ranking papers based on their short-term impact [26].

The Weighted Citation score of paper i, is calculated as follows:

$$s_i = \sum_j A_{i,j} v_{v(j)} e^{-0.117(t_j - t_i)} \tag{5.10}$$

Equation 5.10 implements the method's intuitions by counting all citations of paper i, each one multiplied by factors $v_{v(j)}$ and $e^{-0.117(t_j - t_i)}$. Factor $v_{v(j)}$ is a venue score called the *article influence score* of the venue where paper j is published. This score is calculated based on the formula $v_i = 0.01 \dfrac{v_i^{\{EF\}}}{n_i}$, where:

- $v_i^{\{EF\}}$ is the EigenFactor score [4] of venue i. The EigenFactor score is a PageRank-like score calculated for venues, on a citation graph with venues as nodes.

- n_i denotes the number of articles published in venue i over a five-year target window, normalised by the total number of articles published in all venues over the same time window.

Factor $e^{-0.117(t_j - t_i)}$ is a time decaying function based on citation gap (see Section 5.3.2). The constant factor in the exponent is empirically determined based on the best fit of an exponential function on the empirical distribution of citations which papers receive in the years after their publication, on the dataset where the method was originally evaluated. Based on this distribution, articles receive most of their citations one to two years after their publication, while in the years after this, the number of incoming citations decreases exponentially. Similar empirical data on citations are presented in other works (e.g., [41]).

5.4.5 Yet Rank

"Yet another paper ranking method advocating recent pubs", for convenience referred to as YetRank [22], is another PageRank variation which uses modified landing probabilities. These landing probabilities depend on publication age and venue metadata. The method defines the age-based weights ρ_i, of paper i, as:

$$\rho_i = \frac{e^{-\frac{t_i}{\tau}}}{\tau} \tag{5.11}$$

where τ denotes a characteristic decay time. Further, the method uses the Impact Factor [17] of venues, calculated over the period of the last 5 years as the score of paper i's publication venue (i.e., $v_{v(i)}$ denotes the Impact Factor of paper i's publication venue). The method combines the two weights, for each paper i, as:

$$w_i = \frac{v_{v(i)}\rho_i}{\sum_j v_{v(j)}\rho_j} \tag{5.12}$$

YetRank uses these weights to modify both the landing probabilities and the transition probabilities from dangling nodes of transition matrix P, used by PageRank. Hence, the YetRank of paper i is given by the following formula:

$$s_i = \alpha \sum_j P_{i,j} s_j + (1 - \alpha)w_i \tag{5.13}$$

where the transition matrix $P_{i,j}$ has been modified to incorporate the values of w_j, instead of the uniform value $1/N$ on its rows corresponding to dangling nodes. YetRank's modifications on PageRank have the following effect on the behavior of the random researcher: they prefer to choose recently published papers in high prestige venues, when starting to randomly read papers, as well as when choosing a paper in the citation network after visiting a dangling node.

5.4.6 Time-Aware Ranking – RAM

The *Retained Adjacency Matrix* (RAM) [18] is a modification of the common Adjacency Matrix A, where the non-zero elements of the matrix (i.e., those which denote a citation relationship) depend on the citation age (see Section 5.3.3). In particular the Retained Adjacency Matrix is defined as follows:

$$R_{i,j} = \begin{cases} \gamma^{t_c - t_j} & \text{if } A_{i,j} = 1 \\ 0 & \text{otherwise} \end{cases} \tag{5.14}$$

where t_c denotes the current time (see Section 5.2.1) and $0 < \gamma < 1$. Calculating citation counts based on this matrix has the effect of benefiting papers that have received recent citations compared to those that have accumulated a comparable number of citations, but in the past. The RAM score of paper i, is calculated as follows:

$$s_i = \sum_j R_{i,j} \tag{5.15}$$

5.4.7 TimeAware Ranking – ECM

The *Effective Contagion Matrix* (ECM) [18] is a Katz-centrality based score, which uses the retained adjacency matrix. In particular, the ECM matrix is calculated based on the following formula:

$$E = \sum_i \alpha^i R^i \tag{5.16}$$

where it holds $0 < \alpha < 1$. Essentially, this matrix encodes weights that depend on the length of citation chains, where the weights of individual citations depend on the citation age (see Section 5.3.2). The ECM scores of papers i are calculated using the equation:

$$s_i = \sum_j E_{i,j} \tag{5.17}$$

This calculation has the effect of advantaging papers that are part of many citation chains, where chains that include papers that have been published more recently contribute more to the cited paper scores.

5.4.8 NewRank

NewRank [15] is another PageRank variant, which uses time-aware weights both in the transition matrix and to modify landing probabilities. In both cases the weights are determined by publication age. NewRank uses the aging function defined by CiteRank (i.e., $\rho_i = e^{-\frac{t_i}{\tau}}$). Based on this function, NewRank introduces weights

$w_{i,j}$, into PageRank's transition matrix, as follows:

$$w_{i,j} = \frac{\rho_i}{\sum_i A_{i,j}\rho_i} \tag{5.18}$$

Note that in Eq. 5.18 weights are determined by the cited paper age, normalized based on the weights of all papers i cited by paper j. Based on these weights, the NewRank score of paper i is calculated as:

$$s_i = \alpha \sum_j w_{i,j}P_{i,j}s_j + (1 - \alpha)\frac{1}{N}\frac{\rho_i}{\|\underline{\rho}\|} \tag{5.19}$$

where $\|\rho\|$ equals the sum of ρ values of all papers. This formula modifies the behavior of a random researcher, so that he prefers to start reading recent papers, when initially selecting a paper, and to select the most recently published ones next, among those in the reference list of the paper they are currently reading.

5.4.9 MutualRank

MutualRank [25] is a ranking method that uses a single network, which encodes relations between papers, authors, and venues, and performs iterative calculations on this network to calculate paper, author and venue scores. This heterogeneous network is composed of three different inter-networks (i.e., networks with nodes of the same type) and three intra-networks (i.e., with nodes of different types). The three inter-networks encode author to author, paper to paper, and venue to venue relations. The three intra-networks encode author-paper, venue-paper, and author-venue relations. In the following we examine all network matrices which encode the aforementioned relations.

Stochastic citation matrix P. This is the classic per-column normalized stochastic matrix used by PageRank.

Author co-authorship matrix R. This matrix encodes which authors have written papers together. Its values are set as $R_{x,y} = \sum_{x,y,i:M_{i,x}=1,M_{i,y}=1} M_{i,x}$, and $R_{x,y} = 0$ if authors x and y have not co-authored any papers. Essentially the values of matrix R are equal to the number of papers co-authored by the pair of authors that correspond to its row and column. The method uses a row normalized version of this matrix, $R^{\{N\}}$.

Venue citation matrix J. This matrix encodes the citation relationship between papers published in different venues. Its values are defined as $J_{pq} = \sum_{i,j:v(i)=p,v(j)=q} A_{i,j}$, i.e., each cell has a value equal to the number of citations from papers in venue q to papers in venue p. A row normalized version of this matrix is, $J^{\{N\}}$, is used by the method.

Author-Paper network matrix $M^{\{N\}}$. This is a row normalized version of the classic authorship matrix defined in Section 5.2.1.

Venue-Paper network matrix $V^{\{N\}}$. This is a row normalized version of venue matrix V described in Section 5.2.1.

Author-Venue network matrix W. This matrix encodes which authors published in each venue, where $W_{p,q} = \frac{1}{n_q}$, n_q equals the number of authors that have published in venue q if author p has published in venue q, and $W_{p,q} = 0$ otherwise.

MutualRank defines a number of concepts, which it encodes as scores, and which are recalculated in each of its iterations. These concepts are paper authority and soundness, venue prestige, and author importance. The intuitions that are encoded by MutualRank are the following:

- High soundness papers cite high authority papers, and vice versa.
- High authority and high soundness papers are published in high prestige venues.
- Important authors publish high authority and high soundness papers and vice versa.

MutualRank encodes the above in the following set of equations:

$$
\begin{aligned}
s_i^{\{auth\}} = {} & \xi s_i^{\{auth\}} + \gamma(1-\xi)\left(\lambda \sum_j P_{i,j} s_j^{\{sound\}} + (1-\lambda)\frac{1}{N_p}\right) \\
& + \frac{(1-\gamma)(1-\xi)}{2}\left(\lambda \sum_j M_{i,j}^{\{N\}} a_j + (1-\lambda)\frac{1}{N_p}\right) \\
& + \frac{(1-\gamma)(1-\xi)}{2}\left(\lambda \sum_j V_{i,j}^{\{N\}} v_j + (1-\lambda)\frac{1}{N_p}\right)
\end{aligned}
\tag{5.20}
$$

$$
\begin{aligned}
s_i^{\{sound\}} = {} & \xi s_i^{\{sound\}} + \gamma(1-\xi)\left(\lambda \sum_j P_{i,j} s_j^{\{auth\}} + (1-\lambda)\frac{1}{N_p}\right) \\
& + \frac{(1-\gamma)(1-\xi)}{2}\left(\lambda \sum_j M_{i,j}^{\{N\}} a_j + (1-\lambda)\frac{1}{N_a}\right) \\
& + \frac{(1-\gamma)(1-\xi)}{2}\left(\lambda \sum_j V_{i,j}^{\{N\}} v_j + (1-\lambda)\frac{1}{N_v}\right)
\end{aligned}
\tag{5.21}
$$

$$a_i = \frac{(1-\gamma)(1-\xi)}{2}\left(\lambda \sum_j M_{i,j}^{\{N\}^{\mathsf{T}}} s_j^{\{auth\}} + (1-\lambda)\frac{1}{N_p}\right)$$

$$+ \frac{(1-\gamma)(1-\xi)}{2}\left(\lambda \sum_j M_{i,j}^{\{N\}^{\mathsf{T}}} s_j^{\{sound\}} + (1-\lambda)\frac{1}{N_p}\right)$$

$$+ \xi\left(\lambda \sum_j R_{i,j}^{\{N\}} a_j + (1-\lambda)\frac{1}{N_a}\right) + \gamma(1-\xi)\left(\lambda \sum W^{i,j^{\mathsf{T}}} v_j + (1-\lambda)\frac{1}{N_v}\right)$$

$$(5.22)$$

$$v_i = \frac{(1-\gamma)(1-\xi)}{2}\left(\lambda \sum_j V_{i,j}^{\{N\}^{\mathsf{T}}} s_j^{\{auth\}} + (1-\lambda)\frac{1}{N_p}\right)$$

$$+ \frac{(1-\gamma)(1-\xi)}{2}\left(\lambda \sum_j V_{i,j}^{\{N\}^{\mathsf{T}}} s_j^{\{sound\}} + (1-\lambda)\frac{1}{N_p}\right)$$

$$+ \gamma(1-\xi)\left(\lambda \sum_j W_{i,j}^{\mathsf{T}} a_j + (1-\lambda)\frac{1}{N_a}\right) + \xi\left(\lambda \sum_j J_{i,j}^{\{N\}} v_j + (1-\lambda)\frac{1}{N_v}\right)$$

$$(5.23)$$

where $\xi, \lambda, \gamma \in (0, 1)$ and N_p, N_a, N_v denote the total number of papers, authors, and venues respectively. The set of equations translates as follows: paper authority scores (Eq. 5.20) are determined by the papers own authority score, the PageRank score calculated based on the the soundness scores of other papers that cite it, based on the author importance scores of its authors, and the venue prestige score. The latter two, authors, and venues, contribute to each paper's authority score via a PageRank-like calculation. Similarly, paper soundness (Eq. 5.21) is determined by the paper's own soundness, a PageRank like calculation using paper authority scores, and PageRank-like calculations using its authors' importance and venue prestige scores. Author importance scores are determined by the authority and soundness scores of their papers, the author importance scores of their co-authors and the venue prestige of venues where they published. Finally, venue prestige scores (Eq. 5.23) are determined by the authority and soundness scores of papers they publish, the importance scores of authors that publish in them and the prestige of venues with papers that cite papers they publish.

The set of Eq. 5.20–5.23 can be rewritten in matrix-vector multiplication form, on which the power method is applied, in a fashion similar to PageRank, however on a heterogenenous network of papers, authors and venues.

5.4.10 Wang et al.

Wang et al. [48] rank papers using a framework that can be considered a generalization of FutureRank. Their method defines two types of scores for papers, namely hub and authority scores. Additionally, it defines hub scores for authors and venues. All scores are iteratively calculated, with hub scores being used in the authority score calculations of papers and authority scores of papers being used to calculate all hub scores. In particular, the hub scores of paper, author, and venue i, respectively ($s_i^{\{hub\}}, a_i^{\{hub\}}, v_i^{\{hub\}}$), are calculated based on the authority scores $s^{\{auth\}}$ of papers, according to the following set of formulas:

$$s_i^{\{hub\}} = \frac{\sum_j A_{i,j}^\mathsf{T} s_j^{\{auth\}}}{\sum_j A_{i,j}^\mathsf{T}} \tag{5.24}$$

$$a_i^{\{hub\}} = \frac{\sum_j M_{i,j}^\mathsf{T} s_j^{\{auth\}}}{\sum_j M_{i,j}^\mathsf{T}} \tag{5.25}$$

$$v_i^{\{hub\}} = \frac{\sum_j V_{i,j}^\mathsf{T} s_j^{\{auth\}}}{\sum_j V_{i,j}^\mathsf{T}} \tag{5.26}$$

where A^T is the transpose adjacency matrix of the citation network, M^T is the transpose authorship matrix, and V^T is the transpose venue matrix. Based on Eq. 5.24–5.26, a paper's hub score is the normalized sum of all authority scores of papers it cites. Similarly, the hub score of an author is the normalized sum of the authority scores of their authored papers and the hub score of a venue is the normalized sum of the authority scores of papers it publishes.

The authority scores for papers are given by a combination of author, paper, and venue hub scores, based on the following formula:

$$s_i^{\{auth\}} = \alpha s_i^{\{PR\}} + \beta \frac{1}{Z(M)} \sum_j M_{i,j} a_j^{\{hub\}}$$

$$+ \gamma \frac{1}{Z(V)} \sum_j V_{i,j} v_j^{\{hub\}} + \delta \frac{1}{Z(A)} \sum_j P_{i,j} s_j^{\{hub\}} \tag{5.27}$$

$$+ \theta e^{-\rho(t_c - t_i)} + (1 - \alpha - \beta - \gamma - \delta - \theta) \frac{1}{N}$$

where parameters $\alpha, \beta, \gamma, \delta, \theta \in (0, 1)$. Equation 5.27 calculates paper authority ($s_i^{\{auth\}}$) as a weighted sum of the paper's PageRank score ($s_i^{\{PR\}}$), three scores propagated by its authors', venue's, and citing papers' hub scores, a time-based weight, based on paper i's publication age, and a constant factor that is equal for all papers. $Z(M), Z(V), Z(A)$ are normalization factors, based on the total hub scores, of authors, venues and papers, respectively. Finally, the exponential factor ρ in Eq. 5.27 is set to the same value used in FutureRank.

The method also defines time-aware variations of its hub and authority score calculations. The variation applies time-aware weights for hub scores in a uniform manner. For example time-aware author hub scores are calculated as:

$$a_i^{\{hub\}} = \frac{\sum_j M_{i,j}^{\mathsf{T}} w_{i,j} s_j^{\{auth\}}}{\sum_j M_{i,j}^{\mathsf{T}} w_{i,j}} \tag{5.28}$$

where weights $w_{i,j} = \alpha^{t_c - t_i}$ and $\alpha > 1$. This way older articles have a bigger influence in the calculation of an author's hub score. Time-aware weights for papers and venues are calculated in a similar fashion, by introducing weights $w_{i,j}$ in Eq. 5.24 and 5.26.

Similarly, the factors that depend on hub scores in Eq. 5.27 are modified using time-aware weights. The modified equation becomes:

$$s_i^{\{auth\}} = \alpha s_i^{\{PR\}} + \beta \frac{1}{Z(M)} \sum_j M_{i,j} w_{i,j} a_j^{\{hub\}}$$

$$+ \gamma \frac{1}{Z(V)} \sum_j V_{i,j} w_{i,j} v_j^{\{hub\}} + \delta \frac{1}{Z(A)} \sum_j P_{i,j} w_{i,j} s_j^{\{hub\}} \tag{5.29}$$

$$+ \theta e^{-\rho(t_c - t_i)} + (1 - \alpha - \beta - \gamma - \delta - \theta) \frac{1}{N}$$

where all $w_{i,j}$ are calculated as:

$$w_{i,j} = \frac{1}{1 + b(t_c - t_i)} \tag{5.30}$$

and b is set to 1. Hence, in the authority score calculations, in contrast to the hub score calculations, paper, author, and venue hub scores contribute more, the more recently each paper was published.

An extension of this method, called COIRank [2], focuses on introducing additional weights in the above model, based on the previous or possible collaboration of authors. The aim of this modification is to provide fairer rankings that take into account conflicts of interest.

5.4.11 Non Linear PageRank

Non-Linear PageRank [53] is a modification of PageRank where each element of the transition matrix P, as well as the rank score of each citing paper, is raised to the power of $\theta + 1$, $\theta > 0$. Additionally, Non-Linear PageRank takes the $(\theta + 1)$-root of the scores it aggregates (in the first addend of Eq. 5.1). The effect of raising scores to the power of $\theta + 1$ is that those citing papers with higher scores contribute more to the score of cited papers. At the same time, the $\theta + 1$ root of the sum of scores reduces the effect of citations from low scoring papers. Given the above, Non-Linear PageRank's formula is described as follows:

$$s_i = \alpha \sqrt[\theta+1]{\sum_j P_{i,j}^{\circ(\theta+1)} s_j^{\theta+1}} + (1 - \alpha) \tag{5.31}$$

where $P_{i,j}^{\circ(\theta+1)}$ denotes PageRank's transition matrix with each element raised to the power of $\theta + 1$. As with simple PageRank, α is in the range $[0, 1]$.

5.4.12 SPRank

SPRank [57] is a variation of PageRank, which focuses on differentiating the influence of citing papers on the calculation of rank scores, based on how similar or dissimilar they are to the cited ones. Effectively, the method modifies the behavior of a random researcher, so that they prefer following citations to the most similar among the cited papers of a paper they currently read. The method is not restricted to the use of a particular similarity measure, $f_{i,j}$, between papers i and j. However, its authors make use of the following formula:

$$f_{i,j} = \frac{z(i) \cap z(j)}{\sqrt{k_i^{out} k_j^{out}}} \tag{5.32}$$

where $z(i)$ and $z(j)$ denote the set of papers cited by papers i and j, respectively. Hence, this similarity depends on the overlap of cited papers and the number of papers cited by each of the two papers that are compared. The SPRank score of each paper i is calculated by introducing this similarity measure into PageRank's formula, as follows:

$$s_i = \alpha \sum_j f_{i,j}^\theta P_{i,j} s_j + (1 - \alpha) u_i \tag{5.33}$$

where $\theta \geq 0$ is a tunable parameter, and u_i is set to the uniform landing probability ($u_i = \frac{1}{N}$).

In Eq. 5.32, $\tau(i)$ and $\tau(j)$ denote the set of papers cited by papers p_i and p_j, respectively. k_i and k_j are the numbers of papers cited by p_i and p_j, respectively. Thus, SPRank modifies the random researcher model, changing it into a "focused" researcher model. Here, the researcher does not follow each reference with equal probability. Instead, he prefers following references to papers that cite, to the largest extent possible, the same set of articles cited by the one he currently reads.

5.4.13 EWPR

Ensemble Enabled Weighted Pagerank (EWPR) is an ensemble-based method that ranks papers based on the combination of three scores: a time-aware modification of PageRank, a publication venue score, and a score based on their authors. These scores are rescaled to have common average values and are then combined based on the following formula:

$$s_i = \frac{s_i^{\{T\}} + \alpha v_{v(i)} + \beta l_{1(i)}}{1 + \alpha + \beta} \tag{5.34}$$

where α and β are tunable parameters, $s_i^{\{T\}}$ denotes the time-aware PageRank weight, $v_{v(i)}$ denotes the score of paper i's publication venue, and $l_{1(i)}$ denotes a score derived from paper i's authors.[2] We now detail how these individual scores are calculated.

Time-aware PageRank $(s_i^{\{T\}})$. The time-aware PageRank variation adopted by EWPR defines citation weights $w_{i,j}$, which do not fall into the categorization presented in Section 5.3.2. Here, citing paper citation weights decrease with time, only if the paper in question has already passed its citation peak, i.e., the year in which it gathered its highest number of citations. Weights $w_{i,j}$ are defined as:

$$w_{i,j} = \begin{cases} 1, t_i < p_j \\ \frac{1}{(\ln(e+t_i-p_j))^\tau}, t_i > p_j \end{cases} \tag{5.35}$$

where τ is a decaying factor and p_j denotes the year when paper j peaks in terms of citations. These time based weights are used to modify PageRank, giving the following formula:

$$s_i^{\{T\}} = \alpha \sum_j \frac{A_{i,j} w_{i,j} s_j^{\{T\}}}{A_{i,j} w_{i,j}} + (1 - \alpha) \tag{5.36}$$

Venue based scores $(v_{v(i)})$. Venue scores are calculated based on the paper edge weights of Eq. 5.35, this time applied on a venue-venue graph. Let J denote be the venue-venue network matrix. J is a weighted adjacency matrix defined as:

$$J_{p,q} = \begin{cases} \sum_{i,j:v(i)=p,v(j)=q} A_{i,j} w_{i,j} \\ 0, \text{otherwise} \end{cases} \tag{5.37}$$

The values in each cell of J are therefore equal to the sum of time-based weights of papers in venue q citing papers in venue p. To calculate venue based scores Eq. 5.36 is used, with matrix J replacing matrix P and venue scores v_i replacing paper scores s_i.

Author based scores $(l_{1(i)})$. Author-based scores are calculated as the average scores of their authored papers.

---

[2] Note, that the method additionally proposed defining affiliation based scores, however in the experimental setting examined, these did not lead to better results, and where, thus, omitted. Hence we do not include them in the method's description.

5.4.14 NTUWeightedPR

NTUWeightedPR [13] is a PageRank variation that introduces a compound weight w_i into PageRank's formula. In particular, weights w_i depend on publication venues, authors, author affiliations and yearly citations. NTUWeightedPR's formula is defines as follows:

$$s_i = \alpha \sum_j \frac{w_i}{\sum_i A_{i,j} w_i} P_{i,j} s_j + (1 - \alpha) \frac{w_i}{\sum_i w_i} \qquad (5.38)$$

where both addends of Eq. 5.38 use normalized values of weights w_i. In the case of the first addend, w_i is normalized, per citing paper j, based on the weights of all papers cited by paper j. The second addend normalizes weight w_i based on all papers in the graph. Equation 5.38 modifies the random researcher's behavior both when selecting to read papers from a reference list, as well as when randomly selecting any paper.

The weights w_i themselves are calculated based on the following formula:

$$w_i = w_i^{\{C\}} + w_i^{\{V\}} + w_i^{\{L\}} + w_i^{\{F\}} \qquad (5.39)$$

where $w^{\{C\}}$ denotes a citation-based weight, $w_i^{\{V\}}$ denotes a venue-based weight, $w_i^{\{L\}}$ denotes a weight based on paper i's authors, and $w_i^{\{F\}}$ denotes an affiliation-based weight. In the following we examine each of these weights.

Average Citations Per Year ($w^{\{C\}}$). This weight is defined as $w_i^{\{C\}} = \frac{\sum_j A_{i,j}}{t_c - t_i + 1}$, which equals the average number of yearly citations received by paper i. For papers with zero citations this weight is set to a negligible small value, $w_i^{\{C\}} = \epsilon$.

Venue-based Weight ($w_i^{\{V\}}$). This weight is defined as the sum of average yearly citations of papers in paper i venue, normalized by their number: $w_i^{\{V\}} = \frac{\sum_i V_{i,j} w_i^{\{C\}}}{\sum_i V_{i,j}}$.

Author-based Weight ($w_i^{\{L\}}$). This weight is is calculated as an average of an author-based weight, $w_j^{\{A\}}$, which is in turn based on the previous two weights defined. The author-based weight, $w_j^{\{A\}}$ for author j is given as follows:

$$w_j^{\{A\}} = \frac{1}{\sum_i M_{i,j}} \sum_i M_{i,j} \left(\frac{w_i^{\{C\}} + w_i^{\{V\}}}{\sum_k M_{i,k}} \right) \qquad (5.40)$$

which translates as the sum, for all the papers the author has written, of citation and venue weights, normalized by the number of each paper's authors, and the resulting quantity being normalized by the number of papers written by the author. To produce $w_i^{\{L\}}$, the method normalizes the sum of $w_j^{\{A\}}$, $j \in \ell(i)$, by the number of i's authors, as: $w_i^{\{L\}} = \frac{\sum_j M_{i,j} w_j^{\{A\}}}{\sum_j M_{i,j}}$.

Affiliation Weight ($w_i^{\{F\}}$). This weight is calculated following the same procedure used for the author-based weight, with the difference that in place of authorship matrix

M, an affiliation matrix is used. Let F be the affiliation matrix. Then $F_{i,j} = 1$ if paper i is written by authors with affiliation j, and $F_{i,j} = 0$, otherwise. Equation 5.40 is used with matrix F in place of matrix M, to get an affiliation specific score, which is then normalized by all affiliations of the paper, to produce $w_i^{\{F\}}$.

5.4.15 Bletchleypark

Team bletchleypark's solution [21] to the WSDM 2016 cup challenge is an ensemble method that utilizes various scores calculated based on papers, authors and venues, and institutions. The method ultimately calculates a single score based on the linear combination of these individual scores. Each paper's score is calculated based on the following formula:

$$s_i = 2.5s_i^{\{cit\}} + 0.1s_i^{\{age\}} + s_i^{\{PR\}} + l_{\ell(i)} + 0.1v_{v(i)} + 0.01s_i^{\{inst\}} \qquad (5.41)$$

where $s_i^{\{cit\}}$ is a citation-based score, $s_i^{\{age\}}$ is an age-based score, $s_i^{\{PR\}}$ is the paper's PageRank score, $l_{\ell(i)}$ is a score based on the paper's author list, $s_i^{\{v\}}$ is a score based on paper i's publication venue, and $s_i^{\{inst\}}$ is a score based on the institutions of the paper's authors. The various coefficients were experimentally determined based on the ground truth ranking provided by WSDM's 2016 cup ranking challenge. We now examine the individual scores.

Citation-based score $(s_i^{\{cit\}})$. This score is calculated based on the number of citations, per author of the paper, with a threshold on a maximum value. This is calculated based on:

$$s_i^{\{cit\}} = \begin{cases} \frac{\sum_j A_{i,j}}{\sum_j M_{i,j}}, & \sum_j A_{i,j} < \tau \\ \frac{\tau}{\sum_j M_{i,j}}, & \sum_j A_{i,j} > \tau \end{cases} \qquad (5.42)$$

where τ is a threshold set to $\tau = 5000$.

Age-based score $(s_i^{\{age\}})$. This is a simple score set equal to paper i's publication year, i.e., $s_i^{\{age\}} = t_i$. This essentially means that a paper's $s_i^{\{age\}}$ increases linearly the more recent a paper is.

PageRank score $(s_i^{\{PR\}})$. This is calculated based on the PageRank formula given in Eq. 5.1, with the addition of a "dummy" paper in the transition matrix, which is cited, and cites all other papers, except for itself.

Author list-based score $(l_{\ell(i)})$. This score is based on the mean value of average citations per paper of each of paper i's authors and is given by:

$$l_{\ell(i)} = \frac{\sum_j M_{i,j} \left(\frac{\sum_{p,q:M_{p,j}=1, A_{p,q}=1} A_{p,q}}{\sum_k M_{k,j}} \right)}{\sum_j M_{i,j}} \qquad (5.43)$$

Venue-based score ($v_{v(i)}$). This score is determined based on all citations that papers in venue $v(i)$ receive, i.e., $v_{v(i)} = \sum_{i,j:V_{i,v(i)}=1} A_{i,j}$.

Institution-based score ($s_i^{\{inst\}}$). This score is calculated as the total number of citations which papers of all authors of a particular institution have received, for all institutions the authors of the paper are affiliated with, normalized by the number of these institutions. If F is the set of institutions the authors of paper i are affiliated with and $F_{(f)}$ is the set of authors affiliated with institution f, then this score is given as:

$$s_i^{\{inst\}} = \frac{\sum_{f \in F} \sum_{p,q,z:z \in f, M_{p,z}=1} A_{p,q}}{|F|} \tag{5.44}$$

5.4.16 Age- and Field- Rescaled PageRank

Age-Rescaled PageRank [38] is a method that aims to alleviate PageRank's bias against recent publications. To achieve this it post-processes PageRank scores, based on their average and standard deviation values, calculated for papers published in a particular time frame. The intuition behind this, is that a paper can only be fairly compared in terms of rank scores with other papers of similar publication dates. Hence, the Age-Rescaled PageRank for paper i is given by:

$$s_i = \frac{s_i^{\{PR\}} - \mu_l}{\sigma_i} \tag{5.45}$$

where $s_i^{\{PR\}}$ denotes the PageRank score of paper i and μ_i and σ_i denote the mean and standard deviation, respectively, of PageRank scores for a particular group of papers. Note, that the mean and standard deviation are calculated *per paper*, based on a particular group of papers. In particular, the method initially labels all papers by age and orders them. Following this, a set of n papers, $n/2$ of which were published before paper i and $n/2$ of which were published after paper i, based on the ordering of papers, are selected. Finally, the mean and standard deviation of PageRank scores are based on this set of n papers are calculated for each paper.

Age Rescaled PageRank scores were shown to produce top-ranked sets of papers with more equal distributions of publication years. An extension of Age-Rescaled PageRank, termed Age- and Field- Rescaled PageRank [44], applies the same principles, with the exception that the set of papers n is not only chosen based on publication age, but also based on the scientific field of paper i, i.e., the rescaling of scores is based only on papers of similar age and scientific discipline. In the original work, the respective fields were given by Microsoft Academic Graph, on which the method was evaluated.

5.4.17 SARank

SARank [37] is an ensemble method which uses a linear combination of three different scores, calculated on multiple networks, namely on: the citation graph, the author-paper graph, and the venue-venue graph. These scores are combined, producing the SARank paper score as:

$$s_i = \alpha s_i^{\{C\}} + \beta v_{v(i)} + (1 - \alpha - \beta) l_{\ell(i)} \tag{5.46}$$

where $\alpha, \beta \in [0, 1]$, $s_i^{\{C\}}$ is a citation-based score of paper i, $v_{v(i)}$ is the score of its venue and $l_{\ell(i)}$ is a score derived from its authors. Each of the individual scores, in turn result from a combination of a "popularity" and a "prestige" score for papers, venues, and authors. We now examine each of these individual scores.

Paper citation-based Scores. SARank considers the importance of an article as a combination of its "prestige", which is defined by a time-aware PageRank variant, and its "popularity", which is defined based on how recently it gathered citations. The total citation-based score of paper i is given as:

$$s_i^{\{C\}} = (s_i^{\{T\}})^\lambda (s_i^{\{P\}})^{1-\lambda} \tag{5.47}$$

where $s_i^{\{T\}}$ is the time-aware PageRank variant score, $s_i^{\{P\}}$ is the popularity score for paper i, and $\lambda \in [0, 1]$.

The popularity score $s_i^{\{P\}}$ of a paper is given as a time-weighted citation count, based on citation age:

$$s_i^{\{P\}} = \sum_j A_{i,j} e^{\sigma(t_c - t_j)} \tag{5.48}$$

where σ is a negative decaying factor and t_c is the current year.

The prestige of a paper is given by a time-aware PageRank variant based on a modification of citation gap. The method makes use of a paper citation trajectory categorisation, presented in [9]. From this, it follows that papers exhibit different citation trajectories where, notably, some display an early, and others a late peak in citations.[3] Hence, the time-aware weights are dependent on the year of each paper's citation peak. The weights $w_{i,j}$ employed are calculated as:

$$w_{i,j} = \begin{cases} 1, t_j < p_i \\ e^{\sigma(t_j - p_i)}, t_j > p_i \end{cases} \tag{5.49}$$

where p_i denotes the year when paper i peaks in terms of citations and σ is a negative decaying factor. The citation peak year p_i is calculated in relative terms, as the maximum of the function $\Psi_i = \frac{\Phi_i}{\log Z_i}$, where Φ_i denotes the number of citations of paper i in a particular year and Z_i denotes the total number of citations paper i accumulated until the same year. Intuitively, Eq 5.49 states that a citation is

---

[3] Other citation trajectories (e.g., monotonically increasing citations per year) exist as well. For a full reference see [9].

considered of less importance if citing paper j cites paper i, after the latter has reached its citation peak. The weights $w_{i,j}$ are incorporated into the following Time-aware PageRank formula:

$$s_i^{\{T\}} = \alpha \sum_j \frac{w_{i,j}}{W_{i,j}} P_{i,j} s_j^{\{T\}} + (1 - \alpha) u_i \tag{5.50}$$

where the landing probabilities u_i are defined as uniform, i.e., $1/N$ for all papers, and $W_{i,j}$ is the sum of citation weights $w_{i,j}$ from all papers j citing paper i.

Author Scores. Similarly to papers, SARank calculates author scores which are, per author, a combination of "popularity" and "prestige". The scores are combined using Eq. 5.47, with author scores replacing paper scores. In the case of authors that wrote paper i, their popularity and prestige is defined as the average popularity and prestige of their authored papers. The total author score $l_{\ell(i)}$ of paper i results as the average of all individual author scores.

Venue Scores. SARank's venue scores are calculated similarly to the individual paper scores, i.e., as a combination of venue "prestige" and venue "popularity" on a venue graph, where nodes represent venues and edges between them represent citations from papers in one venue to papers in the other. Venue prestige is calculated using Eq. 5.50 with venues as nodes and weights $w_{p,q}$ given as follows:

$$w_{p,q} = \sum_{i,j:v(i)=p,v(j)=q} A_{i,j} w_{i,j} \tag{5.51}$$

where p, q are publication venues of cited paper i and citing paper j, respectively. Venue popularity is calculated as the average popularity of all papers it publishes. Overall venue scores, $v_{v(i)}$ of the venue publishing paper i, result from applying Eq. 5.47 on venue popularity and prestige.

5.4.18 MR-Rank

MR-Rank [56] (Mutual Reinforcement Rank) is a time-aware method, based on citation gap, which calculates venue and paper scores that mutually reinforce each other. In addition to citation matrix A, and venue matrix V, the method additionally uses a venue-venue network. This network is encoded by a venue network matrix, which we denote with J. Each edge in this network is weighted, based on the fraction of total citations made by papers in each venue to papers published in each other venue, or:

$$J_{p,q} = \frac{\sum_{i,j:v(i)=p,v(j)=q} A_{i,j}}{\sum_{i,j:v(j)=q} A_{i,j}} \tag{5.52}$$

MR-Rank iteratively calculates paper and venue scores based on the following set of formulas:

$$s_i = \alpha v_{v(i)} + (1 - \alpha) \sum_j A_{i,j} s_j e^{-\tau(t_j - t_i)} \tag{5.53}$$

$$v_i = \beta \sum_j J_{i,j} v_j + (1 - \beta) \frac{\sum_j V_{j,i} s_j}{\sum_j V_{j,i}} \tag{5.54}$$

where $\alpha, \beta \in [0, 1]$. The score $v_{v(i)}$ of paper i's publication venue is calculated as the average of the values it takes for the last three years. Parameter τ is tunable and determines how quickly a citation weight decreases, based on citation gap. Essentially, the pair of Eq. 5.53 and 5.54 state: (a) a paper's score is a weighted sum of its publication venue score and it's number of citations, weighted by citation gap, and (b) a venue's score is the weighted sum of the average score of papers it publishes, and the scores of other venues with papers that cite papers it publishes. The scores calculated by Eq. 5.53 and 5.54 are normalized after each iteration and the computations complete when paper and venue scores converge.

5.4.19 ArtSim

Most methods that predict impact-based paper rank rely on the existing citation history of each paper. However, since very limited citation data are available for recent articles, these methods fail to provide good estimations for their impact-based rank. ArtSim [11], is an approach that was introduced to alleviate this issue. The key idea is that in addition to the paper's citation data, it also takes into account the history of similar papers for which more concrete data are available. The intuition is that similar papers are likely to follow a similar trajectory in terms of popularity.

To estimate paper similarity, ArtSim exploits author and topic metadata of each paper. In particular, ArtSim relies on the JoinSim [50] path-based similarity measure for knowledge graphs to capture the similarity of papers based on their common authors and on their common topics.

It is important to highlight that ArtSim initially assigns a popularity score based on another paper ranking method, and then recalculates scores based on the aforementioned similarity measures for all recent papers. Hence, ArtSim is not an independent method, but rather it is applied on top of another to improve its accuracy by better estimating the rank of recently published papers. Based on the above, the ArtSim score of each paper i is calculated based on the following formula:

$$s_i = \begin{cases} \alpha s_i^{\{auth\}} + \beta s_i^{\{top\}} + \gamma s_i^{\{def\}}, & t_i \geq t_c - y \\ s_i^{\{def\}}, & \text{otherwise} \end{cases}$$

where α, β, γ and y are parameters of the method (with $\alpha + \beta + \gamma = 1$ and $y \in \mathbb{R}$), t_c stands for the current year, $s_i^{\{def\}}$ is the popularity score of paper i according to the predetermined (default) rank prediction method, while $s_i^{\{auth\}}$ and $s_i^{\{top\}}$ are the average popularity scores (according to the default ranking method) for all papers that are similar to i based on author-based and topic-based similarity, respectively. To calculate these averages, for each paper, the papers that achieve similarity above a predetermined threshold k according to each of the two similarity measures are considered, and k is a tunable parameter.

5.4.20 AttRank

AttRank [27] is a PageRank variant that uses a comination of a publication age-based weight, similar to the one used by FutureRank, as well as a weight based on each paper's recent attention, in place of PageRank's uniform landing probabilities. The recent attention-based weight is calculated by a time-restricted version of the preferential attachment mechanism. Preferential attachment describes the behavior where nodes newly into a network prefer to connect to other highly connected nodes (e.g., scientists are more likely to cite highly cited papers, for example due to visibility), with probability that is proportional to the latter nodes' in-degree [3]. This behavior is common in many types of real networks, which are characterised by a power law distribution of node degrees, a distribution that well-approximates the real citation distributions observed in citation networks.

AttRank defines a time-restricted form of preferential attachment, where the last y years of the citation network are considered, and paper weights are defined based on their share of total citations in the network during these years. These weights are called the paper's *recent attention*.

Hence, the score of each paper i is calculated based on the following formula:

$$s_i = \alpha \sum_j P_{i,j} s_j + \beta s_i^{\{Att\}} + \gamma s_i^{\{T\}} \tag{5.55}$$

where $\alpha, \beta, \gamma \in [0, 1]$ and $\alpha + \beta + \gamma = 1$. The age-based weight $s_i^{\{T\}}$ is defined as $s_i^{\{T\}} = ce^{-\rho(t_c - t_i)}$, where c is a normalization factor and ρ is determined by fitting an exponential function to the distribution of citations papers receive on average n years after their publication. The recent attention-based score, $s_i^{\{Att\}}$ is calculated as:

$$s_i^{\{Att\}} = \frac{\sum_{j:t_j > t_c - y} A_{i,j}}{\sum_{p,k:t_p > t_c - y, t_k > t_c - y} A_{p,k}} \tag{5.56}$$

and corresponds to paper i's share of citations in the network in the last y years, where y is a tunable parameter.

Hence, AttRank defines the behavior of a random researcher, who starts by reading a paper and then either picks another one at random from the reference list, or chooses any paper in the network, with a preference for those recently published, or recently heavily cited.

5.5 Discussion

In this chapter, we have examined paper ranking methods with focus on the problem of ranking papers based on their expected popularity. While there is an abundance of literature on different methods, there is limited work on directly comparing their effectiveness. Additionally, while some studies, such as [15], do directly compare different ranking methods, they do not focus on their effectiveness in ranking based on the expected popularity. In [26], we experimentally compared a number of methods based on their effectiveness in ranking papers according to short-term impact. We found that the most efficient approaches were those that used time-awareness, based on citation age. Other approaches that performed well in this scenario were those using time-awareness based on publication age, and in some cases those using citation gap. Hence, we conclude that in any case, time-awareness is vital for ranking based on expected popularity by citation network analysis methods. In contrast, more traditional ranking methods, such as PageRank, which are not time-aware, did perform well in different scenario, that of ranking based on the *influence* of papers, i.e., based on their centrality in the overall citation network. Finally, we found that there is still room for improvement with regards to the popularity-based ranking problem, based on the effectiveness of the methods in the literature.

Aside from the problem of ranking based on the expected popularity, as mentioned in Section 5.1, two other problems are closely related to the impact-based paper rank prediction problem: the impact indicator prediction and the impact classification problem. Although a detailed study of these problems is out of scope of this chapter, we briefly discuss them here and the reader may follow the references for more details. Regarding the impact indicator prediction problem, most approaches aiming to solve it attempt to estimate the exact number of citations each paper will receive in a given future period, a problem known as *Citation Count Prediction* (CCP). These approaches are based on regression models (e.g., linear regression, SVR, neural networks) and they utilize a wide range of paper metadata as input features (indicative examples: [1, 31, 32, 34, 49, 52]). Regarding the the impact classification problem, existing work is more limited. In [40] the authors follow a link-prediction-inspired approach and investigate its effectiveness both in solving CCP but also in a relevant classification problem based on a set of predefined classes. Moreover, in [43] and [45] the authors study two impact-based classification problems examining various classification approaches (e.g., logistic regression, random forest, SVM).

5.6 Conclusion

In this chapter, we presented the literature related to the problem of ranking scientific papers according to their expected impact based on citation network analysis-based methods. We formally defined the problem of ranking papers based on their expected short-term impact, presented a high level overview of approaches followed in the literature, and examined in detail those methods that are time-aware, and/or have been evaluated, or shown to perform well in ranking papers based on their expected impact. Finally, we closed the chapter by discussing previous findings regarding the effectiveness of the various approaches in the literature in ranking based on their expected short-term impact. We hope that the reader has found the information presented here a valuable starting point to spark further interest and research in this area.

References

1. Abrishami, A., Aliakbary, S.: Predicting citation counts based on deep neural network learning techniques. Journal of Informetrics **13**(2), 485–499 (2019)
2. Bai, X., Xia, F., Lee, I., Zhang, J., Ning, Z.: Identifying anomalous citations for objective evaluation of scholarly article impact. PloS one **11**(9), e0162364 (2016)
3. Barabási, A.L.: Network Science. Boston, MA: Center for Complex Network, Northeastern University. Available online at: http://barabasi.com/networksciencebook (2014)
4. Bergstrom, C.T., West, J.D., Wiseman, M.A.: The Eigenfactor™ metrics. The Journal of Neuroscience **28**(45), 11433–11434 (2008)
5. Bernstam, E.V., Herskovic, J.R., Aphinyanaphongs, Y., Aliferis, C.F., Sriram, M.G., Hersh, W.R.: Using citation data to improve retrieval from medline. Journal of the American Medical Informatics Association **13**(1), 96–105 (2006)
6. Bollen, J., Van de Sompel, H., Hagberg, A., Chute, R.: A principal component analysis of 39 scientific impact measures. PloS one **4**(6), e6022 (2009)
7. Bornmann, L., Mutz, R.: Growth rates of modern science: A bibliometric analysis based on the number of publications and cited references. Journal of the Association for Information Science & Technology **66**(11), 2215–2222 (2015)
8. Brin, S., Page, L.: The anatomy of a large-scale hypertextual web search engine. Computer Networks & ISDN Systems **30**(1), 107–117 (1998)
9. Chakraborty, T., Kumar, S., Goyal, P., Ganguly, N., Mukherjee, A.: On the categorization of scientific citation profiles in computer science. Communications of the ACM **58**(9), 82–90 (2015)
10. Chang, S., Go, S., Wu, Y., Lee, Y., Lai, C., Yu, S., Chen, C., Chen, H., Tsai, M., Yeh, M., Lin, S.: An ensemble of ranking strategies for static rank prediction in a large heterogeneous graph. WSDM Cup (2016)
11. Chatzopoulos, S., Vergoulis, T., Kanellos, I., Dalamagas, T., Tryfonopoulos, C.: Artsim: Improved estimation of current impact for recent articles. In: Proceedings of the ADBIS-TPDL-EDA Common International Workshops DOING, MADEISD, SKG, BBIGAP, SIM-PDA, AIMinScience and Doctoral Consortium, pp. 323–334 (2020)
12. Chen, P., Xie, H., Maslov, S., Redner, S.: Finding scientific gems with Google's PageRank algorithm. Journal of Informetrics **1**(1), 8–15 (2007)
13. Chin-Chi, H., Kuan-Hou, C., Ming-Han, F., Yueh-Hua, W., Huan-Yuan, C., Sz-Han, Y., Chun-Wei, C., Ming-Feng, T., Mi-Yen, Y., Shou-De, L.: Time-aware weighted PageRank for paper ranking in academic graphs. WSDM Cup (2016)
14. Diodato, V.P., Gellatly, P.: Dictionary of Bibliometrics (Haworth Library and Information Science). Routledge (1994)
15. Dunaiski, M., Visser, W.: Comparing paper ranking algorithms. In: Proceedings of the South African Institute for Computer Scientists & Information Technologists Conference, pp. 21–30 (2012)
16. Feng, M., Chan, K., Chen, H., Tsai, M., Yeh, M., Lin, S.: An efficient solution to reinforce paper ranking using author/venue/citation information - The winner's solution for WSDM Cup. WSDM Cup (2016)

17. Garfield, E.: The history and meaning of the journal impact factor. Jama **295**(1), 90–93 (2006)
18. Ghosh, R., Kuo, T.T., Hsu, C.N., Lin, S.D., Lerman, K.: Time-aware ranking in dynamic citation networks. In: Proceedings of the 11th IEEE International Conference on Data Mining Workshops (ICDMW), pp. 373–380 (2011)
19. Groth, P., Gurney, T.: Studying scientific discourse on the web using bibliometrics: A chemistry blogging case study. In: Proceedings of the WebSci10: Extending the Frontiers of Society On-Line (2010)
20. Haveliwala, T.H.: Topic-sensitive Pagerank: A context-sensitive ranking algorithm for web search. IEEE Transactions on Knowledge & Data Engineering **15**(4), 784–796 (2003)
21. Herrmannova, D., Knoth, P.: Simple yet effective methods for large-scale scholarly publication ranking. CoRR **abs/1611.05222** (2016)
22. Hwang, W.S., Chae, S.M., Kim, S.W., Woo, G.: Yet another paper ranking algorithm advocating recent publications. In: Proceedings of the 19th International Conference on World Wide Web (WWW), pp. 1117–1118 (2010)
23. Ioannidis, J.P.: Why most published research findings are false. PLoS medicine **2**(8), e124 (2005)
24. Jeh, G., Widom, J.: Scaling personalized web search. In: Proceedings of the 12th International World Wide Web Conference (WWW), pp. 271–279 (2003)
25. Jiang, X., Sun, X., Zhuge, H.: Towards an effective and unbiased ranking of scientific literature through mutual reinforcement. In: Proceedings of the 21st ACM International Conference on Information & Knowledge Management (CIKM), pp. 714–723 (2012)
26. Kanellos, I., Vergoulis, T., Sacharidis, D., Dalamagas, T., Vassiliou, Y.: Impact-based ranking of scientific publications: A survey and experimental evaluation. IEEE Transactions on Knowledge & Data Engineering **33**(4), 1567–1584 (2021)
27. Kanellos, I., Vergoulis, T., Sacharidis, D., Dalamagas, T., Vassiliou, Y.: Ranking papers by their short-term scientific impact. In: Proceedings of the 37th IEEE International Conference on Data Engineering (ICDE), pp. 1997–2002 (2021)
28. Kleinberg, J.M.: Authoritative sources in a hyperlinked environment. Journal of the ACM **46**(5), 604–632 (1999)
29. Langville, A.N., Meyer, C.D.: Google's PageRank and Beyond: The Science of Search Engine Rankings. Princeton University Press (2011)
30. Larsen, P., Von Ins, M.: The rate of growth in scientific publication and the decline in coverage provided by science citation index. Scientometrics **84**(3), 575–603 (2010)
31. Li, C., Lin, Y., Yan, R., Yeh, M.: Trend-based citation count prediction for research articles. In: Proceedings of the 19th Pacific-Asia Conference on Advances in Knowledge Discovery & Data Mining (PAKDD), vol. 2, pp. 659–671 (2015)
32. Li, M., Xu, J., Ge, B., Liu, J., Jiang, J., Zhao, Q.: A deep learning methodology for citation count prediction with large-scale biblio-features. In: Proceedings of the IEEE International Conference on Systems, Man & Cybernetics (SMC), pp. 1172–1176 (2019)
33. Liao, H., Mariani, M.S., Medo, M., Zhang, Y., Zhou, M.Y.: Ranking in evolving complex networks. Physics Reports **689**, 1–54 (2017)
34. Liu, L., Yu, D., Wang, D., Fukumoto, F.: Citation count prediction based on neural hawkes model. IEICE Transactions on Information & Systems pp. 2379–2388 (2020)
35. Luo, D., Gong, C., Hu, R., Duan, L., Ma, S.: Ensemble enabled weighted pagerank. CoRR **abs/1604.05462** (2016)
36. Ma, N., Guan, J., Zhao, Y.: Bringing PageRank to the citation analysis. Information Processing & Management **44**(2), 800–810 (2008)
37. Ma, S., Gong, C., Hu, R., Luo, D., Hu, C., Huai, J.: Query independent scholarly article ranking. In: Proceedings of the 34th IEEE International Conference on Data Engineering (ICDE), pp. 953–964 (2018)
38. Mariani, M.S., Medo, M., Zhang, Y.C.: Identification of milestone papers through time-balanced network centrality. Journal of Informetrics **10**(4), 1207–1223 (2016)
39. Page, L., Brin, S., Motwani, R., Winograd, T.: The PageRank citation ranking: Bringing order to the web. Tech. rep., Stanford InfoLab (1999)

40. Pobiedina, N., Ichise, R.: Citation count prediction as a link prediction problem. Applied Intelligence **44**(2), 252–268 (2016)
41. Sayyadi, H., Getoor, L.: Futurerank: Ranking scientific articles by predicting their future Pagerank. In: Proceedings of the 9th SIAM International Conference on Data Mining (SDM), pp. 533–544 (2009)
42. Smith, D.R.: A 30-year citation analysis of bibliometric trends at the Archives of Environmental Health, 1975-2004. Archives of Environmental & Occupational Health **64**(sup1), 43–54 (2009)
43. Su, Z.: Prediction of future citation count with machine learning and neural network. In: Proceedings of the Asia-Pacific Conference on Image Processing, Electronics & Computers (IPEC), pp. 101–104 (2020)
44. Vaccario, G., Medo, M., Wider, N., Mariani, M.S.: Quantifying and suppressing ranking bias in a large citation network. Journal of Informetrics **11**(3), 766–782 (2017)
45. Vergoulis, T., Kanellos, I., Giannopoulos, G., Dalamagas, T.: Simplifying impact prediction for scientific articles. In: Proceedings of the Workshops of the EDBT/ICDT 2021 Joint Conference (2021)
46. Wade, A.D., Wang, K., Sun, Y., Gulli, A.: Wsdm cup 2016: Entity ranking challenge. In: Proceedings of the 9th ACM International Conference on Web Search & Data Mining (WSDM), pp. 593–594 (2016)
47. Walker, D., Xie, H., Yan, K.K., Maslov, S.: Ranking scientific publications using a model of network traffic. Journal of Statistical Mechanics: Theory & Experiment **2007**(06), P06010 (2007)
48. Wang, Y., Tong, Y., Zeng, M.: Ranking scientific articles by exploiting citations, authors, journals, and time information. In: Proceedings of the 27th AAAI Conference on Artificial Intelligence (2013)
49. Wen, J., Wu, L., Chai, J.: Paper citation count prediction based on recurrent neural network with gated recurrent unit. In: Proceedings of the 10th IEEE International Conference on Electronics Information & Emergency Communication (ICEIEC), pp. 303–306 (2020)
50. Xiong, Y., Zhu, Y., Yu, P.S.: Top-k similarity join in heterogeneous information networks. IEEE Transactions on Knowledge & Data Engineering **27**(6), 1710–1723 (2015)
51. Yan, E., Ding, Y.: Weighted citation: An indicator of an article's prestige. Journal of the American Society for Information Science & Technology **61**, 1635–1643 (2010)
52. Yan, R., Tang, J., Liu, X., Shan, D., Li, X.: Citation count prediction: learning to estimate future citations for literature. In: Proceedings of the 20th ACM International Conference on Information & Knowledge Management (CIKM), pp. 1247–1252 (2011)
53. Yao, L., Wei, T., Zeng, A., Fan, Y., Di, Z.: Ranking scientific publications: The effect of nonlinearity. Scientific Reports **4** (2014)
54. Yu, P.S., Li, X., Liu, B.: On the temporal dimension of search. In: Proceedings of the 13th International World Wide Web Conference on Alternate Track Papers & Posters, pp. 448–449 (2004)
55. Yu, P.S., Li, X., Liu, B.: Adding the temporal dimension to search - a case study in publication search. In: Proceedings of the IEEE/WIC/ACM International Conference on Web Intelligence (WI), pp. 543–549 (2005)
56. Zhang, F., Wu, S.: Ranking scientific papers and venues in heterogeneous academic networks by mutual reinforcement. In: Proceedings of the 18th ACM/IEEE on Joint Conference on Digital Libraries (JCDL), pp. 127–130 (2018)
57. Zhou, J., Zeng, A., Fan, Y., Di, Z.: Ranking scientific publications with similarity-preferential mechanism. Scientometrics **106**(2), 805–816 (2016)

Chapter 6
Properties of an Indicator of Citation Durability of Research Articles

Natsuo Onodera and Fuyuki Yoshikane

Abstract Knowledge about citation durability of articles is less accumulated compared with that about citation count itself. One of the essential reasons for it is thought to be that any quantitative index for measuring citation durability is not established yet. In this chapter, we discuss the citation durability of articles focusing on the properties of an index of citation durability. After presenting a review of the main studies on this subject, we describe the authors' research investigating a selected indicator of citation durability ("Citation Delay" introduced by Wang et al. [43]) from the viewpoint of its basic properties, relatedness to the long-term citation count, and dependence on other characteristics of articles. In this research we aimed to examine whether there are some general trends across subject fields regarding the properties of the indicator. Finally, we discuss the results of our research by relating to those reported by existing studies.

6.1 Introduction

The number of citations an academic paper receives is often used as a measure of the scientific impact of the paper. However, papers with the same total citation count can show different time distribution patterns of the count. Many articles are rarely cited for some time after publication, then receive a growing number of citations to arrive at a peak somewhere between two and six years after publication, before the citation count decreases, while some receive most of the citations within a year or

Natsuo Onodera (correspondence author)
University of Tsukuba, 1-2, Kasuga, Tsukuba, Ibaraki 305-8550, Japan
e-mail: nt.onodera@y5.dion.ne.jp

Fuyuki Yoshikane
Faculty of Library, Information and Media Science, the University of Tsukuba, 1-2, Kasuga, Tsukuba, Ibaraki 305-8550, Japan
e-mail: fuyuki@slis.tsukuba.ac.jp

© The Author(s), under exclusive license to Springer Nature Switzerland AG 2021
Y. Manolopoulos, T. Vergoulis (eds.), *Predicting the Dynamics of Research Impact*,
https://doi.org/10.1007/978-3-030-86668-6_6

two, others are cited constantly for a long period, and still others remain unmarked before a sudden wave of citations arrives seven or ten years afterwards.

How citation counts change over time, in other words, ageing (obsolescence) or durability of citations has been studied from various viewpoints (see Section 6.2 for details).

The classification into early-cited ('flash in the pan'), delayed-cited ('sleeping beauty'), and normal is often used [12, 13, 20, 21, 31, 38, 44], but the criteria of the classification differ by the authors. Although the systematic investigations comparing the relation of citation durability with other characteristics of articles are not so many, several works have reported that delayed recognized papers tend to receive more citations in the long run than early recognized ones [3, 6, 12, 28, 29, 35, 42]. It is, however, not clear if this conclusion applies to any subject field because the samples employed in these investigations were either limited to a small number of highly-cited papers (HCPs) or composed of papers from mixed fields. The relation of citation durability with characteristics other than citedness of articles has been addressed by only a few studies [7, 12, 13, 15, 43, 45].

However, knowledge about citation durability of articles is less accumulated compared with that about citation count itself. One of the essential reasons for it is thought to be that any quantitative index for measuring citation durability is not established yet. Some aging parameters such as the cited half-life (median of citation age distribution) do not reflect the entire pattern of life-time citations. In contrast, the Citation Delay introduced by Wang et al. is the measure of citation durability reflecting the entire life-time citation information, but little has been known about its properties [43].

In this chapter, we discuss the citation durability of articles focusing on the properties of an index of citation durability. In the rest of this chapter, Section 6.2 presents a review of the main studies on this subject. In Section 6.3, we describe our study investigating the Citation Delay from the viewpoint of its basic properties, dependence on other characteristics of articles, and relatedness to the long-term citation count. In Section 6.4, we discuss the results of, and findings in, our research by relating to those reported by existing studies.

The main parts of this chapter originate in the study by Onodera [36]. We reused this article with permission from the publishers of the journals in which the article was published, making some additions and modifications. In particular, we added the descriptions on the many studies published after the publication of that article and the discussion of the relation between these added studies and ours.

6.2 Literature Review

6.2.1 Mathematical Models for the Aging of Papers

Studies on the obsolescence function that describes the temporal changes of the citation count have been conducted for a long time. The most simplistic approach consists of fitting the temporal changes to an exponentially decaying curve, but this method is not adequate, even qualitatively, because the citation count is generally recognized as reaching its peak a few years after publication.

Avramescu proposed two types of aging function $c(t)$ describing the citation count after a lapse of time t, which can fairly approximate the citation history of numerous articles by adjusting three parameters in those functions [4]. Egghe and Ravichandra Rao examined the aging factor $a(t) = c(t+1)/c(t)$ and proposed a log-normal model for $c(t)$ based on the empirical observations that, in many cases, $a(t)$ has a minimum at a certain t [17]. Burrell in [9] analyzed citation age function based on the failure rate function in the reliability theory and supported the conclusions of Egghe and Ravichandra Rao.

Glänzel and Schoepflin used a stochastic model for the process to acquire citations from a set of articles and defined indices for the speed of early reception and later aging [22]. Based on applications of these indices to actual cases, they demonstrated that the aging patterns depend on discipline rather than journals and that slow aging does not necessarily mean slow reception.

Wang et al. derived a mechanistic model for the citation dynamics of individual articles [41]. They indicated that, using this model, the citation histories of articles from different journals and disciplines can be collapsed into a single curve and that all articles follow the same universal temporal pattern. This model has been called the WSB model for citation histories.

Della Briotta Parolo et al. investigated the change of citation decay with the publication year for articles published during the 1960-1990 period and found that, the more recently the articles had been published, the faster they reached the citation peak and also the shorter their citation half-life became [14]. In addition, they showed that the citation trajectories after the peak year fit better to exponential decay than power-law decay.

The studies mentioned above suppose the case of ordinary citation histories in which the yearly citations rise for a while after publication until reaching a peak and beginning to decrease. Along with the growing interest in unusual citation histories, such as sleeping beauties (SBs), several models expressing citation histories involving unusual patterns have appeared. Among the three following cases, the first two proposed a model-free expression, and the last one extended the WSB model of Wang et al. [41].

Bornmann et al. proposed a new method to express the citation histories of articles as a sequence of annual field- and time-normalized citation scores based on the four-leveled "characteristic scores and scales" (CSS) [6]. Using all articles published in 2000 indexed by Web of Science Core Collection (WoS9 with their annual citation scores until 2015 as the data set, they compared the sequences of the annual citation scores of six broader fields. Most of the articles classified into the two most highly cited groups by the CSS method showed an increasing tendency in their sequence (increase in the citation score year by year).

Galiani and Gálvez proposed a data-driven approach that estimates citation aging functions based on a quantile regression, not by fitting actual citation history data to a defined mathematical model [19]. Using a data set consisting of citations to nearly

60,000 articles spanning 12 dissimilar research fields, they tested this methodology and presented various types of citation aging curves.

He et al. modeled the citation history patterns of "atypical" articles whose citation trajectories did not follow the single peak pattern subject to the WSB model [25]. Based on the results of the systematic classification of citation patterns of atypical articles, they found that atypical patterns can be characterized as SBs (articles awakened after a long period of non-recognition), "second act" articles (those awakened for a second time), and a combination of both. To accurately describe the citation dynamics of atypical articles, they proposed a model that adds new terms to the WSB citation aging function.

6.2.2 Comparison of the Citation Durability Between Highly-cited Papers and Other Papers

Early studies by Line demonstrated that highly-cited papers (HCPs) have greater citation durability. This finding raised interest in the citation age distribution of HCPs. Aversa [3] and Cano and Lind [11] characterized the citation aging pattern for a long period for the HCPs they selected. Both demonstrated that the papers could be classified into two groups with short and long citation periods. Aversa also observed that the papers in the latter group had higher total citation counts than the former. In addition, Aksnes found that the citation age of the HCPs was somewhat longer than that of the other papers from an analysis of articles published by Norwegian authors from1981 to 1989 [2].

Levitt and Thelwall addressed the issue of the temporal changes of citation counts for HCPs in a more systematic manner [28]. Based on aging patterns for 100 HCPs published in 1974, they claimed that the patterns were highly diverse and far more complicated than dividing them into two groups, as shown by Aversa [3] and Cano and Lind [11]. They further analyzed the citation age distribution during 36 years for 36 HCPs, each from six disciplines published in 1969-1971, concluding that: (i) the ratio of the early citation count (in the initial six years) to the total citation count varied remarkably among the articles regardless of the discipline; (ii) articles with higher total citation counts generally tended to show fewer early citations; and (iii) the citation durability of HCPs was higher than non-HCPs for all disciplines. Levitt and Thelwall also found, for HCPs in the field of information and library science, a moderate correlation indicating a delayed citation tendency for articles ranked as more highly cited [29].

6.2.3 Classification of Papers by their Citation Durability

Some researchers have attempted to classify articles with respect to their citation durability. In particular, articles that attract attention after a prolonged period without citations, called "delayed recognition papers" (DRs) [20, 21] or "sleeping beauties" (SBs) [38], are a focus of interest. Conversely, papers frequently cited immediately after publication and then forgotten are called "flashes in the pan" [13]. Articles that have been regarded as "flashes in the pan" in the past suddenly starting to be cited at some point in time were named "all-elements sleeping beauties" [30, 33].

Glänzel et al. examined the citation history for 21 years for papers published in 1980 and identified 60 DRs [21]. Meanwhile, van Raan defined SBs according to: (a) the maximum citations per year during "sleep", (b) the sleeping period, and (c) the minimum citations per year after awakening, identifying about 360 papers meeting this definition from those published in and after 1980 [38]. Although the DRs (or SBs) identified by these two studies look rather rare, Burrell demonstrated that they appear with a frequency higher than that assuming statistically random events based on calculations using his citation aging model, suggesting that there must be some reasons behind delayed recognition [10].

Li et al. addressed the citation age distribution during the sleeping period, which they called the "heartbeat spectrum" of sleeping papers and revealed that papers with "late heartbeats" have much higher awakening probabilities (more likely to become SBs) than those with "early heartbeats" [32].

Lachance and Larivière extracted SBs from papers published during the 1963-1975 period, as well as papers that received no citation for 10 years or more since publication but had obtained some citations since then (sleepers but not SBs), and compared the citation pattern of the SBs (about 5% of all papers) with that of the non-sleeper papers of the same period (control group) [27]. Although the yearly citation distribution of non-SB sleepers after awakening showed a gradual decline generally similar to that of the control group, the SBs showed no decline but even an increase, differing from the patterns of the control group.

Li and Shi in [31] claimed that the criteria for SBs set by van Raan [38] do not apply for "genius work", that is, articles whose citations grow exponentially in an extended period and proposed two new criteria (according to the length of, and yearly average citations in, sleeping and extending periods) based on a much longer citation history. From articles of Nobel Prize laureates with at least 50 years of citation history, they extracted 25 SBs according to their criteria, finding that only 10 of them met van Raan's criteria. From this case study, they demonstrated that the lower the increase ratio of the exponential function, the longer the sleeping period of the genius work becomes.

Li and Ye pointed out the defects in the three existing kinds of criteria for distinguishing SBs (i.e., average-based criteria, quartile-based criteria, and parameter-free criteria) and proposed four rules that should be noticed when distinguishing SBs: (i) the early citations should be penalized; (ii) the whole citation history should be taken into account; (iii) the awakening time of SBs should not vary over time; and (iv) arbitrary thresholds on sleeping period or awakening intensity should be avoided [34].

Baumgartner and Leydesdorff proposed a method of dividing a set of papers into subsets according to the distribution pattern of citation age using group-based trajectory modeling (GBTM), a non-parametric statistical technique [5]. Applying this method to the 16-year citation history of several sets of papers published in 1996 yielded three to seven groups characterized by not only the citation frequency but citation durability. However, the authors admitted the difficulty of applying this method to large sets of papers from a number of journals because the number of groups obtained depends on subjective judgment, and the model requires simplification by eliminating outliners and defining initial values.

Bornmann et al. in [7] identified "hot papers" (HPs), articles that receive a boost of citations shortly after publication, and "DRs", articles that scarcely receive citations over a long period before a considerable citation boost occurs, based on the method proposed by Bornmann et al. in [6] (see Section 6.2.1) and that by Ye and Bornmann in [44] (see Section 6.2.5). They identified 2,130 HPs and 315 DRs from a data set with the citation data until 2015 of about five million articles published between 1980 and 1990. While scholars in the fields "biochemistry & molecular biology", "immunology", and "cell biology" published significantly more HPs than DRs, the opposite result was seen for "surgery" and "orthopedics".

Fang proposed a derivative analysis to comprehensively characterize the citation curves of SBs [18]. This analysis can differentiate among periods with different variation tendencies (ascending, declining, or unchanged) from successive derivatives in an array of yearly citations. They demonstrated that this method helps us precisely perform the identification of SBs, determine awaking times, classify SB types, and analyze other complicated behaviors of citation curves.

In recent years, studies have appeared identifying articles maintaining high numbers of citations for a very long time somewhat differently from the traditional type of SBs, as the observation of citation histories over much longer time periods becomes possible compared to the days when the idea of DRs was first advocated by Glänzel et al. [21] or van Raan [38].

One such study was conducted by Zhang et al., who identified articles that display a continual rise in annual citations without a decline within a sufficiently long time period (they named such articles "evergreens") by a functional data analysis method [46]. Applying this method to the citation trajectories over 30 years of a sample of 1,699 articles published in 1980 in the American Physical Society (APS) journals, they extracted 45 evergreens. From the results, they concluded that evergreens exist as one of the general types of citation trajectories and that it is impossible to predict whether an article would become an evergreen from its other attributes, such as the number of authors or the number of references.

Another case is the studies by Bornmann et al. proposing an approach for identifying "landmark papers", which maintain a very high level of citation impact across many years [8]. Using a cluster analysis of citation history data expressed by a sequence of field- and year-normalized citation scores [6] (see Section 6.2.1), they identified 1,013 landmark papers from the citation data across 25 years of about five million articles between 1980 and 1990.

6.2.4 Change in the Citation Ranking of Articles with Altering the Citation Window Length

Abramo et al. estimated the differences that occur when decreasing the citation window length based on the citation data until the end of 2008 of papers published in 2001 by Italian authors [1]. For the three-year citation window including the publication year, for example, the cumulative citation counts amounted to 65%-88% (depending on the discipline) of the final (nine-year citation window) counts, and Spearman's rank correlation coefficients for the citation rankings between the three- and nine-year citation windows were in the range [0.79..0.96].

When studying the correlation between short and long citation impacts of papers, Wang introduced "Citation Speed" as an indicator to measure how fast an article acquires its citations [42]. Let the publication year of an article be 0 and the cumulative citation count from year 0 to year t be $C(t)$; then, the Citation Speed of the article at year T is given by:

$$\text{Citation Speed} = \frac{1}{T} \sum_{t=0}^{T-1} \frac{C(t)}{C(T)}$$

The Citation Speed lies between 0 and 1, and the earlier an article is cited, the more closely it approaches 1. From the observation of the citation data during 31 years of articles published in 1980, Wang showed that the distribution of the Citation Speed of articles is heterogeneous not only across subject fields but also within a subject field or even within a journal, meaning the existence of a considerable difference in the citation rankings between the short- and the long-term citation windows. He also indicated that the more highly-cited articles tend to receive more delayed citations.

Both Abramo et al. in [1] and Wang in [42] indicated the problem of evaluating the impact of articles using citation data in a short period after publication and claimed that the citation window in research assessment should be carefully chosen considering the trade-off between accuracy and timeliness in measurement. Moreover, Levitt and Thelwall reported in [28] that the correlation between the citation rank within HCPs in the first six years and that in the entire period (36 years) was insignificant in five out of six disciplines they surveyed (see Section 6.2.2).

6.2.5 Indicators for Measuring the Citation Durability and Citation Delay of Articles

Wang et al. defined "Citation Delay" by subtracting the Citation Speed (see Section 6.2.4) from one, and analyzed how this measure calculated using long-term (13 years) citation data from articles published in 2001 is influenced by the interdisciplinarity of the articles and some other article features [43]. Besides Citation Delay, no quantitative indicator of citation durability reflecting information on the entire citation age distribution has been proposed. Although the index G_s introduced by Li et al. is very similar to Citation Delay, they only used that concept for measuring the inequality of the "heartbeat spectrum" (see Section 6.2.3) and not as a general indicator of citation durability [32].

Ke et al. introduced the "Beauty Coefficient" indicator, which measures quantitatively the extent of sleeping [26]. This indicator measures the gap between the yearly citations estimated from the citations at the year of the citation peak and the actual yearly citations until the peak year. Therefore, the longer the sleeping period is, and the steeper the growth of citations at the year of awaking is, the larger the indicator value becomes. The distributions of this indicator for citation data from two data sets for a long period were subject to the power law. They concluded, from these distributions, that there are no clear demarcation values to separate SBs from "normal" papers. The Beauty Coefficient indicator has drawn attention as a simple way to find SBs, and some studies have tried to modify it.

Dey et al. in [15] tried to identify SBs in the domain of computer science using a variant of the methodology proposed in [26]. They extracted more than 5,000 SBs from about 180,000 articles included in Microsoft Academic Search that were published in and after 1950 and received more than 20 citations until 2012, characterizing them based on their subfield and citation profile after awakening.

Ye and Bornmann in [44] introduced the "dynamic citation angle" (β) as a simple way to identify quantitatively not only SBs but also "smart girls" (SGs; articles gaining high citations immediately after publication, the same as "hot papers" described in Section 6.2.3). The indicator β is the slope angle of the straight line connecting the original point and the citation peak of a citation curve, complementing the Beauty Coefficient. The angle β for the first and the second peaks is related to SG and SB, respectively. The calculation of β for nearly 170,000 articles published in 1980 in natural sciences revealed that about 3.4% of the articles are typical SGs and about 0.08% typical SBs.

Du and Wu proposed a modified Beauty Coefficient denoted as Bcp, which substituted yearly citations in the Beauty Coefficient with a yearly cumulative percentage of citations [16]. They also redefined the awakening year, sleeping length, and sleeping depth to avoid arbitrary thresholds as much as possible. The results of testing the new index using the data of articles from *Science* and *Nature* showed that, compared to Beauty Coefficient, Bcp is more sensitive in identifying SBs whose citations are not so high.

Zhang and Glänzel in [45] investigated diachronous aging (distribution of citations according to the age of articles published in a certain year) at different levels of aggregation (i.e., major fields, subfields, journals, and the individual paper level). The share of fast citations (measured by the prospective Price index introduced by Glänzel et al. in [23]) was positively correlated with the share of recent references (measured by Price index) at different levels of aggregation, although the consistency varies among different aggregations.

6.2.6 Relationships Between Citation Durability and Other Characteristics of Articles

The systematic comparison of the relationship between the citation durability and other characteristics of articles has been investigated only in a few studies [7, 12, 13, 15, 43, 45]. These studies will be discussed later again (see Section 6.4.2) in relation to the authors' work.

Van Dalen and Henkens classified the 1, 371 articles published in 17 demographic journals in 1990-92 according to the citation pattern in 10 years after publication into four classes: [I] few cited, [II] early cited, [III] delayed cited, and [IV] normal [13]. They discussed the effects of the characteristics of individual articles on their attribution to the classes using a logit polynomial regression analysis with class I as a reference. Among the explanatory variables, publication in top-ranked journals, the author's reputation (higher citation count in the past), and the number of pages were found to increase the probability of an article belonging to class II, III, or IV compared to class I; the number of authors and their national affiliations were not significant. The data suggest that the three significant variables are correlated most strongly with class IV. As for classes II and III, publication in top-ranked journals favors allocation to class II, but the other two variables show no difference.

A similar classification into four classes (but with a different criterion) was performed by Costas et al. for articles published in 1983-2003 and indexed in WoS according to citation data up to 2008 [12]. A mutual comparison of classes II, III, and IV revealed that the articles in class III have a significantly smaller number of authors, affiliated institutions, national affiliations, and cited references, but they had a greater number of pages than those in class II.

Using the citation data of 13 years received by articles published in 2001, Wang et al. carried out a multiple regression analysis with Citation Delay (see Section 6.2.5) as the dependent variable and several measures of the interdisciplinarity of articles and also some control variables as the independent variables [43]. Their main interest lay in the relationship between interdisciplinarity and Citation Delay, but we address here the control variables that reflect more general article characteristics. Wang et al. showed that, among the control variables, the number of authors, the internationality of co-authors, and the number of references had a significant negative effect on Citation Delay, while the number of pages had a significant positive effect.

Zhang and Glänzel analyzed the correlation of a measure of citation immediateness of an article to a measure of the recency of references of the article [45]. They used the prospective Price index (PPI; the percentage of citations up to five years after publication) as the citation immediateness measure and Price index (PI) as the reference recency measure. Based on the citation data until 2015 of articles published in 1992, they compared average PPI to average PI, as well as the ratio of articles with PPI > 0.5 to the ratio of articles with PI > 0.5 within each of 14 major fields and 64 subfields, finding a generally positive correlation ($r \in [0.75..0.85]$).

Dey et al., using the articles sampled by the method shown in Section 6.2.5, attempted to predict whether an article is likely to become an SB using a machine learning-based classification approach [15]. The features for classification included both those determined at the time of publication (e.g., the number of authors, the number of keywords, etc.) and those determined a few years after publication (e.g., citations received by the article or by the author(s), etc.). The more important features for predicting the likelihood of becoming to an SB were: (1) the diversity (measured by entropy) of the subfields of the citing articles, (2) the number of articles published in the venue, (3) the type of venue, and (4) the average number of citations received by the author.

Bornmann et al., using 315 DRs identified by the method shown in Section 6.2.3, examined whether the distribution of values of some article features statistically differs or not between the group of DRs identified and that of other articles [7]. Although the distributions of the number of references, the number of pages, and the interdisciplinarity of the publishing journal differ significantly (distributions of the number of authors and the number of countries of the authors are not significant), they suggested that it is difficult to predict whether an article becomes a DR based on its particular properties.

6.3 Authors' Research on Citation Durability

In this section, the research conducted by the authors (called "our research" or "this research") are described, mainly based on [36].

6.3.1 Objective of this Research

In this research, using the Citation Delay mentioned in Section 6.2.5 as an index of the citation durability of articles, we investigate the following matters:

1. elucidating the characteristics of the distribution of the Citation Delay,
2. examining relationships between this durability index and the citation count of articles to reveal whether or not there is any difference in the long-range citation counts between early-cited and delayed-cited articles, and
3. examining relationships between the durability index and other characteristics of articles.

For this research, we took the approach similar to that used in our research described in the Chapter "Extrinsic Factors Affecting Citation Frequencies of Research Articles". Namely, we applied a same analytical method to several selected subject fields, aiming to find any common tendencies across fields. For this, we analyzed citation data of 15 years after publication of original articles published in the same year in the several journals (all in English only) selected from each of the six different fields in natural science.

6.3.2 Definition of the Index of Citation Durability

The Citation Delay (referred to symbol D hereinafter), which was introduced in [43] and was used as the index of citation durability in this research, was described briefly in Section 6.2.5. It is defined again as follows.

Let the publication year of an article be $j = 0$, and the citation count within each of the years $j = 0, 1, 2, \ldots, T$ be $c(j)$, where T is the last year in which the citations were observed. Then, the cumulative citation count $C(t)$ from the publication year to $j = t$ is given by:

$$C(t) = \sum_{j=0}^{t} c(j) \tag{6.1}$$

The relative cumulative citation count $x(t)$ up to $j = t$ is:

$$x(t) = \frac{C(t)}{C(T)} \tag{6.2}$$

where $C(T)$ is the total number of citations received between $j = 0$ and $j = T$. D of the article at $j = T$ is defined as:

$$D = 1 - \frac{1}{T} \sum_{t=0}^{T-1} x(t) \tag{6.3}$$

This definition is illustrated in Fig. 6.1, where the horizontal axis represents t ($0 \leq t \leq T - 1$), and the vertical axis refers to $x(t)$. The area of the shaded part relative to the total area of the rectangle corresponds to D.

If an article receives citations in the year of publication ($t = 0$) only and none subsequently, $x(t)$ reaches 1 at $t = 0$ and remains there, which means that the area of the shaded part (D) is 0. If, conversely, the citation count remains at 0 from $t = 0$ to $t = T - 1$ and assumes a certain value at $t = T$, the total area of the rectangle of

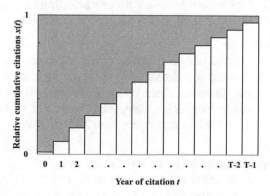

Fig. 6.1 Illustration of the Citation Delay D (reproduced from [36]). When the area of the rectangle is 1, the area of slant line part is D.

Fig. 6.1 is shaded, and therefore $D = 1$. Therefore, D of an article lies between 0 and 1; earlier citations lead to smaller areas of the shaded part (i.e., smaller D), and later citations result in larger D values. Thus, D can be considered as an index of citation durability reflecting information on the entire citation age distribution.

D depends on the relative shape of the graph shown in Fig. 6.1 but not on the total number of citations. It should be noted that D is not defined for $C(T) = 0$ and has little meaning for very low $C(T)$. The present work is principally concerned with articles with $C(T)$ of 5 or more for 15 years.

6.3.3 Data and Methods Used

Data sources
We selected the following six subject fields from the WoS Subject Categories in order to find any common or different tendencies about citation durability across fields:

- Condensed-matter physics (CondMat)
- Inorganic and nuclear chemistry (Inorg)
- Electric and electronic engineering (Elec)
- Biochemistry and molecular biology (Biochem)
- Physiology (Physiol)
- Gastroenterology and hepatology (Gastro)

The abbreviations shown in parentheses will be used hereinafter.

Four journals were chosen from each field and all of the normal articles (assigned the document type "article" in WoS) published in 2000 in the 24 journals were extracted for analysis, excluding those simultaneously classified as "proceedings papers", those shorter than three pages, and those without author names. The journals and the numbers of the articles from each journal are shown in Table 6.1.

The subject fields and journals selected are same as those selected in our research described in the Chapter "Extrinsic Factors Affecting Citation Frequencies of Research Articles". For the reasons of the selection refer to Section 4 of that chapter.

Obtaining citation data
Citation data from a considerably long period of time are necessary for the purposes of this research. Moreover, it is desirable to compare the citation data of articles published in the same year. For these reasons, articles published in 2000 were selected and the citation data were downloaded from WoS in March 2015. The citation counts were recorded for every year from 2000 to 2014.

Calculation of D
Out of 18,702 articles contained in the 24 journals (shown in Table 6.1), 331 were never cited in the whole 2000-2014 period; that is, $C(T) = 0$. D was calculated using Eqs. 6.1-6.3 for the remaining 18,371 articles, here $T = 15$. It should be noted that 1,556 articles among them have $C(T)$ values lower than 5 and consequently

Table 6.1 Selected subject fields and journals (modified from [37]).
(a) The journal titles at the time of 2000, although some were changed after that.

Subject field	Journal title[a]	Publishing country	Normal articles	Articles in sample
Condensed Matter Physics (CondMat)	Physical Review B	USA	4738	47
	Journal of Physics Condensed Matter	GBR	813	42
	European Physical Journal B	DEU	538	49
	Physica B	NLD	148	39
Inorganic & Nuclear Chemistry (Inorg)	Inorganic Chemistry	USA	931	50
	Journal of the Chemical Society - Dalton Transactions	GBR	682	49
	Inorganica Chimica Acta	CHE	546	53
	Transition Metal Chemistry	NLD	139	40
Electric & Electronic Engineering (Elec)	IEEE Transactions on Microwave Theory & Techniques	USA	295	37
	IEEE Transactions on Circuits & Systems I - Fundamental Theories & Applications	USA	218	50
	Signal Processing	NLD	178	33
	IEE Proceedings - Circuits, Devices & Systems	GBR	52	25
Biochemistry & Molecular Biology (Biochem)	Journal of Biological Chemistry	USA	5504	60
	Journal of Molecular Biology	USA	875	60
	European Journal of Biochemistry	GBR	788	57
	Journal of Biochemistry (Tokyo)	JPN	275	54
Physiology (Physiol)	Journal of General Physiology	USA	110	34
	Journal of Physiology - London	GBR	472	55
	Pflugers Archive European Journal of Physiology	DEU	238	50
	Japanese Journal of Physiology	JPN	72	60
Gastro-enterology (Gastro)	Gastroenterology	USA	259	56
	Gut	GBR	277	52
	American Journal of Gastroenterology	USA	430	47
	Journal of Gastroenterology	JPN	124	44

are not useful for discussing citation durability. The analysis was therefore chiefly concerned with the remaining 16,815 articles with $C(T) \geq 5$.

Multiple regression analysis

To find the characteristics of articles significantly correlated with D, a linear multiple regression (LMR) analysis was performed for each of the six subject fields. In each field, 60 or somewhat less articles are sampled out from each of the four journals and the articles of which $C(T)$ is at least 5 are chosen for the analysis. The numbers of articles selected for LMR are also shown in Table 6.1.

The explanatory variables are the characteristics shown in Table 6.2. Logarithmic transforms were used for variables $C(T)$, *Eqs, Age, RatePubl,* and *MedCite* because they showed a skewness of 2 or higher or a ratio of the mean to the median of 1.5 or higher in more than half of the subject fields. Since each of these variables may have a value of 0 for certain articles, 1 was added to $C(T)$, *Eqs*, and *Age*; 0.1 to *RatePubl*; and 0.01 to *MedCite* before transformation. The values added were chosen considering the distribution of each variable.

The explanatory variables shown in Table 6.2 are same as those selected in the Chapter "Extrinsic Factors Affecting Citation Frequencies of Research Articles" except for $C(T)$ and *IF*. The reasons of the selection and the procedure to obtain the data for these variables are explained in Sections 2 and 4 of that chapter.

Table 6.2 Explanatory variables for LMR analysis predicting D (reproduced from [36])

Variable	Definition	Log transform
Authors	Number of authors of the article	No transform
Insts	Number of institutions with which the authors are affiliated	No transform
Countries	Number of countries where the institutions are located	No transform
Refs	Number of references cited in the article	No transform
Price	Price index (percentage of the references whose publication year is within 5 years before the publication year of the article)	No transform
Length	Number of normalized pages of the article	No transform
Figures	Number of figures in the article	No transform
Tables	Number of tables in the article	No transform
Eqs	Number of numbered equations in the article	Logarithm of the value plus 1
Age	Active years (elapsed years from the year of the first article publication to the year 2000) of the first author	Logarithm of the value plus 1
RatePubl	Number of articles published per annum by the first author during their active years	Logarithm of the value plus 0.1
MedCite	Median of the number of citations received per annum by each published article	Logarithm of the value plus 0.01
$C(T)$	Number of citations received by the article till the end of 2014	Logarithm of the value plus 1
IF	Impact Factor of the journal in which the article was published (average of the values of 2001 and 2002)	No transform

SPSS Statistics Base 17.0 was used for the LMR analysis. Statistically significant explanatory variables were selected in a stepwise process.

6.3.4 Results

Distribution of the total number of citations $C(T)$
Since it is well known that the distribution of citation counts is highly skewed, we consider logarithmic-transformed citation counts here. Table 6.3 shows the distribution statistics of $\log[C(T) + 1]$ for all articles (including those with $C(T) = 0$). It is known from this table that the mean and the median of $\log[C(T) + 1]$ are almost identical and the skewness is very low. These facts suggest a nearly symmetrical distribution of $\log[C(T) + 1]$ around the mean. Fig. 6.2 displays Q-Q plots to examine whether the distribution of $\log[C(T) + 1]$ is close to a normal one. The distribution curves for the six fields seem to be nearly linear in the range of a normal theoretical quantile [-2, 2] (roughly corresponding to the percentile range [2.5..97.5]), except that slight curving is seen for the "Gastro" field. Therefore, statistical analyses of the supposition of normality (Pearson's product-moment correlation analysis and LMR analysis) are applied to $\log[C(T) + 1]$ in the following.

Distribution of D
Little is known how D values distribute in a given set of articles. Wang et al. claimed that "Citation Delay is roughly normally distributed" but did not show any data supporting this idea [43]. In this research, the distribution of D was examined and compared to the normal distribution for each of the six fields.

Fig. 6.3 shows the distributions of D calculated from Eq. 6.3 for articles with $C(T) \geq 5$ in the six fields. These distributions are approximately symmetrical around $D = [0.4..0.5]$. More specifically, Tables 4a and 4b show the statistics of D of articles with $C(T) \geq 1$ and $C(T) \geq 5$, respectively:
The tables indicate that:

- The standard deviations are clearly lower for the $C(T) \geq 5$ group than for the $C(T) \geq 1$ group (by 3%-24%, depending on the subject field). Similarly, the quartile differences (P50-P25 and P75-P50) are smaller for the $C(T) \geq 5$ group

Table 6.3 Statistics on $\log[C(T) + 1]$ (reproduced from [36])

Field	#Articles	Mean	S.D.	Median	Skewness
CondMat	6,237	1.20	0.51	1.20	−0.1
Inorg	2,298	1.29	0.42	1.30	−0.3
Elec	743	1.00	0.51	1.00	0.1
Biochem	7,442	1.64	0.39	1.65	−0.3
Physiol	892	1.38	0.46	1.43	−0.5
Gastro	1,090	1.52	0.54	1.59	−0.6

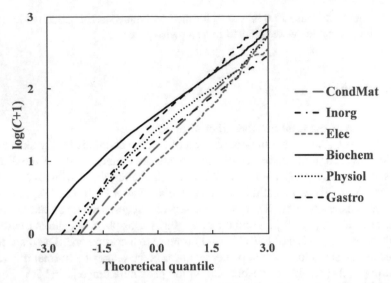

Fig. 6.2 Q-Q plot for $\log[C(T) + 1]$ of the articles with $C(T) \geq 1$ (reproduced from [36])

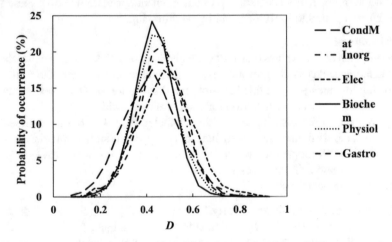

Fig. 6.3 Distribution of D of articles with $C(T) \geq 5$ (reproduced from [36])

in most fields. In other words, the distribution of D is narrower if only articles with a greater total number of citations are considered.

•   The means and medians of D are considerably close, particularly for the $C(T) \geq 5$ group, in any field. The differences between P50 and P25 and those between P75 and P50 are also close to each other. The visual impression of symmetrical distribution displayed in Fig. 6.3 is thus supported numerically.

The ratios of P50-P25 or P75-P50 to the SD of the distribution of D, as shown in Table 6.4, are in the range [0.52..0.73], which is close to the theoretical ratio of

Table 6.4 Statistics on the distribution of D (reproduced from [36])

Field	n	Mean	S.D.	P25	P50	P75	P50-P25	P75-P50
CondMat	6,028	0.411	0.139	0.321	0.411	0.499	0.090	0.088
Inorg	2,278	0.450	0.119	0.380	0.452	0.523	0.073	0.070
Elec	698	0.462	0.157	0.359	0.473	0.555	0.114	0.081
Biochem	7,432	0.431	0.087	0.376	0.434	0.488	0.058	0.054
Physiol	876	0.441	0.102	0.382	0.443	0.500	0.061	0.057
Gastro	1,059	0.462	0.106	0.400	0.465	0.531	0.065	0.066

(a) Articles with $C(T) \geq 1$

Field	n	Mean	S.D.	P25	P50	P75	P50-P25	P75-P50
CondMa	5,097	0.420	0.119	0.338	0.419	0.498	0.081	0.079
Inorg	2,061	0.456	0.105	0.388	0.457	0.522	0.069	0.065
Elec	517	0.480	0.119	0.399	0.484	0.554	0.085	0.069
Biochem	7,327	0.432	0.084	0.378	0.435	0.488	0.057	0.054
Physiol	808	0.443	0.088	0.385	0.443	0.496	0.058	0.053
Gastro	1,005	0.464	0.095	0.402	0.466	0.528	0.064	0.062

(b) Articles with $C(T) \geq 5$

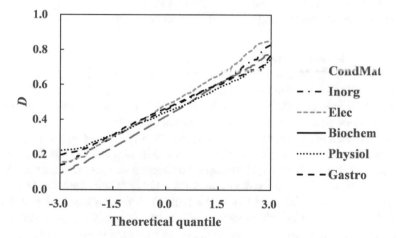

Fig. 6.4 Q-Q plot for D of the articles with $C(T) \geq 5$

0.674 for the normal distribution. The Q-Q plots shown in Fig. 6.4 comparing the distributions of D of articles with $C(T) \geq 5$ to normal ones are almost linear for all the six fields. The Q-Q plots for articles with $C(T) \geq 1$ (not shown) are also nearly linear, but the slope gradually becomes steeper in the region $D > 0.6$. It is thus concluded that D follows the normal distribution fairly closely, particularly for articles with citation counts greater than a certain level.

It is well known that the citation distribution of articles largely differs from field to field; does this hold for the distribution of D? Distributions of D and $\log[C(T) + 1]$ (both for articles $C(T) \geq 1$) are shown using box plots in Fig. 6.5, where the

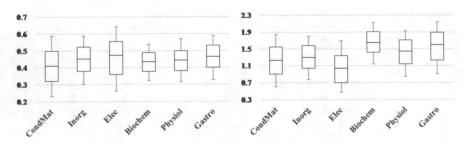

Fig. 6.5 Difference among fields of: (left) distribution of D and (right) distribution of $\log[C(T)+1]$ (reproduced from [36]).
The bar in the middle of the box is the median (the second quartile), the lower and upper boundary of the box indicate the first and third quartile, respectively, and the lower and upper bar outside the box are the 10 and 90 percentiles, respectively.

difference among fields can be visually confirmed. Compared to $\log[C(T) + 1]$, the distribution of D considerably overlaps among the different fields. Thus, citation durability is not as dependent on the subject field as the citation count. Although many previous studies have revealed that the citation count level among fields follows the general trend of biomedical > physical/chemical > engineering, D does not reflect such a trend.

Relationship between D and $\log[C(T) + 1]$

Table 6.5 shows the Pearson's product-moment correlations of D with $\log[C(T)+1]$ of articles with $C(T) \geq 5$ in the six fields. All coefficients lie in the range [0.16..0.33] and do not include 0 in their 95% confidence intervals. This suggests that a greater total number of citations means a higher citation durability.

A closer examination reveals, however, that the relationship is not linear. Table 6.6 shows the mean of $\log[C(T) + 1]$ values for D in intervals of 0.05 for each field and indicates that $\log[C(T) + 1]$ assumes a maximum at a certain value of D. This finding is illustrated as scatter diagrams for the fields "CondMat" and "Biochem" in Fig. 6.6. The mean values (dots in the figures) show nonlinear relationships of $\log[C(T) + 1]$ with D. In other words, citations of a frequently cited article tend to spread over the entire period instead of being concentrated in very early or very late phases.

Table 6.5 Pearson's correlation coefficients r between D and $\log[C(T) + 1]$ (reproduced from [36])

Field	n	r	95% CI
CondMat	5,097	0.325	[0.300, 0.350]
Inorg	2,061	0.167	[0.124, 0.208]
Elec	517	0.259	[0.177, 0.338]
Biochem	7,327	0.288	[0.267, 0.309]
Physiol	808	0.291	[0.227, 0.353]
Gastro	1,005	0.198	[0.138, 0.256]

Table 6.6 The relationship between D and the mean of $\log[C(T)+1]$ (reproduced from [36]). The values are shown only when at least two articles are included in the division.

Range of D	Mean of $\log[C(T)+1]$ [a]				
	CondMat	Inorg	Biochem	Physiol	Gastro
$0.05 \leq D < 0.1$	0.88		0.78		
$0.1 \leq D < 0.15$	0.96	0.92	0.95		
$0.15 \leq D < 0.2$	1.06	0.99	1.11		1.02
$0.2 \leq D < 0.25$	1.08	1.07	1.21	1.08	1.17
$0.25 \leq D < 0.3$	1.18	1.14	1.34	1.24	1.26
$0.3 \leq D < 0.35$	1.25	1.28	1.49	1.31	1.44
$0.35 \leq D < 0.4$	1.34	1.35	1.63	1.42	1.57
$0.4 \leq D < 0.45$	1.42	1.42	1.69	1.48	1.62
$0.45 \leq D < 0.5$	1.47	1.45	1.75	1.57	1.69
$0.5 \leq D < 0.55$	1.49	1.42	1.75	1.60	1.66
$0.55 \leq D < 0.6$	1.53	1.46	1.77	1.63	1.71
$0.6 \leq D < 0.65$	1.48	1.39	1.71	1.59	1.70
$0.65 \leq D < 0.7$	1.50	1.31	1.56	1.10	1.50
$0.7 \leq D < 0.75$	1.63	1.26	1.55		1.59
$0.75 \leq D < 0.8$	1.13	0.91	0.85		
$0.8 \leq D < 0.85$	0.86	0.88			
$0.85 \leq D < 0.9$	1.18				

It is seen in Fig. 6.6 that the range of D is widely spread for lower citation counts (in the lower part of the graph) but becomes narrower for higher counts (in the upper part). This finding is consistent with what was stated above in this subsection, namely, the distribution of D is narrower if only articles with a greater total number of citations are considered.

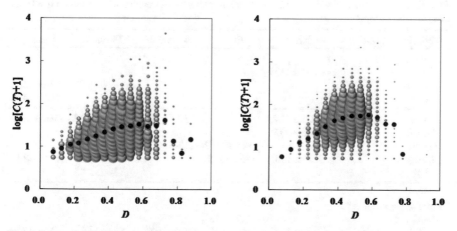

Fig. 6.6 Non-linear relations of $\log[C(T)+1]$ with D for: (left) "CondMat" field and (right) "Biochem" field (reproduced from [36]). Bubble sizes show the number of articles falling under each division. Dots (•) show the mean of $\log[C(T)+1]$ in each D region.

Table 6.7 The results of the multiple regression analysis: the coefficient of determination (R^2), adjusted R-squared (R_c^2), and variance ratio (F) (reproduced from [36])

	CondMat	Inorg	Elec	Biochem	Physiol	Gastro
R^2	0.136	0.131	0.177	0.153	0.130	0.350
R_c^2	0.116	0.108	0.154	0.138	0.112	0.330
F	6.76	5.61	7.54	10.20	7.23	17.23

LMR analysis of the relationship of citation durability with other characteristics of articles

The LMR analysis with D as the response variable and the characteristics shown in Table 6.2 as the explanatory variables resulted in significant correlations (with a probability of significance $p < 10^{-4}$ in all the six fields. Coefficients of determination R^2, adjusted coefficients of determination R_c^2, and variance ratios F are summarized in Table 6.7. R_c^2 values are not particularly high and lie in the range of $[0.1..0.3]$.

Table 6.8 shows the standardized partial correlation coefficients β and their 95% confidence intervals for the explanatory variables. The criteria for the selection of significant variables in the stepwise approach were 10% for the upper limit for selection and 20% for the lower limit for elimination. The table therefore indicates βs of variables that are significant at the 10% level, hence there are a few βs that include 0 in the 95% confidence interval. The relatively small sample size (about 200) was the reason for setting a moderate level of significance. In the field "Gastro", the variable Eqs (number of equations) was not included in the analysis because none of the articles in this field included equations.

Table 6.8 The results of the LMRs: the standardized partial regression coefficients β^a (reproduced from [36]). The estimated value and 95% confidence interval of coefficients are shown if selected in a stepwise process.

Variable	CondMat	Inorg	Elec	Biochem	Physiol	Gastro
Authors						
Insts		−.164 ± .139				
Countries						.136±.122
Refs						−.163 ± .132
Price	−.184 ± .149	−.223 ± .140		−.302 ± .143	−.232 ± .145	−.380 ± .123
Length		-.185±.171				
Figures			−.128 ± .152	−.136 ± .125	−.207 ± .157	
Tables		.226±.154	.185 ± .153			.111±.125
Eqs					−.136 ± .143	
Age						
RatePubl	.160±.142					
MedCite			−.212 ± .152			
C(T)	.308±.152	.241±.153	.340±.154	.423±.148	.354±.156	.464±.150
IF	−.183 ± .147			−.144 ± .141		−.191 ± .154

The following five variables were selected at least in three subject fields out of the six.

Price. Negatively correlated with D in five subject fields other than "Elec", meaning that articles that cite more recent references (within five years after publication) tend to be cited earlier.

Figures and *Tables.* Positive and negative correlations, respectively, with D in three fields each. This finding means that articles containing more figures tend to be cited earlier, and those with more tables later. These relationships are, however, not very strong.

$C(T)$ and *IF.* D has a relatively strong positive correlation with $C(T)$ in all the fields and a somewhat weak negative correlation with *IF* in three fields.

The correlation with the variables other than the five variables mentioned above was not significant.

6.3.5 Discussion

Characteristics of the distribution of the citation durability index D
The distribution of D, used as the citation durability index of an article in this research, was found to show the following characteristics across six different subject fields:

- follows the normal distribution closely, without high skewness, showing a remarkable contrast to the distribution of the citation count;
- becomes higher in normality and narrower in variance if only articles with greater citation counts are considered; and
- is not as dependent on the subject field as the distribution of the citation count.

D is considered to be an appropriate index of citation durability from those properties, together with reflecting information on the entire citation lifetime of an article.

Relationship between the total number of citations and D
The following results were demonstrated in this research based on the data for the citation window of fifteen years.

- D is positively correlated with $C(T)$ (see Table 6.5). The multiple regression analysis also showed that the partial regression coefficient of $C(T)$ was significantly positive in all the fields investigated (see Table 6.8).
- However, $C(T)$ does not linearly increase with D, but the mean of $C(T)$ has a maximum value at a certain value of D (see Table 6.6 and Fig. 6.6).

The relationship of these findings with other studies described in Section 6.2 will be discussed in Section 6.4.1.

Relationships between the citation durability and other characteristics of articles - comparison to the characteristics affecting the citation count

The NBMR (negative binomial multiple regression) analysis predicting citation frequency C in the Chapter "Extrinsic Factors Affecting Citation Frequencies of Research Articles" and the LMR analysis predicting citation durability D in this chapter are based on practically the same sample articles and use the similar explanatory variables. These results are now drawn on for a comparison of the characteristics closely related to the citation count and those closely related to the citation durability.

Price shows a strong correlation with both C and D but with opposite signs: An article with a high *Price* value (citing more recent reference materials) will receive more total citations but with a bias toward early periods. While this is not contrary to the definition of *Price*, articles with high *Price* values represent somewhat special cases since articles with high citation counts are more likely to receive delayed citations (see Section 6.3.4).

Figures and *Tables* are also significantly correlated with D, but their relationship with C was not significant. A few other studies have included the number of figures and tables in an article among the explanatory variables in the multiple regression analysis to explain citation counts [24, 39], but they also did not show a significant correlation. The present work is the first to demonstrate that the number of figures and tables in an article is correlated with citation durability but not the citation count. The general tendency (though not too strong) that articles containing more figures are cited earlier and those containing more tables are cited over a longer period may suggest the behavior of citing authors, that is, they are attracted by the articles including more figures with strong visual impact when browsing recent articles while they place importance on the articles including more tables with rich information when searching articles with a longer view.

Limitations of the present work and future issues
The value of D of an article varies with the observed citation period. The changing pattern differs among articles. Therefore, the distribution shape and other properties of D obtained from the citation history of 15 years may change when a much longer citation period is used. Or, the ranking of articles according to D may considerably change. How these phenomena change longitudinally constitutes future issues.

Two articles whose patterns of citation history are the same have the same value of D even if their total citations greatly vary. For example, if the citation count of an article is double that of another article every year, the D values of the two articles are the same. However, the reverse is not always true. That is, the patterns of citation history may be quite different even if they have the same D values. Recently, Sun et al. introduced an "obsolescence vector" as an indicator that can distinguish the different patterns to some extent [40]. This indicator is a two-dimensional vector consisting of a parameter Gs, which is proposed by Li et al. [32] and is similar to D, as well as a parameter A^- detecting drastic fluctuations of citation curves. Further development of citation durability indicators allowing us to discriminate a fine difference in the structure of citation history is another issue for future investigation.

6.4 Concluding Remarks

6.4.1 Studies on the Relationship Between the Long-term Citation Rate and the Citation Durability of Articles

Here, the results of our research shown in Section 6.3 are compared to several previous studies.

Line [35], Aversa [3] and Levitt and Thelwall [28, 29] have reported that higher cited papers (HCP) in the long run tend to receive their citations more lately than fewer cited ones (see Section 6.2.2). Wang in [42], investigating the correlation between the short and long citation impacts of papers, also found that more highly cited articles in their life-time tend to receive more delayed citations, as described in Section 6.2.4 Bornmann et al. showed that, for very highly-cited papers, the longer time elapsed since their publication, the higher their CSS (a yearly- and field-normalized citation score) tended to rise (see Section 6.2.1) [6]. Our finding that D is positively correlated with $C(T)$ corroborates those results mentioned above by deeper analysis over several different subject fields.

Our finding that the positive relationship between D and $C(T)$ is not linear but the mean of $C(T)$ has a maximum value at a certain value of D can be compared with those of [12] (see Section 6.2.6). They examined the field-normalized citation counts (CPP/FCSm) among three classes of citation durability: early cited [II], delayed cited [III], and normal [IV]. For citation windows of five, ten and twenty years after publication, the orders of mean CPP/FCSm were IV > II > III, IV > III > II, and III > IV > II, respectively. The non-linear relationship of the citation count to the citation durability found in our work is consistent with the trend seen for the 10-year window employed by Costas et al. [12].

6.4.2 Studies on the Relationship Between the Citation Durability and other Properties of Articles

Here, we discuss the studies concerning the relationship of various attributes of articles to citation durability including our research, which are summarized in Table 6.9. These previous studies were mentioned in Section 6.2.6 and our study was described in Section 6.3. There are considerably fewer studies on this subject than those on the relationships to the citation rate. Moreover, the studies adopting approaches considering interactions between variables such as multiple regression analysis are only those by Wang et al. [43], Dey et al. [15] and ours. It is therefore difficult to make definite statements, but the followings are supposed from Table 6.9.

Work	Sample size (n)	Attributes that may influence on citation durability										
		#Authors	#Institutions	#Countries	#References	Price index	Interdisciplinarity of references	Article length	#Figures	#Tables	Author performance	Venue prestige
vanDalen & Henkens (2005)	Demography	1,371	Not						Not			Not	Neg
Costas et al. (2010)	Multidiscipline	8.34M	Neg	Neg	Neg	Neg			Pos				
Wang et al. (2015)	Multidiscipline	646,669	Neg		Neg	Neg			Pos				
Zhang & Glänzel (2017)	Multidiscipline	0.542M					Neg						
Dey et al. (2017)	Computer science		Not			Not		Pos				Pos	
Bornmann et al. (2018)	10 Subject Categories of WoS	961	Not		Not	Pos		Pos	Pos				
This research	Six fields	145 - 231 for a field	Not	Not	Not	Not	Neg	Not	Not	Neg	Pos	Not	Neg

Table 6.9 Correlatedness of various attributes of articles with citation durability (neg for negative, pos for positive).

- Factors having a high possibility of a significant relation with the citation durability: Price index (negative correlation); the interdisciplinarity of references (positive correlation)
- Factors having some possibility of a significant relation with the citation durability: the article length (positive correlation); the number of figures (negative correlation); the number of tables (positive correlation); the authors' performance (positive correlation); the citation impact of the publishing journal (negative correlation)
- Factors having little possibility of a significant relation with the citation durability: the number of authors; the number of affiliated institutions; the number of affiliated countries.

It is suggested that the attributes of references highly correlate with the citation durability while those of the authors' collaboration do not. The results of our research are compared to those found by other studies below. The results of van Dalen and Henkens [13] are not contradictory with what our analysis shows: Citation durability has no definite relationship with the number of authors, the length of the article, or the author's reputation but is negatively correlated with publication in top-ranked journals. The results of Zhang and Glänzel that an article with higher Price index tends to show a lower citation durability (i.e., higher prospective Price index) [45] are also consistent with ours.

We compare the results of the studies of Costas et al. [12] and Wang et al. [43] with ours. Both these studies suggest significant relationships between the citation durability and the degree of collaboration, the number of references, and article length, which are not recognized as significant in our work. The reasons for the discrepancy are considered as follows.

- The sample size used is greatly different. That used in our research is of the order of 10^2 or 10^3, while Costas et al. had a sample size of about 8 million, and Wang et al. analyzed 0.3 million articles. This leads to considerably different effect sizes being statistically significant (although it is impossible to compare the effect sizes because different analytic methods were adopted for the respective studies).
- The relationships of some variables with citation durability found by Costas et al. and Wang et al. might be controlled by other variables used in the LMR analysis by ours.

To examine the second possibility above, we tried to carry out an LMR analysis excluding $C(T)$, which is the variable most strongly related to D among the explanatory variables. Contrary to expectations, when $C(T)$ was excluded, the cases resulting in a rise in significance or an increase in the effect size (standardized partial correlation coefficient) were rather few. The variables such as *Authors*, *Insts*, *Refs*, and *Length* were still not significant in most fields. These results suggest that the correlations between D and these variables become clearer by the controlling variable $C(T)$.

References

1. Abramo G., C.T., D'Angelo, C.A.: Assessing the varying level of impact measurement accuracy as a function of the citation window length. Journal of Informetrics **5**(4), 659–667 (2011)
2. Aksnes, D.W.: Characteristics of highly cited papers. Research Evaluation **12**(3), 159–170 (2003)
3. Aversa, E.: Citation patterns of highly cited papers and their relationship to literature aging: A study of the working literature. Scientometrics **7**(3–6), 383–389 (1985)
4. Avramescu, A.: Actuality and obsolescence of scientific literature. Journal of the American Society for Information Science **30**(5), 296–303 (1979)
5. Baumgartner, S.E., Leydesdorff, L.: Group-based trajectory modeling (GBTM) of citations in scholarly literature: Dynamic qualities of "Transient" and "Sticky Knowledge Claims". Journal of the Association for Information Science & Technology **65**(4), 797–811 (2014)
6. Bornmann, L., Ye, A.Y., Ye, F.Y.: Sequence analysis of annually normalized citation counts: An empirical analysis based on the characteristic scores and scales (CSS) method. Scientometrics **113**(3), 1665–1680 (2017)
7. Bornmann, L., Ye, A.Y., Ye, F.Y.: Identifying 'hot papers' and papers with 'delayed recognition' in large-scale datasets by using dynamically normalized citation impact scores. Scientometrics **116**(2), 655–674 (2018)
8. Bornmann, L., Ye, A.Y., Ye, F.Y.: Identifying landmark publications in the long run using field-normalized citation data. Journal of Documentation **74**(2), 278–288 (2018)
9. Burrell, Q.L.: Age-specific citation rates and the Egghe–Rao function. Information Processing & Management **39**(5), 761–770 (2003)
10. Burrell, Q.L.: Are "Sleeping Beauties" to be expected? Scientometrics **65**(3), 381–389 (2005)
11. Cano, V., Lind, N.: Citation life cycles of ten citation classics. Scientometrics **22**(2), 297–312 (1991)
12. Costas, R., van Leeuwen, T.N., van Raan, A.F.J.: Is scientific literature subject to a Sell-By-Date? A general methodology to analyze the durability of scientific document. Journal of the American Society for Information Science & Technology **61**(2), 329–339 (2010)
13. van Dalen, H.P., Henkens, K.: Signals in science - On the importance of signaling in gaining attention in science. Scientometrics **64**(2), 209–233 (2005)
14. Della Briotta Parolo, P., Pan, R.K., Ghosh, R., Huberman, B.A., Kaski, K., Fortunato, S.: Attention decay in science. Journal of Informetrics **9**(4), 734–745 (2015)
15. Dey, R., Roy, A., Chakraborty, T., Ghosh, S.: Sleeping beauties in Computer Science: Characterization and early identification. Scientometrics **113**(3), 1645–1663 (2017)
16. Du, J., Wu, Y.: A parameter-free index for identifying under-cited sleeping beauties in science. Scientometrics **116**(2), 959–971 (2018)
17. Egghe, L., Ravichandra Rao, I.K.: Citation age data and the obsolescence function: Fits and explanations. Information Processing & Management **28**(2), 201–217 (1992)
18. Fang, H.: Analyzing the variation tendencies of the numbers of yearly citations for sleeping beauties in science by using derivative analysis. Scientometrics **115**(2), 1051–1070 (2018)
19. Galiani, S., Gálvez, R.H.: An empirical approach based on quantile regression for estimating citation ageing. Journal of Informetrics **13**(2), 738–750 (2019)
20. Garfield, E.: Premature discovery or delayed recognition. Why? Essays of an Information Scientist **4**, 488–493 (1980)
21. Glänzel, W., Schlemmer, B., , Thijs, B.: Better late than never? On the chance to become highly cited only beyond the standard bibliometric time horizon. Scientometrics **58**(3), 571–586 (2003)
22. Glänzel, W., Schoepflin, U.: A bibliometric study on ageing and reception process of scientific literature. Journal of Information Science **21**(1), 37–53 (1995)
23. Glänzel, W., Thijs, B., Chi, P.S.: The challenges to expand bibliometric studies from periodical literature to monographic literature with a new data source: The book citation index. Scientometrics **109**(3), 2165–2179 (2016)

24. Haslam, N., Ban, L., Kaufmann, L., Loughnan, S., Peters, K., Whelan, J., Wilson, S.: What makes an article influential? Predicting impact in social and personality psychology. Scientometrics **76**(1), 169–185 (2008)
25. He, Z., Lei, Z., Wang, D.: Modeling citation dynamics of 'atypical' articles. Journal of the Association for Information Science & Technology **69**(9), 1148–1160 (2018)
26. Ke, Q., Ferrara, E., Radicchi, F., Flammini, A.: Defining and identifying sleeping beauties in science. Proceedings of the National Academy of Sciences **112**(24), 7426–7431 (2015)
27. Lachance, C., Larivière, V.: On the citation lifecycle of papers with delayed recognition. Journal of Informetrics **8**(4), 863–872 (2014)
28. Levitt, J.M., Thelwall, M.: Patterns of annual citations of highly cited articles and the prediction of their citation raking: A comparison across subject categories. Scientometrics **77**(1), 41–60 (2008)
29. Levitt, J.M., Thelwall, M.: Citation levels and collaboration within library and information science. Journal of the American Society for Information Science & Technology **60**(3), 434–442 (2009)
30. Li, J.: Citation curves of 'all-elements-sleeping-beauties': 'flash in the pan' first and then 'delayed recognition'. Scientometrics **100**(2), 595–601 (2014)
31. Li, J., Shi, D.: Sleeping beauties in genius work: When were they awakened? Journal of the Association for Information Science & Technology **67**(2), 432–440 (2016)
32. Li, J., Shi, D., Zhao, S.X., Ye, F.Y.: A study of the 'heartbeat spectra' for 'sleeping beauties'. Journal of Informetrics **8**(3), 493–502 (2014)
33. Li, J., Ye, F.Y.: The phenomenon of all-elements-sleeping-beauties in scientific literature. Scientometrics **92**(3), 795–799 (2012)
34. Li, J., Ye, F.Y.: Distinguishing sleeping beauties in science. Scientometrics **108**(2), 821–828 (2016)
35. Line, M.B.: Citation decay of scientific papers: Variation according to citations received. Journal of Information Science **9**(2), 90–91 (1984)
36. Onodera, N.. Properties of an index of citation durability of an article. Journal of Informetrics **10**(4), 981–1004 (2016)
37. Onodera, N., Yoshikane, F.: Factors affecting citation rates of research articles. Journal of the American Society for Information Science & Technology **66**(4), 677–690 (2015)
38. van Raan, A.F.J.: Sleeping beauties in science. Scientometrics **59**(3), 467–472 (2004)
39. Snizek, W.E., Oehler, K., Mullins, N.C.: Textual and non-textual characteristics of scientific papers. neglected science indicators. Scientometrics **20**(1), 25–35 (1991)
40. Sun, J., Min, C., Li, J.: A vector for measuring obsolescence of scientific articles. Scientometrics **107**(2), 745–757 (2016)
41. Wang, D., Song, C., Barabási, A.L.: Quantifying long-term scientific impact. Science **342**(6154), 127–132 (2013)
42. Wang, J.: Citation time window choice for research impact evaluation. Scientometrics **94**(3), 851–872 (2013)
43. Wang, J., Thijs, B., Glänzel, W.: Interdisciplinarity and impact: Distinct effects of variety, balance, and disparity. PLOS ONE **10**(5), e0127298 (2015)
44. Ye, F.Y., Bornmann, L.: 'Smart girls' versus 'sleeping beauties' in the sciences: The identification of instant and delayed recognition by using the citation angle. Journal of the Association for Information Science & Technology **69**(3), 359–367 (2018)
45. Zhang, L., Glänzel, W.: A citation-based cross-disciplinary study on literature ageing: Part II—diachronous aspects. Scientometrics **111**(3), 1559–1572 (2017)
46. Zhang, R., Wang, J., Mei, Y.: Search for evergreens in science: A functional data analysis. Journal of Informetrics **11**(3), 629–644 (2017)

Chapter 7
Wider, or Deeper! On Predicting Future of Scientific Articles by Influence Dispersion Tree

Tanmoy Chakraborty, Sumit Bhatia, Annu Joshi, Partha Sarathi Paul

Abstract The number of citations received by a scientific article has been used as a proxy for its influence/impact on the research field since the long past. Raw citation count, however, treats all the citations received by a paper equal and ignores the evolution and organization of follow-up studies inspired by the article, thereby failing to capture the depth and the breadth of the paper's influence on the research field. To address this issue, we propose *Influence Dispersion Tree* (IDT), a novel data structure to model the organization of follow-up papers and their dependencies through citations. IDT organizes the follow-up papers in several branches such that each branch represents an independent body of research in the field. The number of branches of IDT and their length captures the breadth and the depth, respectively, to compute the influence of the paper on the corresponding research field. We present a suite of metrics, viz., *Influence Dispersion Index* (IDI) and *Normalized Influence Divergence* (NID), to quantify the influence of a paper. We posit that a highly influential paper should increase the knowledge of a research field vertically and horizontally. To capture this, we develop the notion of an ideal IDT for every paper representing the configuration that maximizes the article's impact. Our theoretical analysis shows that an ideal IDT configuration should have equal depth and breadth and minimize the NID value. We establish the utility of our proposed framework using two large real-world datasets from two research domains, viz., Computer Science and Physics. We present an empirical study of different characteristics of IDTs and establish the superiority of NID as a better measure of influence when compared with raw citation counts.

Tanmoy Chakraborty, Annu Joshi, Partha Sarathi Paul
Department of Computer Science & Engineering, Indraprastha Institute of Information Technology (IIIT), Delhi, India,
e-mail: {tanmoy,annuj,parthap}@iiitd.ac.in

Sumit Bhatia
IBM Research AI, Delhi, India
e-mail: sumitbhatia@in.ibm.com

7.1 Introduction

The total number of citations received by a paper (citation count) is one of the most prevalent metrics to measure the paper's impact or influence on the research field. Despite various shortcomings [7], it is widely used in assessing the research quality of researchers, institutions and research groups. There are a number of compelling reasons to study the properties of a scientific article apart from its citation count, which can help us understand more obvious and subtle interconnections between the articles.

The authors in [40] provided a PageRank-based approach of citation analysis to rank scientific publications and to assist scientific award winner prediction. In [25] they attempted to predict the Nobel Laureates using the connections between cited and citing papers. Various works attempted to discover star scientists by considering it as a multidimensional ranking problem. They exploited some uncorrelated state-of-the-art scientometric indices to comprise different dimensions of the problem space, and introduced new ranking schemes to rank the scientists for this purpose [38, 41].

These tasks was not possible simply by considering the citation count of these articles. Further, citation count has often been criticised due to unknown citing motivations – some articles that are influential are cited, while some other articles exhibiting similar influence are not cited. Most of the times, citations are added without proper context [5, 27, 37]. These observations go against mere looking at the raw citation counts and make it more imperative to study the organisation of citing articles of the paper.

In this chapter, we hypothesize that the effect of citations can better be captured by understanding the relationships between the citing articles of a paper – the interconnections between the citing articles will provide us with a better perception of their impact compared to the aggregated citation count of a paper. We can analyse the propagation of influence in different directions using these interconnection patterns. For instance, a specific paper p_i has influenced deeply another paper p_j which in turn has played a leading role in the development of some new ideas in the domain, and thereby has increased the depth of the corresponding research field. The above may be considered to carry a higher impact than simply increasing the breadth of citations, i.e., influencing a number of papers from the same domain.

We compare two articles having the same citation counts in Fig. 7.1; citation network subgraphs for these articles are presented in the left side of Fig. 7.1(a) and Fig. 7.1(b), and their IDTs on the right side of the same diagrams. We notice that, even if these articles have the same citation count (a value of 8 in the present example), they have different research impact. Analysing the depth and the breadth of the citing articles can capture the impact mentioned above. In other words, we can study the influence of a research article from two broad perspectives: (i) *breadth* – the number of novel ideas a particular research article gives direction to, and (ii) *depth* – the degree to which each of these novel ideas grows in its direction.

In our previous study [30], we proposed a metric to quantify a new facet of influence that a paper has on its follow-up papers and made the following contributions:

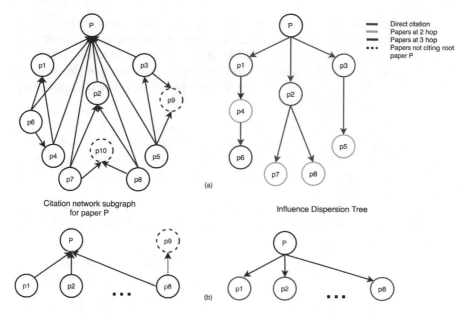

Fig. 7.1: Schematic diagrams illustrating the importance of understanding the organization of citing papers of a scholarly article P. In either scenarios, the article P has the same citation count (= 8). However, in scenario (a), citing articles $p1$, $p2$ and $p3$ give rise to new articles like $p6$, thus increasing the depth of the field; on the other hand, in scenario (b), there is no further contribution in the new fields originating from P, it only increases the breadth of the tree. The difference mentioned is highlighted using IDTs shown on the right.

(a) We defined a framework to model the depth and the breadth of a scientific paper. We developed a novel network structure that we termed the *Influence Dispersion Tree (IDT)* (Section 7.3). These IDTs are used to model the depth and the breadth of a paper and to study the paper's influence dispersion. We further presented a theoretical analysis of the properties of the IDT structure and their relationships with the citation count of the paper. (b) We proposed a novel metric, called *Influence Dispersion Index* (IDI) to quantify the influence of a scientific paper (Section 7.3.2). We argued that, in an ideal scenario, the influence of a paper should be dispersed to maximize the depth and the breadth of its influence. We proved that such an optimal IDT configuration would have equal depth and breadth (and is equal to $\lceil \sqrt{n} \rceil$, where n is the number of citations of a given paper). We also proposed a metric, called *Influence Divergence* (ID) to measure how far the original IDT configuration is from the optimal IDT configuration. We further derived a normalized version of ID, called *Normalized Influence Divergence* (NID), that normalizes the value of the influence divergence in the range $[0, 1]$ to allow fair comparison between different papers with different citation counts based on their NID values. We evaluated the results in the said article on a computer science bibliographic dataset.

In this chapter, we build upon our previous work and provide a detailed analysis of the impact of depth and breadth in quantifying the quality of a paper. We perform our analysis here on two scientific domains, viz., (i) the domain of computer science as the representative of conference-driven domains, and (ii) the domain of physics as the representative of journal-driven domains. Our significant (additional) contributions in this chapter are the following:

1. We use two large bibliographic datasets, one consisting of 3.9 million articles from the domain of computer science and the other consisting of 384K articles from the field of physics, to study the properties of the IDT structure and to test the effectiveness of the influence metrics. We preprocess the datasets and construct IDTs for all the articles in the datasets.

2. Following the preprocessing step, we perform analysis on IDTs. Our rigorous analysis reveals exciting observations (Section 7.5) that we find true for both the domains of our study.

   a. Our observations suggest that when citation count increases, the breadth of an IDT tends to grow much faster than the depth. The maximum breadth of an IDT is significantly higher than the maximum depth of the same. The above holds both for the physics and the computer science datasets. We infer from this observation that acquiring more citations over time often contributes more to breadth than depth.

   b. We empirically find that the value of NID decreases with an increase in citation count. This finding strengthens our hypothesis that the IDT of a highly influential paper tends to reach its optimal configuration by enhancing both the depth and the breadth of its research field.

   c. We show that NID performs better than the raw citation count as an early predictor to forecast the impact of a scientific article (Section 7.6.1)[1]. More importantly, we show that, for a paper to be influential, it is unnecessary to have many follow-up papers (i.e., high citation count); rather, the organization of the follow-up papers matters more. A highly influential paper tends to have citations spread in both directions to increase its corresponding IDT's breadth and depth.

For reproducibility of our results, we make available to the community the code and the datasets used in this study at `https://github.com/LCS2-IIITD/influence_dispersion_extension`.

---

[1] In our previous work, we manually curated 40 papers recognized as the most influential papers by the scientific communities. We made the above selection of the articles based on the 'Test of Time award' or the '10 years influential paper award' for the computer science community. We find that NID outperforms the raw citation count in identifying these influential papers. However, we cannot make a similar comparison for the physics dataset because, in the physics community, there is no such award given to a paper after a specific time interval upon publication.

7.2 Related Work

Measuring the impact of scientific articles by analysing the citations received has been a prominent research area in Scientometrics [17, 18]. We now study the relevant works done in this domain. We conducted our analysis in this context from two perspectives – (i) studies dealing with citation count and its variants for measuring the article impact, and (ii) studies exploring the structure of citations of an article and its effect on article impact.

7.2.1 Quantifying the Scientific Impact

Citation data is frequently used to measure scientific impact as mentioned earlier. Most citation indicators are based on citation count – *Journal Impact Factor* [19], *h-index* [23], *Eigenfactor* [16], *i10-index* [14], *c-index* [32], etc. are few of them, all of these have their limitations. There have been several attempts to overcome the drawbacks of these indices such as *m-quotient* [23, 42], *g-index* [15], *e-index* [47], C^3-*index* [33, 34], and *hg-index* [1]. The authors in [31] extended the idea of Journal Impact Factor to author level to define a metric for author impact and proposed author impact factor (AIF), whereas in [48] they discussed in details about different author-centric metrics and may be consulted for further reading.

The authors in [22] they argued against considering only the citation count while measuring the impact and proposed *Paper Impact Effectiveness* (PIE) that combines the citation count and the usage pattern (in terms of downloads, reads and views) [44]. In [2] they measured the impact of an article as the number of lead authors that have been influenced by the article. In [21] they reviewed the strengths and limitations of six established approaches for measuring research impact. In [26] they examined the impact of the ease of access of an article (open-access publication versus subscription-based publication) on its citations pattern.

In [29] the authors focused on the citations, but primarily on the dynamic behaviour of citations over the life time of a paper and showed how we can predict the future citations based on previous citations. In [39] they discussed how a couple of special regions of an author's citation curve, they coined as the *penalty areas* of the curve, can be exploited to distinguish the *influential authors* from *mass producers*. They proposed a number of indices, viz., *Perfectionism Index* and *eXtreme Perfectionism Index*, using the said regions of the citation curve for the purpose mentioned. In [9] they showed that the change in yearly citation count of articles published in journals is different from articles published in conferences.

Even the evolution of yearly citation count of papers varies across disciplines [10, 35] that often leads to a new direction of designing domain-specific impact measurement metrics. In [8] they categorized the evolution of citation count of articles into six classes – (i) *Early Risers* (ER) having a single peak in citation count within the first five years after publication; (ii) *Frequent Risers* (FR) having distinct multiple peaks in citation count; (iii) *Late Risers* (LR) having a single peak after five years of the publication (but not in the last year); (iv) *Steady Risers* (SR) with monotonically increasing citation count; (v) *Steady Dropper* (SD) with monotonically decreasing citation count, and (vi) *Other* (OT) which does not follow any of the patterns mentioned above.

The authors further adopted a *stratified learning approach* to predict the future citation count of scientific articles. In the stratified learning approach, one divides the elements of the stratified space into homogeneous subgroups (*aka* strata) before sampling. The authors exploited the previously-mentioned categorization of the articles as strata in their two-stage prediction model. They observed that the above categorization of articles helped in better predicting the future citation count (and hence, the future impact) of a paper.

7.2.2 Citation Expansion of Individual Papers

Despite extensive criticism on the shortcomings of using citation count, it is widely used in the scientific community. The analysis of citation structure to articles is unexplored mostly. There have been a few recent studies which attempted to understand the organization of citations around a scientific entity (paper, author, venue, etc.). These works focused particularly on the topology of the graph constructed from the induced subgraph of the papers citing the seed paper.

In [45] the authors took an evolutionary perspective to propose an algorithm for constructing genealogical trees of scientific papers on the basis of their citation count evolution over time. This is useful to trace the evolution of certain concepts proposed in the seed paper. In [46] they studied the importance and the differences in the breadth and depth of citation distribution in scientific articles. They proposed indices that capture the distribution of citations in terms of breadth and depth. In [20] they proposed to construct for each paper a rooted citation graph (RCG). RCG is an induced subgraph of all the nodes that can be reachable from the root paper via directed paths. They then derived two metrics, viz., *average degree* and *core influence*, from the RCG for the root paper. They further extended the proposed measures to authors and journals.

The authors in [6] proposed a multidimensional framework to capture the impact of a scientific article. The authors studied different dimensions of citations, including the depth and the breadth. However, they used different indicators to quantify the depth and the breadth of a publication. They considered the citations of paper p along with their relationship with references of p. They introduced two variables, viz. *R[citing pub]* and *R[cited pub]*, for each publication citing a target paper. The distribution of these variables yields two sets of metrics to characterize the target publication's citation impact, viz. *depth* and *breadth* from the former distribution, and *dependence* and *independence* from the latter distribution. However, the notion of *depth* and *breadth* in our approach is different from their approach. In our approach, we introduce a tree representation of the citations of a given article and coin it as an influence dispersion tree. We then define the depth and the breadth of a paper as parameters of that tree representation.

In [13] the authors claimed that the highly-cited papers might not always be the most influential articles in a domain, as academic citations often occur out of compulsions; the authors often cite papers without reading them. To reduce these 'citation bias', they applied a transitive reduction approach on the network and claimed that the combination of transitive reduction with citation count could reveal new and exciting features about the publications.

The authors in [43] they applied transitively reduction technique on citation networks to highlight influential papers based on the degree distribution of the articles. They claimed that this approach could become an alternative to text analysis techniques for retrieving influential papers in a scientific domain. In [4] they computed the centrality of a paper in the citation network as a measure of its impact. The authors introduced Eigenfactor Metrics for this purpose. The notion of the Eigenfactor algorithm is similar to Google's PageRank algorithm for ranking webpages.

7.3 Influence Dispersion Tree

In this section, we define the idea of Influence Dispersion Tree of a scholarly paper and how it is different from other similar classical structures. We further describe some of the properties of IDTs and some metrics that we use to measure the influence of a scientific article.

7.3.1 Influence Dispersion Tree: Modelling Citations of an Article

Modeling the citations of a scholarly article in the form of a tree is a traditional practice. In these representations, the article P is positioned as the root of the tree and the articles citing P form the branches of the tree. All the citing articles are considered to be equal; the relationships (if any) between these articles are not taken care of in these models. However, in real life, not all citations are equal; they can have different motivations, and hence can have a different impact of these citations to the cited article. To understand better the impact of the citing articles, we propose a new citation representation model, called the *Influence Dispersion Tree* (IDT).

Let us consider a scholarly article P and $C_P = \{p_1, p_2, \ldots, p_n\}$ be the set of articles citing P. We first create a Influence Dispersion Graph defined as follows:

Definition 1 (Influence Dispersion Graph) The Influence Dispersion Graph (IDG) of an article P is a directed and rooted graph $\mathcal{G}_P(\mathcal{V}_P, \mathcal{E}_P)$ with $\mathcal{V}_P = C_P \cup \{P\}$ as the vertex set, P as the root of the graph, and the edge set $\mathcal{E}_P$ consists of directed edges of the form $\{p_u \rightarrow p_v\}$ such that $p_u \in \mathcal{V}_P, p_v \in C_P$, and the article p_v cites the article p_u.

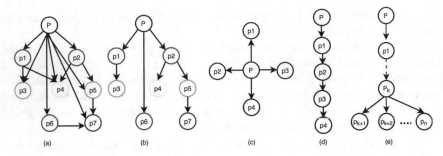

Fig. 7.2: (a)-(b) Illustration of the construction of (b) IDT from (a) IDG of paper
P. Papers circled red cite only the paper P; papers circled green cite the paper P as
well as one of the other papers; paper circled blue cites the paper P and more than
one other papers. In case of the paper circled yellow, due to equal possibility of p_4
being connected from p_1 and p_2, a tie situation occurs. In order to break the tie,
we select the node that can rise the depth of the IDT. Further ties are resolved by
randomly connecting an option node, like p_4 from p_2 in the example IDT. (c)-(d)
Two corner cases to illustrate the lower bound – minimum and maximum number
of leaf nodes. (e) A configuration of a P-rooted IDT with (n) non-root nodes that
results in maximum IDI value.

Fig 7.2 shows an illustration of an IDG for an article P and its citing article set
$\{p_1, p_2, p_3, p_4, p_5, p_6, p_7\}$. Observe that, the IDG of the article P is same as the
induced subgraph of the citation graph consisting of P and all its citing articles, and
with directed edges in the opposite direction to indicate the propagation of influence
from the cited paper to the citing paper. Note further that, the IDG is restricted strictly
to the one-hop citation neighborhood of P (i.e., among the papers that are directly
influenced by the paper P). Thus, an IDG takes only into account those follow-up
papers that are *directly influenced* by a given paper. For example, if p_1 cites P, and
if p_2 cites p_1, but it does not cite P, it is not always clear whether p_2 is influenced
both by P and p_1, or only by p_1. Therefore, in such a scenario, IDG considers only
p_1 to avoid any possibility of ambiguity.

We convert the IDG to Influence Dispersion Tree (IDT) to analyse the impact
of citations in terms of depth and breadth. A tree data structure, by definition, is
used to represent the hierarchy in data. We use IDT to study the influence of an
article P on its citing articles and provide an easy-to-understand representation of
the organisation of articles citing P and their contribution to the depth and breadth
of the field. The IDT of the article P is a directed and rooted tree $\mathcal{T}_P = \{\mathcal{V}_P, \mathcal{E}'_P\}$
with P as the root. The vertex set is the same as that of IDG of P, and the edge set
$\mathcal{E}'_P \subset \mathcal{E}_P$ is derived from the edge set of IDG as described next.

Note that, an article $p_v \in C_P$ can cite more than one article in $\mathcal{V}_P$, which gives
rise to the following three possibilities:

1. p_v cites only the root article P. In this case, we add the edge $P \to p_v$ in $\mathcal{E}'_P$, thus creating a new branch in the tree emanating from the root node (e.g., the edges $P \to p_1$ and $P \to p_2$ in Fig. 7.2(b)).

2. p_v cites the root article P and an article $p_u \in C_P \setminus \{p_v\}$. In this case, we say that the article p_v is influenced by P as well as by the article p_u. There are two possible edges here: $P \to p_v$ and $p_u \to p_v$. However, since p_u is also influenced by P, the edge $p_u \to p_v$ captures indirectly the influence that P has on p_v. Therefore, we discard $P \to p_v$ and retain only the edge $p_u \to p_v$. This choice leads to the addition of a new leaf node in IDT capturing the chain of impact starting from P up to the leaf node p_v (e.g., the edges $p_1 \to p_3$ and $p_2 \to p_5$ in Fig. 7.2(b)).

3. p_v cites the root article P, and to each of a set of other articles $P_u \subseteq C_P \setminus \{p_v\}$, $|P_u| \geq 2$. Note that, by definition, each article $p \in P_u$ cites the root article P. The possible candidate edges to add here are $E = \{p \to p_v | p \in P_u\}$. We add to $\mathcal{E}'_P$ the edges $e = p \to p_v$ that satisfy the following:

$$p = \arg\max_{p' \in P_u} \ shortestPathLength(P, p') \tag{7.1}$$

The edge $p_5 \to p_7$ in Fig. 7.2(b) is such an edge.

The intuition behind adding edges in $\mathcal{E}'_P$ in the way mentioned above is to maximize the depth of IDT. If there are more than one candidate edges that can maximize the depth of the IDT, then we choose one of them at random. For example, between $p_1 \to p_4$ and $p_2 \to p_4$ in Fig. 7.2(b), we randomly choose the later one. The above edge construction mechanism is motivated by the *citation cascade graph* [24, 28]. We add a new citing article in $\mathcal{T}_P$ in such a way that the impact of the article P to its citing articles is maximally preserved. Impact maximization can be thought of either as maximizing the breadth or as maximizing the depth of the IDT. We choose the latter version in our approach in order to capture the cascading effect in the resultant IDT.

Definition 2 (Influence Dispersion Tree) The Influence Dispersion Tree (IDT) of an article P is a tree $\mathcal{T}_P(\mathcal{V}_P, \mathcal{E}'_P)$ with vertex set $\mathcal{V}_P$ as the set consisting of P and all the articles citing P, and edge set $\mathcal{E}'_P$ constructed as follows: if a paper p_v cites only P and no other articles in $\mathcal{V}_P$, we add $P \to p_v$ into the edge set $\mathcal{E}'_P$; if p_v cites other papers $P_u \in \mathcal{V}_P \setminus \{P\}$ along with P, we add only one edge $p_x \to p_v$ (where $p_x \in P_u$) according to Eq. 7.1.

7.3.2 Metrics to Measure Influence

Definition 3 (Influence Dispersion Index) The Influence Dispersion Index (IDI) of an article P is defined by the equation:

$$IDI(P) = \sum_{p_l \in P_L} distance(P, p_l) \tag{7.2}$$

where p_L is the set of leaf nodes in $\mathcal{T}_P(\mathcal{V}_P, \mathcal{E}_P)$, the IDT of P.

The IDI of the article P in Fig. 7.2(b) is 8. Intuitively, each leaf node in P's IDT corresponds to a separate branch emanating from the seed article P. IDI refers to the sum of distances of all the leaf nodes from the root node of the IDT. We can interpret IDI as a measure of the *ability* of an article to distribute its influence to other articles.

Note: It has been proved that the optimal configuration for an IDT having n non-root nodes should satisfy [30]:

$$depth = breadth = \lfloor \sqrt{n} \rceil \qquad (7.3)$$

The authors showed that there exists an ideal configuration of the IDT that optimizes the influence dispersion of the paper. They coined the same as the *ideal configuration* of IDT for the article. They claimed further that the IDI of an ideal IDT with n non-root nodes is n.

Definition 4 (Influence Divergence) Influence Divergence (ID) of an article is defined as the difference between the IDI of its original IDT, IDI(P) and that of its corresponding ideal IDT configuration, $I\bar{D}I(P)$. Formally:

$$ID(P) = IDI(P) - I\bar{D}I(P) \qquad (7.4)$$

We further normalize the IDI value using max-min normalization.

Definition 5 (Normalized Influence Divergence) Normalized Influence Divergence (NID) of an article P is defined as the difference between the IDI value of its IDT and the IDI value of its corresponding ideal IDT configuration, normalized by the difference between the maximum and the minimum IDI values of the IDTs having the size same as that of the IDT of P. Formally written:

$$NID(P) = \frac{IDI(P) - I\bar{D}I(P)}{IDI_{|P|}^{max} - IDI_{|P|}^{min}} \qquad (7.5)$$

The normalization is needed to compare two papers with different IDI values. NID ranges between 0 and 1. Clearly, a highly influential paper will have a low $NID(P)$ (i.e., lower deviation from its ideal dispersion index).

For an IDT of article P with n non-root nodes, the value of IDI^{min} will be n, as each node will be encountered at least once while traversing the tree. For an IDT of article P with n non-root nodes, the value of IDI^{max} will be [30]:

$$IDI(P)^{max} = (1 + \lfloor \frac{n-1}{2} \rfloor)(n - \lfloor \frac{n-1}{2} \rfloor) \qquad (7.6)$$

7.4 Dataset Description

We use two publicly available datasets of scholarly articles from two different domains – computer science and physics.

1. **MAS Dataset:** This dataset contains about 4 million articles in computer science domain indexed by Microsoft Academic Search (MAS)[2]. For each paper in the dataset, additional metadata such as, the title of the paper, authors of the paper and their affiliations, year and venue of publication, are included in the dataset. The publication years of the articles present in the dataset span over half a century, allowing us to investigate diverse types of articles in terms of their IDTs. A unique ID is assigned to each author as well as each publication venue upon resolving the named-entity disambiguation by MAS itself.
2. **PRJ Dataset:** This dataset contains all published articles in *Physical Review* (PR) journals [36] from 1950 through 2009[3]. We use only those entries in this dataset which include information about index, title, name of the author(s), year of publication, and references. The filtered dataset contains $384,289$ valid articles and $235,533$ authors.

We consider computer science as a representative domain of the research areas where authors prefer conference publications. In contrast, we choose the physics domain to represent research areas where authors prefer journal publications. Although each of the datasets can be viewed as a proxy for monodisciplinary research activities in science, we believe that analysis of these datasets jointly allows us to support multidisciplinary features which characterize citation dynamics in general [10].

We pass the above datasets through a series of pre-processing steps such as removing papers that do not have any citation and reference, removing papers that have forward citations (i.e., citing a paper that is published after the citing paper; this may happen due to public archiving the paper before actually publishing it), etc. Table 7.1 shows the summary statistics of the filtered datasets we use in all the analysis discussed in the following sections.

Table 7.1: Summary statistics of MAS and PRJ datasets after filtration.

Statistics	MAS	PRJ
Number of papers	3,908,805	384,289
Number of unique venues	5,149	412
Number of unique authors	1,186,412	235,533
Avg. number of papers per author	5.21	10.47
Avg. number of authors per paper	2.57	5.33
Min. (max.) number of references per paper	1 (2,432)	1 (581)
Min. (max.) number of citations per paper	1 (13,102)	1 (4757)

---

[2] https://academic.microsoft.com/

[3] https://journals.aps.org/datasets

7.5 Empirical Observations

In this section, we report various empirical observations about the IDTs of the papers
in the two datasets that provide a holistic view of the topological structure of the
trees in two domains. We also show how depth and breadth of the IDTs, the IDI, and
the NID values vary with the citation count of the papers in these domains.

7.5.1 Structural Properties of IDTs: Comparison of Depth, Breadth, and Citations Patterns

Fig. 7.3 plots the frequency distributions of depth and breadth of the IDTs for all
the papers in the two datasets. We observe that, for both PRJ and MAS datasets, the
values for breadth follow a long tail distribution, with roughly 80% of the papers
having a breadth less than 5 in the MAS dataset, and about 65% of the papers with
breadth less than 5 in the PRJ dataset (notice the logarithmic scale on x-axes in

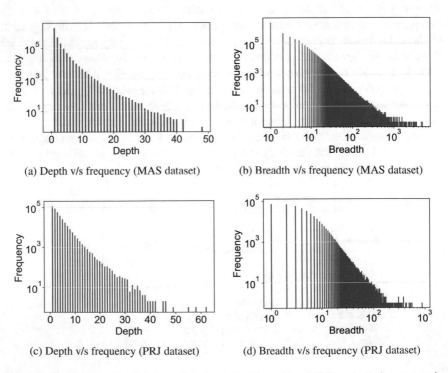

(a) Depth v/s frequency (MAS dataset)   (b) Breadth v/s frequency (MAS dataset)

(c) Depth v/s frequency (PRJ dataset)   (d) Breadth v/s frequency (PRJ dataset)

Fig. 7.3: Frequency distributions for depth and breadth of IDTs of all the papers in
both datasets. The x-axis in plots for breadth is in logarithmic scale.

Figs. 7.3b and 7.3d). On the other hand, the range of the depth values for IDTs is much smaller compared to the range of breadth values. For the MAS dataset, the maximum value of depth is 48 compared to the maximum breadth of 4,892. Likewise, for the PRJ dataset, the maximum value of depth is 62, and the maximum value of breadth is 857. We also observe that the maximum depth in physics is more than that of computer science despite fewer citations. The above indicates the nature of these research fields: the computer science domain receives many citations from various application domains. On the other hand, the physics discipline is more theoretical; a new concept in physics is often explored and enhanced by scientists in several subsequent papers. This might be the reason why the IDTs of the articles from the domain of computer science have larger breadth values, whereas the same for the pieces from physics have larger depth values.

Fig. 7.4 shows the variation of breadth and depth with citations (Fig. 7.4a and Fig. 7.4b for the MAS dataset, and Fig. 7.4d and Fig. 7.4e for the PRJ dataset, respectively) and the correlation between depth and breadth (Fig. 7.4c and Fig. 7.4f). We note that, both the datasets have similar characteristics. We observe that, while breadth is strongly correlated with citation count ($\rho = 0.90$ for MAS and $\rho = 0.83$ for PRJ), the correlation between depth and citation count is relatively weak ($\rho = 0.50$ for MAS and $\rho = 0.63$ for PRJ). (We have used *Spearman's rank-order correlation coefficient* in all the above computations.) These observations indicate that *incremental citations received by a paper often lead to the development of*

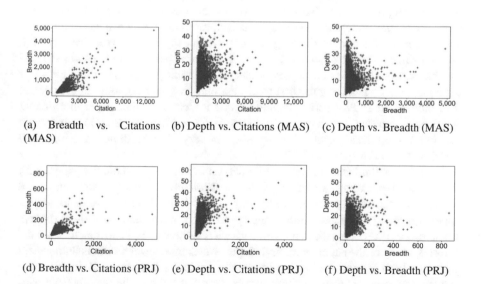

(a) Breadth vs. Citations (MAS) (b) Depth vs. Citations (MAS) (c) Depth vs. Breadth (MAS)

(d) Breadth vs. Citations (PRJ) (e) Depth vs. Citations (PRJ) (f) Depth vs. Breadth (PRJ)

Fig. 7.4: Scatter plots showing variations of breadth with citations [(a) & (d)], depth with citations [(b) & (e)], and spearman's rank order correlation between depth and breadth [(c) & (f)].

new branches in the IDT of the paper rather than increasing the depth of existing branches. This happens because most citations to a paper use the cited paper as a background reference (thus these get added to the IDT as a new branch), rather than extending a body of work represented by an already formed branch (increasing the depth) [3, 11]. Fig. 7.4c and Fig. 7.4f show that the variation in breadth values reduces with increasing depth. Especially for IDTs with depth greater than 30, the values of breadth lie in a relatively narrow band. Almost all IDTs with depth greater than 30 have breadth less than 300 for the MAS dataset and less than 346 for the PRJ dataset. These IDTs corresponds to the highly influential articles from the domains that have produced multiple directions of follow-up works. The incremental citations found in these connections correspond to continuation of these independent directions (thus increasing depth of the IDTs).

7.5.2 Qualitative Analysis of Depth and Breadth

We now present a qualitative analysis to illustrate the types of articles whose IDTs achieve very high breadth and depth values. Tables 7.2 and 7.3 list the top five papers having maximum breadth (Papers 1-5) and maximum depth (Papers 6-10) for the MAS and PRJ datasets, respectively. We note in Table 7.2 that, among the articles with the highest breadth values, the papers 1,2,3, and 5 are famous textbooks of computer science, and the paper 4 is the paper describing the notion of *Fuzzy Sets*. When a citing paper cites a book or a survey paper, it usually does it as a background reference before elaborating a concept. Such citations often lead to a large number of short branches in the corresponding IDTs, and hence these articles receive very high breadth values. On the other hand, papers 6-10 correspond to some of the seminal and breakthrough papers in the field that spawned multiple research directions. For example, Paper 6 is among the first to discuss and propose a solution for the control flow problem in TCP/IP networks, and Paper 7 is Codd's seminal paper introducing relational databases. These groundbreaking works led to multiple follow-up papers that build upon these papers, thus resulting in very high depth and relatively low breadth of the corresponding IDTs.

We note similar behavior for the PRJ dataset in Table 7.3, where the papers with the highest depth values are amongst the most influential ones (papers 6 and 8 are Nobel Prize papers, Paper 9 featured in Physics spotlight). On the other hand, review papers (paper 3) and papers describing specific techniques and algorithms (papers 1,4, and 5) are the papers that achieve very high breadth values. Note that, for both the datasets, many of the papers with high depth values have fewer citations than papers achieving high breadth values. We observe in Table 7.4 that, out of 10 highly cited papers, only two are award winning papers. Note further that the famous Feynman-publication is not present in this list. Thus, using only citations as a measure of scholarly influence would not reveal the impact of these highly influential papers fully. The IDT, on the other hand, enables us to understand the *depth* and *breadth* of

Table 7.2: List of papers with maximum values of breadth (Papers 1-5) and depth (Papers 6-10) in MAS dataset. Papers with high breadth are invariably review/survey papers whereas papers with high depth are seminal papers indicating the *deep* influence on the field.

No.	Paper	# citations	breadth	depth	Remarks
1	Michael Randolph Garey, David S. Johnson. 1979. Computers and Intractability: A Guide to the Theory of NP-Completeness.	13102	4892	34	A book on the theory of NP-Completeness.
2	Thomas H. Cormen, Charles E. Leiserson, Ronald L. Rivest, Clifford Stein. 2001. Introduction to Algorithms, Second Edition.	6777	4596	8	Highly referred text book on Algorithms.
3	William H. Press, S. A. Teukolsky, W. T. Vetterling, B. P. Flannery. 1990. Numerical recipes in c: the art of scientific computing.	5320	3429	17	A Book on scientific computing
4	Lotfi A. Zadeh. 1965. Fuzzy Sets.	7455	3317	17	The notion of an infinite-valued logic was introduced in Zadeh's seminal work Fuzzy Sets where he described the mathematics of fuzzy set theory, and by extension fuzzy logic.
5	David F. Mayers, Gene H. Golub, Charles F. van Loan. 1986. Matrix Computations.	5520	3152	17	A book on Matrix computations.
6	Van Jacobson. 1988. Congestion avoidance and control.	2577	259	48	Highly influential paper describing Jacobson's algorithm for control flow in TCP/IP networks
7	Edgar Frank Codd. 1970. A Relational Model for Large Shared Data Banks.	2141	437	42	Codd's Seminal paper on Relational Databases
8	Ashok K. Chandra, Philip M. Merlin. 1977. Optimal implementation of conjunctive queries in relational data bases.	469	34	42	One of the influential papers in the domain of conjunctive queries in database theory.
9	Roger M. Needham, Michael D. Schroeder. 1978. Using encryption for authentication in large networks of computers (CACM 1979).	1022	101	42	Initiated the field of cryptographic protocol design led to Kerberos, IPsec, SSL, and all modern protocols
10	Leslie Lamport. 1978. Time, clocks, and the ordering of events in a distributed system.	3170	505	40	Lamport's seminal work on concept of and an implementation of the logical clocks.

the impact of these articles on their citing papers and understand how these articles have influenced their respective fields.

Table 7.3: List of papers with maximum values of breadth (Papers 1-5) and depth (Papers 6-10) in PRJ dataset. Papers with high breadth are invariably review/survey papers whereas papers with high depth are seminal papers indicating the *deep* influence on the field. Many of the papers with high depth values describe Nobel Prize winning work.

No.	Paper	# citations	Breadth	Depth	Remarks
1	John P. Perdew, Kieron Burke, and Matthias Ernzerhof. 1996. Generalized Gradient Approximation Made Simple	3083	857	23	A book on Interdisciplinary Applied Mathematics.
2	G. Kresse and D. Joubert. 1999. From ultrasoft pseudopotentials to the projector augmented-wave method	1418	544	10	Overall Google Scholar citation as on date 49,029
3	K. Hagiwara et al.(Particle Data Group). 2002. Review of Particle Properties	1077	511	8	The Review summarizes much of particle physics and cosmology.
4	G. Kresse and J. Furthmüller. 1996. Efficient iterative schemes for ab initio total-energy calculations using a plane-wave basis set	2384	508	14	Phenomenal in experimental solid state physics.
5	John P. Perdew and Yue Wang. 1992. Accurate and simple analytic representation of the electron-gas correlation energy	1443	479	16	Overall Google Scholar citation as on date is 24, 837
6	W. Kohn and L. J. Sham. 1965. Self-Consistent Equations Including Exchange and Correlation Effects	4757	284	62	Nobel Prize, Seminal Paper in the density-functional theory.
7	L. Wolfenstein. 1978. Neutrino oscillations in matter	761	32	58	This work led to an eventual understanding of the MSW effect, which acts to enhance neutrino oscillation in matter.
8	Steven Weinberg. 1967. A Model of Leptons	1391	132	55	Nobel Prize in 1979 and it is the most highly cited research paper in particle physics, yet it was not cited in a journal at all in the first year after its publication, and was referred to only twice in the following two years.
9	P. Hohenberg and W. Kohn. 1964. Inhomogeneous Electron Gas	3710	254	49	Featured in Physics : spotlighting exceptional research, Seminal Paper in the density-functional theory.
10	S. L. Glashow, J. Iliopoulos, and L. Maiani. 1970. Weak Interactions with Lepton-Hadron Symmetry	796	92	46	Quantum field theory, Electroweak unification paper.

Table 7.4: List of papers with maximum values of citation in PRJ dataset. Nobel Prize papers are labelled with an asterisk (*).

No.	Paper Name	Citations	Breadth	Depth
1	W. Kohn and L. J. Sham. 1965. Self-Consistent Equations Including Exchange and Correlation Effects*	4757	284	62
2	P. Hohenberg and W. Kohn. 1964. Inhomogeneous Electron Gas	3710	254	49
3	J. P. Perdew and Alex Zunger. 1981. Self-interaction correction to density-functional approximations for many-electron systems	3182	222	34
4	John P. Perdew, Kieron Burke, and Matthias Ernzerhof. 1996. Generalized Gradient Approximation Made Simple	3083	857	23
5	D. M. Ceperley and B. J. Alder. 1980. Ground State of the Electron Gas by a Stochastic Method	2647	197	30
6	Hendrik J. Monkhorst and James D. Pack. 1976. Special points for Brillouin-zone integrations	2566	308	26
7	G. Kresse and J. Furthmüller. 1996. Efficient iterative schemes for ab initio total-energy calculations using a plane-wave basis set	2384	508	14
8	N. Troullier and Josè Luìs Martins. 1991. Efficient pseudopotentials for plane-wave calculations	1949	343	16
9	David Vanderbilt. 1990. Soft self-consistent pseudopotentials in a generalized eigenvalue formalis	1947	295	23
10	J. Bardeen, L. N. Cooper, and J. R. Schrieffer. 1957. Theory of Superconductivity*	1722	296	35

7.5.3 Comparing IDI and NID vs. Citations

We now study how IDI and NID values vary with the citation counts across papers of the two chosen scientific domains. Fig. 7.5 shows the scatter plots of the IDI and NID values for the papers in the two datasets. We observe that IDI values, in general, increase with the number of citations of an article (Fig. 7.5a and Fig. 7.5c). This observation is well-expected. The reason is that the IDI for a paper is bounded by the number of citations of that paper according to Eq.n 7.6. A more interesting observation that we can make from the plots for NID values (Fig. 7.5b and Fig. 7.5d) is the following: the value of NID, in general, decreases with an increase in citations; papers having a high number of citations tend to have low NID values. Recall that, for a given article, NID captures the 'distance' of the IDI of that paper from its corresponding ideal IDT. Thus, highly influential papers tend to have their IDTs close to their ideal IDT configurations (as depicted by the low NID value). This empirical observation strengthens our hypothesis that *highly influential papers will, in general, lead to a considerable amount of follow-up work (high depth) in multiple directions (high breadth)*. Note that, on the contrary, it is not necessary that a paper will have a low NID value if it has high citation count. Papers with similar citation counts can have different NID values, based on their influence.

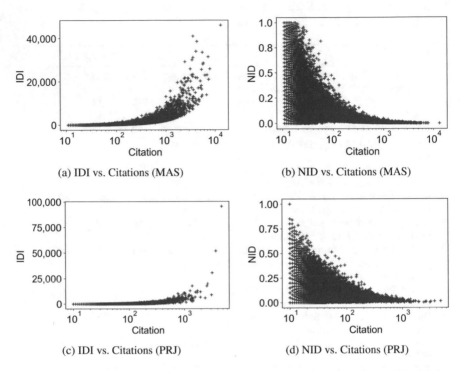

(a) IDI vs. Citations (MAS)   (b) NID vs. Citations (MAS)

(c) IDI vs. Citations (PRJ)   (d) NID vs. Citations (PRJ)

Fig. 7.5: Scatter plots showing variations of (a) IDI and (b) NID values with citation counts.

7.5.4 Depth and Breadth Profiles of Top-cited Papers

Next, we study the depth and breadth profiles of top-cited papers in our datasets. We select the top 200 highly-cited papers after ignoring the top 10 papers. The reason behind this is somewhat tricky. The acquired citation volume of the top 10 articles from a scientific domain often overwhelms the rest of the articles from the field. For example, the range of citations received by the top 10 papers in the MAS dataset is 5715–13102, compared to the same for the following 200 articles from the same dataset, which is 1500 – 5000. Similarly, the range of citations received by top 10 papers in the PRJ dataset is 1722–4757, compared to the same for the following 200 articles from the same dataset, which is 450–1400. Such discrepancies in received citations often yield a skewed behaviour in the dataset. To avoid this skewness, we ignore the top 10 articles in the current analysis.

We compute the depth and breadth values for the selected papers and plot the values, as shown in Fig. 7.6 for the two datasets. The blue lines in the figure correspond to the mean values of depth and breadth for the selected papers. Each paper is represented by a point in the figure and its color corresponds to the number of citations received by the paper. Observe from the figure that papers with similar citation values

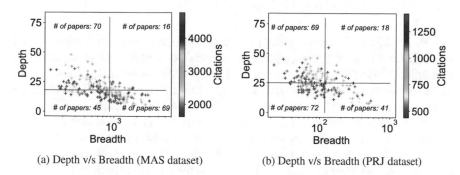

(a) Depth v/s Breadth (MAS dataset)   (b) Depth v/s Breadth (PRJ dataset)

Fig. 7.6: Depth vs Breadth distribution of papers with similar citations for MAS (Fig. 7.6a) and PRJ (Fig. 7.6b) dataset. The x-axis in the plots for breadth is in logarithmic scale. The blue lines indicate the mean depth and breadth values.

have very different characteristics in terms of their depth and breadth values, thus indicating the difference in impact of the papers have had on their respective fields. Note further that most of the papers in the bottom-left quadrant (less than average depth and breadth values) have relatively lower citation count (red points). On the other hand, most of the papers in the top-right quadrant (higher than average depth and breadth values) have relatively higher citation count (blue and violet points). Once again, we observe that analyzing the citation profiles of the papers in terms of their depth and breadth distributions can help us better understand the types of influence of the papers. This way, we can also differentiate the papers with similar citation counts in terms of the depth and breadth of their impact in the scientific field.

7.6 NID as an Indicator of Influence

In our previous work, we show that the highly influential papers produce IDTs that are close to their corresponding ideal IDT configurations. In Section 7.5.3, we have noticed that highly-cited papers have very low NID values. Here, we ask a complementary question – *will a paper having its IDT close to the ideal configuration at a given time be an influential paper in future?* We perform two experiments to answer the above question. In Section 7.6.1, we study if NID can predict how many citations a paper will get in future. In Section 7.6.2, we show how the ranking of the articles from different citation band (and NID band) changes with time.

7.6.1 NID as a Predictor of Future Citations

Let P_v be the set of papers published in a publication venue v (a conference or a journal). Let y_v be the year of organization of v. Over the next t years, papers in P_v will influence the follow-up works and will gather citations accordingly. Let $I(p)$ be an influence measure under consideration. Let $R(v, t, I)$ be the ranked list of papers in P_v ordered by the value of $I(\cdot)$ at t. Thus, the top ranked paper in $R(v, t, I)$ is considered to have maximum influence at t. If $I(\cdot)$ is able to capture the impact correctly, we expect the papers with high influence scores to have more incremental citations in future compared to papers having low influence scores. Let $C(v, t_1, t_2)$ be the ranked list of papers in P_v ordered by the increment in citations from time t_1 to t_2. Thus, the papers that received highest fractional increment in citations in the time period (t_1, t_2) will be ranked at the top. Note that, we have chosen fractional increment in citation count rather than the absolute count to account for papers that are early risers and receive most of their lifetime citations in first few years after publication [9]. Also, we consider only those papers published in a venue (v here) rather than all the papers in our dataset to nullify the effect of diverse citation dynamics across fields and venues [10]. Intuitively, if $I(\cdot)$ is a good predictor of a paper's influence, the ranked lists $R(v, t_1, I)$ and $C(v, t_1, t_2)$ should be very similar – influential papers at time t_1 should receive more incremental citations from t_1 to t_2. Thus, the similarity of the two ranked list could be used as a measure to evaluate the potential of $I(\cdot)$ to be able to capture the influence of papers. We use the normalised version of *Kendall Tau rank distance for sequences*, $\mathcal{K}$ [12], to measure the similarity of the two ranked lists $R(v, t_1, I)$ and $C(v, t_1, t_2)$ as follows.

$$z(v, I) = \mathcal{K}(R(v, t_1, I), C(v, t_1, t_2)) \tag{7.7}$$

The normalised version of Kendall tau distance ranges between 0 and 1; it attains a minimum value of 0 when both sequences are identical and a maximum value of 1 when one sequence is just reverse of the other. Hence, according to the definition of z score (Eq. 7.7), a lower z-value would indicate that the two ranked lists are highly similar. The above, in turn, would show that $I(\cdot)$ has high predictive power in forecasting the future incremental citations. We use this framework to compare NID with the raw citation count of a paper (substituting for $I(\cdot)$ in Eq. 7.7 one by one), and check their respective potential as an early predictor of future incremental citations of a paper. We use the citation count of an article as a competitor of NID, as it is the most common and most straightforward way of judging the influence of a paper [17, 18].

Next, we group all the papers in the two datasets by their venues and compute the values of the influence metrics (NID and citation count) after five years following the publication year (i.e., $t_1 = 5$). A venue is uniquely defined by the year of publication and the conference/journal series. For example, *JCDL 2000* and *JCDL 2001* are considered as two separate venues. Likewise, *Physical Review C-1990*, *Physical Review C-1991* are considered as two different venues. We then compute the incremental citations gathered by the papers ten years after the publication

($t_2 = 10$). For the MAS dataset, we only consider venues with the publication year in the range 1995 and 2000, because we need citation information 10 years after publication (i.e., up to 2010), and the coverage of papers published after year 2010 is relatively sparse in the MAS dataset [10]. This filtering resulted in 1, 219 unique venues and 30, 556 papers in total. Likewise, for the PRJ dataset, we consider venues with the publication year in the range 1990 and 2000, resulting in 77 unique venues and 134, 506 papers in total. Note that, we increase the time range for the PRJ as the number of venues is less in the range 1995-2000 (only 45 venues in 5 years).

With the group of papers published together in a venue and their citation information available, we compute the following three ranked lists:

1. $R_{v,c} = R(v, 5, c)$; the ranked lists of papers in venue v ordered by their citation counts five years after the publication.
2. $R_{v,NID} = R(v, 5, NID)$; the ranked lists of papers in venue v ordered by their NID scores five years after the publication.
3. $C_v = C(v, 5, 10)$; the ranked lists of papers in venue v ordered by the normalized incremental citations received beginning of 5th years after the publication till 10th years after publication.

For each venue v, these lists can be used to compute $z(v, NID)$ and $z(v, c)$ – i.e., the z scores with NID and citation count as influence measures, respectively. For the 1, 219 venues identified as above in the MAS dataset, the average value of z score using citations and NID as the influence measure is found to be 0.5125 and 0.3703, respectively. For the 77 venues identified in the PRJ dataset, the average value of z-score using citations and NID as the influence measure is found to be 0.31566 and 0.2773, respectively. Thus, for both the datasets, we find that, on an average, the Z score is lower when using NID as the influence measure compared to that with citation count. In other words, more papers identified as influential by NID received more incremental future citations compared to the papers identified as influential by citation count.

Fig. 7.7 provides a fine-grained illustration of the difference of z scores achieved by the two influence measures for all the venues in both the datasets. For each venue, we compute the difference in z-scores achieved by NID and citation count. We note that, for most of the venues, the z-score achieved by NID is lower than the z-score achieved by the citation count (positive bars). These observations indicate that, when compared with the raw citation count, NID is a much stronger predictor of the future impact of a scientific paper. As opposed to the raw citation count, the IDT of a paper provides a fine-grained view of the impact of the paper in terms of its depth and breadth as succinctly captured by the IDT of the paper. These results provide compelling evidence for the utility of IDT (and the consequent measures such as IDI and NDI derived from it) for studying the impact of scholarly papers in domains of computer science and physics.

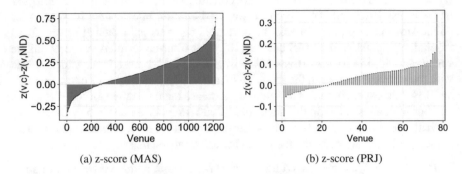

(a) z-score (MAS)   (b) z-score (PRJ)

Fig. 7.7: Papers in a venue are ranked using the NID metric and citations count metric separately. The plot shows the relative gain in resulting z-score values. The horizontal axis represents venues ordered as numbered sequence, and the vertical axis indicates the difference in two obtained z-score values.

7.6.2 Temporal Analysis of Papers Based on NID

In this section, we explore how the ranking of the papers changes over time with respect to citation count. We argue that considering the relationship between the citations received will be a better measure to capture the influence over the years as compared to the raw citation count. We group the set of papers published in the year 2000 – these papers have gathered citations over the years based on their influence. We rank these papers based on their NID and citation count till year 2005. We group these papers into five equal-sized bins, with Bin 1 representing top papers by citation count and Bin 5 having the lowest-ranked ones. However in case of NID, Bin 1 represent papers with low NID value and Bin 5 with high NID value. Recall that highly influential papers have low NID value as they are closer to their ideal configuration and sometimes papers with low citations can also lead to low NID value. We only considered papers having citation count $\geq$ 10 for our analysis. We again find the ranks of the same set of papers following exactly the same approach after 5 years, i.e., in the year 2010. Our objective now is to track down the change in ranks of these articles during this five year time span (i.e., movement of an article form a rank bin in 2005 to a different rank bin in 2010) and to reveal if there is any correlation of the above changes in rank with the article quality metric used.

Fig. 7.8 shows the change in ranks of articles during the chosen time interval [2005, 2010]. We notice that significantly less number of articles deviates one rank bin to a different rank bin during the chosen time span when we use citation count to measure the quality of the articles. This indicates that most of the articles that have received high citations till 2005 continue to get high citations in till 2010. On the other hand, when we use NID to measure the quality of the articles, we find that a significant fraction of articles deviate to different bins during the chosen time interval. This happens only if there is a significant change in the quality of citations

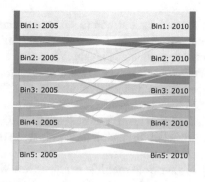

(a) Ranked based on citation count (MAS dataset)

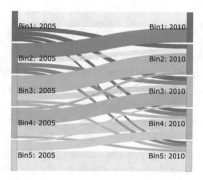

(b) Ranked based on NID (MAS dataset)

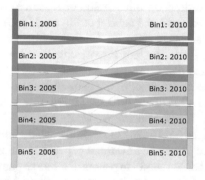

(c) Ranked based on citation count (PRJ dataset)

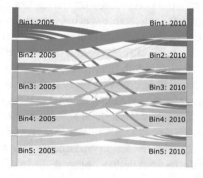

(d) Ranked based on NID (PRJ dataset)

Fig. 7.8: Change in ranking of papers between the years (2005 and 2010) when ranked based on citation count and NID.

received by these articles that change the breadth or depth of IDTs of these articles, which in turn, incur significant change in their NID values. If the citations received by an article bring the tree closer to the optimal IDT, the ranking of the article may improve. The same phenomena is visible for articles from both the datasets. This indicates that maintaining the citation count does not imply that NID will be maintained. In fact, with the increase of citations, papers may move to the lower NID-based bins. This further strengthens our claim that increasing citation count does not guarantee the increase in the depth and the breadth of the IDT, resulting in the change of the NID value.

7.7 Conclusion

In our previous study [30], we introduced a novel data structure that we coined as 'Influence Dispersion Tree' (IDT). IDT can organize the follow-up papers of a given article and find their citation-based inter-dependencies nicely. We can use it to analyze these articles and quantify their scientific influence. This data structure helps to study the structural properties of the above citation relationships like the breadth and depth of impact of a given article on follow-up articles. We introduced a couple of metrics, 'Influence Dispersion Index' (IDI) and 'Normalized Influence Divergence' (NID), capable of capturing the impact of these structural properties on the articles' present and future influence.

This chapter has extended the above study further from multiple perspectives. We have performed both fundamental and advanced analysis on the IDTs to understand their relations with the raw citation count. Fundamental analyses on the dataset refer to the citation count-based experiments performed in the paper. On the contrary, advanced studies on the dataset take into account the tree-based examinations. To verify the consistency of our earlier observations on other fields of research, we have replicated our experiments on a second dataset, the dataset of Physical Review journals, and we find that these results are consistent across research domains.

One striking observation that we made in our earlier study is that, with the increase in citation count, the depth of an IDT grows at a much slower rate than the breadth. We find in our present study that the same holds for either of the datasets we used in our research. We have noticed that, as the citation count grows, the IDT of a paper moves closer to its ideal IDT configuration. The conclusion we would like to draw here is – how to quantify the impact of a scientific article. One should consider how to organize the follow-up papers of the source paper and how those papers contribute to its research field, along with the total number of follow-up papers of a source paper (i.e., citation count). An article can be treated as highly influential only when it has enriched a field equally in both vertical (deepening the knowledge further inside the research area) and horizontal (allowing the emergence of new sub-fields) directions. The NID metric turns out to be superior to the raw citation count on either of the datasets – to predict how many new citations a paper will receive within a particular time window after publication. Our final remark in this chapter is – all the above observations can be generalized across fields of research.

References

1. Alonso, S., Cabrerizo, F., Herrera-Viedma, E., Herrera, F.: hg-index: A new index to characterize the scientific output of researchers based on the h- and g-indices. Scientometrics **82**(2), 391–400 (2009)
2. Aragón, A.M.: A measure for the impact of research. Scientific Reports **3**, 1649 (2013)
3. Balaban, A.T.: Positive and negative aspects of citation indices and journal impact factors. Scientometrics **92**(2), 241–247 (2012)

4. Bergstrom, C.T., West, J.D., Wiseman, M.A.: The Eigenfactor^{TM} metrics. Journal of Neuroscience **28**(45), 11433–11434 (2008)
5. Bornmann, L., Daniel, H.D.: What do citation counts measure? A review of studies on citing behavior. Journal of documentation **64**(1), 45–80 (2008)
6. Bu, Y., Waltman, L., Huang, Y.: A multi-dimensional framework for characterizing the citation impact of scientific publications. Quantitative Science Studies **2**(1), 155–183 (2021)
7. Cavalcanti, D.C., Prudêncio, R.B.C., Pradhan, S.S., Shah, J.Y., Pietrobon, R.S.: Good to be Bad? Distinguishing between positive and negative citations in scientific impact. In: Proceedings of the 23rd IEEE International Conference on Tools with Artificial Intelligence (ICTAI), pp. 156–162 (2011)
8. Chakraborty, T., Kumar, S., Goyal, P., Ganguly, N., Mukherjee, A.: Towards a stratified learning approach to predict future citation counts. In: Proceedings of the 14th ACM/IEEE-CS Joint Conference on Digital Libraries (JCDL), pp. 351–360 (2014)
9. Chakraborty, T., Kumar, S., Goyal, P., Ganguly, N., Mukherjee, A.: On the categorization of scientific citation profiles in computer scienc. Communications of the ACM **58**(9), 82–90 (2015)
10. Chakraborty, T., Nandi, S.: Universal trajectories of scientific success. Knowledge & Information Systems **54**(2), 487–509 (2018)
11. Chubin, D.E., Moitra, S.D.: Content analysis of references: Adjunct or alternative to citation counting? Social Studies of Science **5**(4), 423–441 (1975)
12. Cicirello, V.: Kendall tau sequence distance: Extending Kendall tau from ranks to sequences. EAI Endorsed Transactions on Industrial Networks & Intelligent Systems **7**(23), 163925 (2020)
13. Clough, J.R., Gollings, J., Loach, T.V., Evans, T.S.: Transitive reduction of citation networks. Journal of Complex Networks **3**(2), 189–203 (2015)
14. Connor, J.: Google Scholar citations open to all. https://scholar.googleblog.com/2011/11/google-scholar-citations-open-to-all.html (2011)
15. Egghe, L.: An improvement of the h-index: The g-index. ISSI Newsletter **2**(1), 8–9 (2006)
16. Fersht, A.: The most influential journals: Impact Factor and Eigenfactor. Proceedings of the National Academy of Sciences **106**(17), 6883 (2009)
17. Garfield, E.: "Science Citation Index" - A new dimension in indexing. Science **144**(3619), 649–654 (1964)
18. Garfield, E.: Citation analysis as a tool in journal evaluation. Science **178**(4060), 471–479 (1972)
19. Garfield, E.: The history and meaning of the journal Impact Factor. Jama **295**(1), 90–93 (2006)
20. Giatsidis, C., Nikolentzos, G., Zhang, C., Tang, J., Vazirgiannis, M.: Rooted citation graphs density metrics for research papers influence evaluation. Journal of Informetrics **13**(2), 757–768 (2019)
21. Greenhalgh, T., Raftery, J., Hanney, S., Glover, M.: Research impact: A narrative review. BMC medicine **14**(1), 78 (2016)
22. Halaweh, M.: Paper Impact Effectiveness (PIE): A new way to measure the impact of research papers. Procedia Computer Science **132**, 404–411 (2018)
23. Hirsch, J.E.: An index to quantify an individual's scientific research output. Proceedings of the National Academy of Sciences **102**(46), 16569–16572 (2005)
24. Huang, Y., Bu, Y., Ding, Y., Lu, W.: Number versus structure: Towards citing cascades. Scientometrics **117**(3), 2177–2193 (2018)
25. Klosik, D.F., Bornholdt, S.: The citation Wake of publications detects Nobel Laureates' papers. PLoS One **9**(12), e113184 (2014)
26. Kurtz, M.J., Eichhorn, G., Accomazzi, A., Grant, C., Demleitner, M., Henneken, E., Murray, S.S.: The effect of use and access on citations. Information Processing & Management **41**(6), 1395–1402 (2005)
27. MacRoberts, M.H., MacRoberts, B.R.: Problems of citation analysis: A study of uncited and seldom-cited influences. Journal of the American Society for Information Science & Technology **61**(1), 1–12 (2010)
28. Min, C., Sun, J., Ding, Y.: Quantifying the evolution of citation cascades. Proceedings of the Association for Information Science & Technology **54**(1), 761–763 (2017)

29. Mingers, J., Burrell, Q.L.: Modeling citation behavior in management science journals. Information Processing & Management **42**(6), 1451–1464 (2006)
30. Mohapatra, D., Maiti, A., Bhatia, S., Chakraborty, T.: Go Wide, Go Deep: Quantifying the impact of scientific papers through influence dispersion trees. In: Proceedings of the 19th ACM/IEEE Joint Conference on Digital Libraries (JCDL), pp. 305–314 (2019)
31. Pan, R.K., Fortunato, S.: Author Impact Factor: tracking the dynamics of individual scientific impact. Scientific Reports **4**, 4880 (2014)
32. Post, A., Li, A.Y., Dai, J.B., Maniya, A.Y., Haider, S., Sobotka, S., Choudhri, T.F.: c-index and subindices of the h-index: New variants of the h-index to account for variations in author contribution. Cureus **10**(5) (2018)
33. Pradhan, D., Paul, P.S., Maheswari, U., Nandi, S., Chakraborty, T.: C^3-index: Revisiting author's performance measure. In: Proceedings of the 8th ACM Conference on Web Science, pp. 318–319 (2016)
34. Pradhan, D., Paul, P.S., Maheswari, U., Nandi, S., Chakraborty, T.: C^3-index: A PageRank based multi-faceted metric for authors' performance measurement. Scientometrics **110**(1), 253–273 (2017)
35. Ravenscroft, J., Liakata, M., Clare, A., Duma, D.: Measuring scientific impact beyond academia: An assessment of existing impact metrics and proposed improvements. PloS One **12**(3), e0173152 (2017)
36. Redner, S.: Citation characteristics from 110 years of Physical Review. Physics Today Online **58**(6), 49–54 (2005)
37. Savov, P., Jatowt, A., Nielek, R.: Identifying breakthrough scientific papers. Information Processing & Management **57**(2), 102168 (2020)
38. Sidiropoulos, A., Gogoglou, A., Katsaros, D., Manolopoulos, Y.: Gazing at the skyline for star scientists. Journal of Informetrics **10**(3), 789–813 (2016)
39. Sidiropoulos, A., Katsaros, D., Manolopoulos, Y.: Ranking and identifying influential scientists versus mass producers by the Perfectionism Index. Scientometrics **103**(1), 1–31 (2015)
40. Sidiropoulos, A., Manolopoulos, Y.: A citation-based system to assist prize awarding. ACM SIGMOD Record **34**(4), 54–60 (2005)
41. Stoupas, G., Sidiropoulos, A., Gogoglou, A., Katsaros, D., Manolopoulos, Y.: Rainbow ranking: An adaptable, multidimensional ranking method for publication sets. Scientometrics **116**(1), 147–160 (2018)
42. Thompson, D.F., Callen, E.C., Nahata, M.C.: New indices in scholarship assessment. American Journal of Pharmaceutical Education **73**(6), 111 (2009)
43. Vasiliauskaite, V., Evans, T.S.: Diversity from the topology of citation networks. CoRR **abs/1802.06015** (2018)
44. Watson, A.B.: Comparing citations and downloads for individual articles at the Journal of Vision. Journal of Vision **9**(4), i–i (2009)
45. Waumans, M.C., Bersini, H.: Genealogical trees of scientific papers. PloS one **11**(3), e0150588 (2016)
46. Yang, S., Han, R.: Breadth and depth of citation distribution. Information Processing & Management **51**(2), 130–140 (2015)
47. Zhang, C.T.: The e-index, complementing the h-index for excess citations. PLoS One **4**(5), e5429 (2009)
48. Zhang, F., Bai, X., Lee, I.: Author impact: Evaluations, predictions, and challenges. IEEE Access **7**, 38657–38669 (2019)

Chapter 8
Can Author Collaboration Reveal Impact? The Case of *h*-index

Giannis Nikolentzos, George Panagopoulos, Iakovos Evdaimon, and Michalis Vazirgiannis

Abstract Scientific impact has been the center of extended debate regarding its accuracy and reliability. From hiring committees in academic institutions to governmental agencies that distribute funding, an author's scientific success as measured by *h*-index is a vital point to their career. The aim of this work is to investigate whether the collaboration patterns of an author are good predictors of the author's future *h*-index. Although not directly related to each other, a more intense collaboration can result into increased productivity which can potentially have an impact on the author's future *h*-index. In this paper, we capitalize on recent advances in graph neural networks and we examine the possibility of predicting the *h*-index relying solely on the author's collaboration and the textual content of a subset of their papers. We perform our experiments on a large-scale network consisting of more than 1 million authors that have published papers in computer science venues and more than 37 million edges. The task is a six-months-ahead forecast, i.e., what the *h*-index of each author will be after six months. Our experiments indicate that there is indeed some relationship between the future *h*-index of an author and their structural role in the co-authorship network. Furthermore, we found that the proposed method outperforms standard machine learning techniques based on simple graph metrics along with node representations learnt from the textual content of the author's papers.

Giannis Nikolentzos
Athens University of Economics and Business, Athens, Greece,
e-mail: nikolentzos@aueb.gr

George Panagopoulos
École Polytechnique, Palaiseau, France,
e-mail: george.panagopoulos@polytechnique.edu

Iakovos Evdaimon
Athens University of Economics and Business, Athens, Greece,
e-mail: p3130059@aueb.gr

Michalis Vazirgiannis
École Polytechnique, Palaiseau, France,
e-mail: mvazirg@lix.polytechnique.fr

© The Author(s), under exclusive license to Springer Nature Switzerland AG 2021
Y. Manolopoulos, T. Vergoulis (eds.), *Predicting the Dynamics of Research Impact*,
https://doi.org/10.1007/978-3-030-86668-6_8

8.1 Introduction

Citation counts is undoubtedly the most widely used indicator of a paper's impact and success. It is commonly believed that the more citations a paper has received, the higher its impact. When it comes to authors, measuring impact becomes more complicated since an author may have published multiple papers in different journals or conferences. Still the publication record of an author is in many cases the most important criterion for hiring and promotion decisions, and for awarding grants. Therefore, institutes and administrators are often in need of quantitative metrics that provide an objective summary of the impact of an author's publications. Such indicators have been widely studied in the past years [8, 14, 54, 60]. However, it turns out that not all aspects of an author's scientific contributions can be naturally captured by such single-dimensional measures. The most commonly used indicator is perhaps the h-index, a measure that was proposed by Jorge Hirsch [23]. The h-index measures both the productivity (i. e., the number of publications) and the impact of the work of a researcher. Formally, the h-index is defined as the maximum value of h such that the given author has published h papers that have each been cited at least h times. Since its inception in 2005, the h-index has attracted significant attention in the field of bibliometrics. It is not thus surprising that all major bibliographic databases such as Scopus, the Web of Science and Google Scholar compute and report the h-index of authors. One of the main appealing properties of the h-index is that it is easy to compute, however, the indicator also suffers from several limitations which have been identified in the previous years [7, 41, 52].

It is clear from the definition of h-index that it mainly depends on the publication record of the author. On the other hand, there is no theoretical link between the h-index of an author and the collaborations the author has formed. For instance, it is not necessary that the h-index of an author that has collaborated with many other individuals is high. However, the above example is quite reasonable, and empirically, there could be some relation between h-index and co-authorship patterns [27]. In fact, it has been reported that co-authorship leads to increased scientific productivity (i. e., number of papers published) [4, 12, 25, 30, 62]. For instance, some studies have investigated the relationship between productivity and the structural role of authors in the co-authorship network, and have reported that authors who publish with many different co-authors bridge communication and tend to publish larger amounts of papers [13, 29]. Since research productivity is captured by the h-index, co-authorship networks could potentially provide some insight into the impact of authors. Some previous works have studied if centrality measures of vertices are related to the impact of the corresponding authors. Results obtained from statistical analysis or from applying simple machine learning models have shown that in some cases, the impact of an author is indeed related to their structural role in the co-authorship network [1, 6, 58].

The goal of this paper is to uncover such connections between the two concepts, i. e., h-index and the co-authorship patterns of an author.Specifically, we leverage machine learning techniques and we study whether the collaboration patterns of an author are good indicators of the author's future impact (i. e., their future h-index).

Predicting the future h-index of an author could prove very useful for funding agencies and hiring committees, who need to evaluate grant proposals and hire new faculty members, since it can give them insights about the future success of authors. We treat the task as a regression problem and since the collaboration patterns of an author can be represented as a graph, we capitalize on recent advances in graph neural networks (GNNs) [55]. These models have been applied with great success to problems that arise in chemistry [18], in social networks [16], in natural language processing [35], and in other domains. GNNs use the idea of recursive neighborhood aggregation. Given a set of initial vertex representations, each layer of the model updates the representation of each vertex based on the previous representation of the vertex and on messages received from its neighbors. In this paper, we build a co-authorship network whose vertices correspond to authors that have published papers in computer science journals and conferences. We then train a GNN model to predict the future h-index of the vertices (i.e., the authors). More precisely, we propose two GNN variants tailored to the needs of the h-index prediction task. Our results demonstrate that the structural role of an individual in the co-authorship network is indeed empirically related to their future h-index. Furthermore, we find that local features extracted from the neighborhood of an author are very useful for predicting the author's future h-index. On the other hand, the textual features generated from the author's published papers do not provide much information. Overall, the proposed architectures achieve low levels of error, and can make relatively accurate predictions about the scientific success of authors.

The rest of this paper is organized as follows. Section 8.2 provides an overview of the related work. In Section 8.3, we present our dataset, while Section 8.4 introduces some preliminary concepts and gives details about representation learning in natural language processing and about graph neural networks. Section 8.5 provides a detailed description of our methodology. Section 8.6 evaluates the proposed graph neural network in the task of h-index prediction. Finally, Section 8.7 summarizes the work and presents potential future work.

8.2 Related Work

In the past years, a substantial amount of research has focused on determining whether information extracted from the co-authorship networks could serve as a good predictor of the authors' scientific performance. For instance, McCarty et al. investigated in [33] which features extracted from the co-authorship network enhance most the h-index of an author. The authors found that the number of co-authors and some features associated with highly productive co-authors are most related to the increase in the h-index of an author. In another study [22], Heiberger and Wieczorek examined whether the structural role of researchers in co-authorship networks is related to scientific success, measured by the probability of getting a paper published in a high-impact journal. The authors found that the maintenance of a moderate number of persistent ties, i.e., ties that last at least two consecutive years, can lead to scientific success. Furthermore, they report that authors who connect otherwise unconnected parts of the network are very likely to publish papers in high-impact journals. Parish et al. employed in [40] the R index, an indicator that captures the co-authorship patterns of individual researchers, and studied the association between R and the h-index. The authors found that more collaborative researchers (i. e., lower values of R) tend to have higher values of h-index, and the effect is stronger in certain fields compared to others.

Centrality measures identify the most important vertices within a graph, and several previous works have studied the relationship between these measures and research impact. Yan and Ding investigated in [58] if four centrality measures (i. e., closeness centrality, betweenness centrality, degree centrality, and PageRank) for authors in the co-authorship network are related to the author's scientific performance. They found that all four centrality measures are significantly correlated with citation counts. Abbasi et al. also studied the same problem in [1]. Results obtained from a Poisson regression model suggest that the g-index (an alternative of the h-index) of authors is positively correlated with four of the considered centrality measures (i. e., normalized degree centrality, normalized eigenvector centrality, average ties strength and efficiency). Sarigöl ct al. examined in [42] if the success of a paper depends on the centralities of its authors in the co-authorship network. The authors showed that if a paper is represented as a vector where most of the features correspond to centralities of its authors, then a Random Forest classifier that takes these vectors as input can accurately predict the future citation success of the paper. Bordons et al. explored in [6] the relationship between the scientific performance of authors and their structural role in co-authorship networks. The authors utilized a Poisson regression model to explore how authors' g-index is related to different features extracted from the co-authorship networks. The authors find that the degree centrality and the strength of links show a positive relationship with the g-index in all three considered fields. Centrality measures cannot only predict scientific success, but can also capture the publication history of researchers. More specifically, Servia-Rodríguez ct al. studied in [45] if the research performance of an author is related to the number of their collaborations, and they found that if two authors have similar centralities in the co-authorship network, it is likely that they also share similar publication patterns.

Some other related studies have investigated the relationship between the number of authors and the success of a paper. For instance, Figg et al. analyzed in [17] the relationship between collaboration and the citation of a paper. The authors found that there exists a positive correlation between the number of authors and the number of citations received by a paper, for papers published in six leading journals. Hsu and Huang studied in [24] whether collaboration leads to higher scientific impact. They computed the correlation between the number of citations and the number of co-authors for papers published in eight different journals, and drew the same conclusions as Figg et al. above. Within each journal, there exists a positive correlation between citations and the number of co-authors, while single-authored articles receive the smallest number of citations. In a different study [2], Abramo et al. examined how the international collaborations of Italian university researchers is associated with their productivity and scientific performance. The authors found that the volume of international collaboration is positively correlated with productivity, while the correlation between the volume of international collaboration and the average quality of performed research is not very strong.

Our work is also related to mining and learning tasks in academic data such as collaboration prediction [20, 59] and prediction of scientific impact and success [3, 53]. When the scientific impact is measured by the author's h-index, the learning task boils down to predicting the author's future h-index [11], while a similar problem seeks to answer the question of whether a specific paper will increase the primary author's h-index [10]. In addition, some attempts have been made to predict the impact of an author based on its early indications and the use of graph mining techniques [38].

Finally, GNNs, as learning models, have also been applied to other types of graphs extracted from bibliographic data. More specifically, some of the most common benchmark datasets to evaluate GNN methods are the following paper citation networks: Cora, PubMed and CiteSeer [44]. These datasets are used to predict the field of the paper given the network and the paper's textual information (i. e., abstract), in small scale. In a larger scale, a recent graph transformer has been developed to perform link prediction on a heterogeneous network that includes papers, textual information extracted from papers, venues, authors, citations and fields [26]. The edges represent different types of relations and thus an attention mechanism is utilized to separate between them and aggregate the representations of the nodes. The final tasks evaluated are paper-field, paper-venue and author-paper matching which correspond to link prediction tasks.

8.3 Data

The Microsoft Academic Graph (MAG) is one of the largest publicly available bibliographic databases. It contains more than 250 million authors and 219 million papers [46]. We decided to use MAG instead of other potential databases such as DBLP and AMiner for two reasons. The first is that it has been found that MAG has a quite large coverage [32], and the h-index values of authors estimated from the data contained in MAG were closer to those provided by services like Google Scholar. The second is that MAG is more well-curated compared to other databases (e. g., less conference names and scientific fields were missing).

It should be noted that MAG required some pre-processing before predictive models could be applied, and due to its very large scale, this task turned out to be particularly challenging. The initial pre-processing is delineated in [39] and it included among others removing duplicate authors that are associated with a single, removing stopwords from the text using the NLTK library [5], and determining the scientific field of each author based on the "field of study" of their papers[1]. The "field of study" of each paper is determined based on a machine learning pipeline that uses

---

[1] https://docs.microsoft.com/en-us/academic-services/graph/reference-data-schema

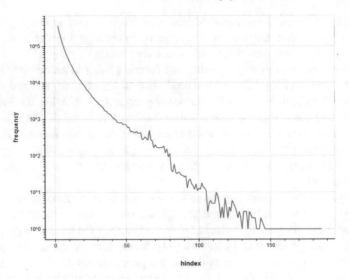

Fig. 8.1 Distribution of h-index values of the authors contained in our dataset

features such as the venues and the papers' textual content as described in [47]. This allowed us to extract subgraphs of the network induced by authors that belong to a specific field of study. In this work, we focused on authors that have published papers which belong to the computer science field. This graph was extracted and already used in another study on influence maximization [37]. It should be noted that the emerging dataset was missing textual information of some papers. If none of an author's papers contained text, we removed the corresponding node from the graph as we would not be able to leverage the heterogeneous information effectively. The final graph consists of 1,503,364 nodes and 37,010,860 edges. The weight of an edge is set equal to the number of papers two scholars have co-authored together. To investigate if the proposed model can accurately predict the future h-index of authors, we extracted the h-index values of the authors from a more recent snapshot of MAG compared to the one we used to construct the co-authorship network (June 2019 vs. December 2018). The h-index distribution of the authors is illustrated in Fig. 8.1. Note that the values on the vertical axis are in logarithmic scale. As we see, it verifies the well-known power law distribution inherent in many real-world networks [15].

8.4 Preliminaries

In this section, we first define our notation, we then present the different graph metrics that we employed, and we finally introduce the concepts of word embeddings and graph neural networks.

8.4.1 Notation

Let $G = (V, E)$ be an undirected and unweighted graph consisting of a set V of vertices and a set E of edges between them. We will denote by n the number of vertices and by m the number of edges. The neighbourhood $\mathcal{N}(v)$ of vertex v is the set of all vertices adjacent to v. Hence, $\mathcal{N}(v) = \{u : (v, u) \in E\}$ where (v, u) is an edge between vertices v and u of V. We denote the degree of vertex v by $d_v = |\mathcal{N}(v)|$. A graph G can be represented by its adjacency matrix $\mathbf{A}$. The $(i, j)^{\text{th}}$ entry of $\mathbf{A}$ is w_{ij} if the edge (v_i, v_j) between vertices v_i and v_j exists and its weight is equal to w_{ij}, and 0 otherwise. An attributed graph is a graph whose vertices are annotated with continuous multidimensional attributes. We use $\mathbf{X} \in \mathbb{R}^{n \times d}$ where d is the number of attributes to denote the graph's vertex information matrix with each row representing the attribute of a vertex.

8.4.2 Graph Metrics

One of the most prominent ways to estimate the scientific impact of an author is based on the author's position in the academic collaboration network [38]. The position can be estimated through multiple network science centralities and metrics developed to capture different dimensions of a node's impact on the graph. In this work, we utilize a number of them to compare them with the proposed approach and evaluate their usefullness in our framework.

- **Degree**: The sum of the weights of edges adjacent to a vertex. Since the co-authorship network is undirected, we can not compute the in-degree and out-degree.
- **Degree centrality**: It is the normalized degree of a vertex, i.e., the number of neighbors of the vertex divided by the maximum possible number of neighbors (i.e., $n - 1$). This measure does not take into account the weights of the edges.
- **Neighbor's average degree**: The average degree of the neighborhood of a vertex:

$$neig\_avg\_deg(v) = \frac{1}{|\mathcal{N}(v)|} \sum_{u \in \mathcal{N}(v)} deg(u) \tag{8.1}$$

where $deg(u)$ is the degree of vertex u. This measure also ignores edge weights.
- **PageRank**: Pagerank is an algorithm that computes a ranking of the vertices in a graph based on the structure of the incoming egdes. The main idea behind the algorithm is that a vertex spreads its importance to all vertices it links to:

$$PR(v) = \sum_{u \in \mathcal{N}(v)} \frac{PR(u)}{deg(u)} \tag{8.2}$$

In practice, the algorithm computes a weighted sum of two matrices, i.e., the column stochastic adjacency matrix and an all-ones matrix. Then, the pagerank scores are contained in the eigenvector associated with the eigenvalue 1 of that matrix [36].

- **Core number**: A subgraph of a graph G is defined to be a k-core of G if it is a maximal subgraph of G in which all vertices have degree at least k [31]. The core number $c(v)$ of a vertex v is equal to the highest-order core that v belongs to.
- **Onion layers**: The onion decomposition is a variant of the k-core decomposition [21].
- **Diversity coefficient**: The diversity coefficient is a centrality measure based on the Shannon entropy. Given the probability of selecting a node's neighbor based on its edge weight $p_{u,v} = \frac{w_{v,u}}{\sum_{u \in N(v)} w_{v,u}}$, the diversity of a vertex is defined as the (scaled) Shannon entropy of the weights of its incident edges.

$$D(v) = \frac{-\sum_{u \in N(v)} (p_{v,u} \log(p_{v,u}))}{\log(|N(v)|)} \tag{8.3}$$

- **Community-based centrality**: This centrality measure calculates the importance of a vertex by considering its edges towards the different communities [50]. Let $d_{v,c}$ be the number of edges between vertex v and community c and n_c the size of community c retrieved by modularity optimization. The metric is then defined as follows:

$$CB_v = \sum_{c \in C} d_{v,c} \frac{n_c}{n} \tag{8.4}$$

- **Community-based mediator**: The mediator centrality takes into consideration the role of the vertex in connecting different communities [50], where the communities are again computed by maximizing modularity. The metric relies on the percentages of the node's weighted edges that lie in its community relative to its weighted degree and the corresponding percentage for edges on different communities. To compute it, we first calculate the internal density of vertex v as follows:

$$p_v^{c_v} = \frac{\sum_{u \in N(v) \cup c_v} w_{v,u}}{deg(v)} \tag{8.5}$$

where c_v is the community to which v belongs and can be replaced with the other communities to obtain the respective external densities. Given all densities, we can calculate the entropy of a vertex as follows:

$$H_v = -p_v^{c_v} \log(p_v^{c_v}) - \sum_{c' \in C \backslash c_v} p_v^{c'} \log(p_v^{c'}) \tag{8.6}$$

where C is the set that contains all the communities. Finally, we compute the community mediator centrality:

$$CM_v = H_v \frac{deg(v)}{\sum_{u \in N(v)} deg(u)} \tag{8.7}$$

8.4.3 Text Representation Learning

In the past years, representation learning approaches have been applied heavily in the field of natural language processing. The Skip-Gram model [34] is undoubtedly one of the most well-established methods for generating distributed representations of words. Skip-Gram is a neural network comprising of one hidden layer and an output layer, and can be trained on very large unlabeled datasets. In our setting, let P_v denote a set that contains the abstracts of some papers published by author v. Let also W denote the vocabulary of all the abstracts contained in $\bigcup_{v \in V} P_v$. Then, to learn an embedding for each word $w \in W$, our model is trained to minimize the following objective function:

$$\mathcal{L} = \sum_{d \in P_v} \sum_{w_i \in d} \sum_{\substack{w_j \in \{w_{i-c}, \ldots, w_{i+c}\} \\ w_j \neq w_i}} \log\left(p(w_j | w_i)\right) \tag{8.8}$$

$$p(w_j | w_i) = \frac{\exp(\mathbf{v}_{w_i}^\top \mathbf{v}'_{w_j})}{\sum_{w \in W} \exp(\mathbf{v}_{w_i}^\top \mathbf{v}'_w)} \tag{8.9}$$

where c is the training context, $\mathbf{v}_{w_i}^\top$ is the row of matrix $\mathbf{H}$ that corresponds to word w_i, and $\mathbf{v}'_{w_j}$ is the column of matrix $\mathbf{O}$ that corresponds to w_j. Matrix $\mathbf{H} \in \mathbb{R}^{|W| \times d}$ is associated with the hidden layer, while matrix $\mathbf{O} \in \mathbb{R}^{d \times |W|}$ is associated with the output layer. Both these matrices are randomly initialized and are trained using the loss function defined above. The embeddings of the words are contained in the rows of matrix $\mathbf{H}$. The main intuition behind the representations learnt by this model is that if two words w_i and w_j have similar contexts (i.e., they both co-occur with the same words), they obtain similar representations. In real scenarios, the above formulation is impractical due to the large vocabulary size, and hence, a negative sampling scheme is usually employed [19]. The training set is built by generating word-context pairs, i.e., for each word w_i and a training context c, we create pairs of the form $(w_i, w_{i-c}), \ldots, (w_i, w_{i-1}), (w_i, w_{i+1}), (w_i, w_{i+c})$.

Note that since the abstract of a paper contains multiple words, we need to aggregate the embeddings of the words to derive an embedding for the entire abstract. To this end, we simply average the representations of the words. Furthermore, an author may have published multiple papers, and to produce a vector representation for the author, we again compute the average of the representations of (the available subset of) their papers:

$$\mathbf{h}_v = \frac{1}{|P_v|} \sum_{d \in P_v} \frac{1}{|d|} \sum_{w \in d} \mathbf{v}_w \tag{8.10}$$

8.4.4 Graph Neural Networks

In the past years, GNNs have attracted a lot of attention in the machine learning community and have been successfully applied to several problems. Most of the proposed GNNs share the same basic idea, and can be reformulated into a single common framework [18]. These models employ a message passing procedure, where vertices update their representations based on their previous representation and on messages received from their neighbors. While the concept of message passing over graphs and that of GNNs have been around for many years [43, 48, 61], GNNs have only recently become a topic of intense research mainly due to the recent advent of deep learning.

A GNN model consists of a series of neighborhood aggregation layers. As already mentioned, each one of these layers uses the graph structure and the node feature vectors from the previous layer to generate new representations for the nodes. The feature vectors are updated by aggregating local neighborhood information. Let $\mathbf{h}_v^{(0)} \in \mathbb{R}^d$ denote the initial feature vector of vertex v, and suppose we have a GNN model that contains T neighborhood aggregation layers. In the t-th neighborhood aggregation layer ($t > 0$), the hidden state $\mathbf{h}_v^{(t)}$ of a node v is updated as follows:

$$
\begin{aligned}
\mathbf{m}_v^{(t)} &= \text{AGGREGATE}^{(t)}\left(\left\{\mathbf{h}_u^{(t-1)} | u \in \mathcal{N}(v)\right\}\right) \\
\mathbf{h}_v^{(t)} &= \text{COMBINE}^{(t)}\left(\mathbf{h}_v^{(t-1)}, \mathbf{m}_v^{(t)}\right)
\end{aligned}
\tag{8.11}
$$

By defining different $\text{AGGREGATE}^{(t)}$ and $\text{COMBINE}^{(t)}$ functions, we obtain a different GNN variant. For the GNN to be end-to-end trainable, both functions need to be differentiable. Furthermore, since there is no natural ordering of the neighbors of a node, the $\text{AGGREGATE}^{(t)}$ function must be permutation invariant. The node feature vectors $\mathbf{h}_v^{(T)}$ of the final neighborhood aggregation layer are usually passed on to a fully-connected neural network to produce the output.

8.5 Methodology

In this section, we give details about the two graph neural networks that we employed and we present their exact architecture.

8.5.1 Proposed Architecture

We propose two different GNN variants which differ from each other only in terms of the employed message passing procedure. Both models merge the AGGREGATE and COMBINE functions presented above into a single function. A high-level illustration

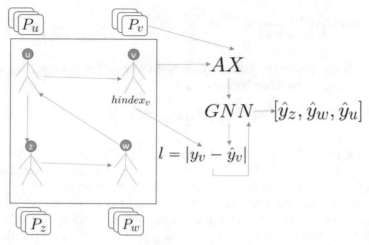

Fig. 8.2 An overview of the proposed model for a graph of 4 nodes (u, v, y, w)

of the proposed approach is shown in Fig. 8.2 via a graph which is represented by its adjacency matrix, whereas the textual and graph-related features of the nodes are stored in a matrix **X**. These two matrices are passed on to a GNN which predicts the authors' future h-index which are then compared against the actual h-index values to compute the loss.

GNN.
Given the aforementioned co-authorship graph $G = (V, E)$ where vertices are annotated with feature vectors $\mathbf{h}_v^{(0)} \in \mathbb{R}^d$ stemming from the learnt representations of the author's papers and/or the graph metrics, each neighborhood aggregation layer of the first model updates the representations of the vertices as follows:

$$\mathbf{h}_v^{(t)} = \text{RELU}\left(\mathbf{W}^{(t)} \mathbf{h}_v^{(t-1)} + \sum_{u \in N(v)} \mathbf{W}^{(t)} \mathbf{h}_u^{(t-1)}\right) \qquad (8.12)$$

where $\mathbf{W}^{(t)}$ is the matrix of trainable parameters of the tth message passing layer. In matrix form, the above is equivalent to:

$$\mathbf{H}^{(t)} = \text{RELU}\left(\tilde{\mathbf{A}} \mathbf{H}^{(t-1)} \mathbf{W}^{(t)}\right) \qquad (8.13)$$

where $\tilde{\mathbf{A}} = \mathbf{A} + \mathbf{I}$. Note that in both Eqs. 8.12 and 8.13, we omit biases for clarity of presentation. The above message passing procedure is in fact similar to the one of the GIN-0 model [56].

GCN.
With regards to the second variant, its node-wise formulation is given by:

$$\mathbf{h}_v^{(t)} = \text{ReLU}\left(\mathbf{W}^{(t)}\frac{1}{1+d_v}\mathbf{h}_v^{(t-1)} + \sum_{u\in\mathcal{N}(v)}\mathbf{W}^{(t)}\frac{1}{\sqrt{(1+d_v)(1+d_u)}}\mathbf{h}_u^{(t-1)}\right) \quad (8.14)$$

where $\mathbf{W}^{(t)}$ is the matrix of trainable parameters of the tth message passing layer. In matrix form, the above is equivalent to:

$$\mathbf{H}^{(t)} = \text{ReLU}\left(\hat{\mathbf{A}}\,\mathbf{H}^{(t-1)}\,\mathbf{W}^{(t)}\right) \quad\quad\quad (8.15)$$

where $\hat{\mathbf{A}} = \tilde{\mathbf{D}}^{-\frac{1}{2}}\tilde{\mathbf{A}}\tilde{\mathbf{D}}^{-\frac{1}{2}}$ and $\tilde{\mathbf{D}}$ is a diagonal matrix such that $\tilde{\mathbf{D}}_{ii} = \sum_{j=1}^{n}\tilde{\mathbf{A}}_{ij}$. Note that in both Eqs 8.14 and 8.15, we omit biases for clarity of presentation. The above message passing procedure is in fact the one employed by the GCN [28].

It should be mentioned that matrix $\hat{\mathbf{A}}$ corresponds to the symmetrically normalized adjacency matrix with self-loops. Normalization is mainly applied to avoid numerical instabilities and exploding/vanishing gradients associated with deep neural networks. One should note that by normalizing, we capture the distribution of the representations of the neighbors of a vertex, but not the exact multiset of representations. Therefore, we could potentially lose the ability to distinguish between authors that are connected with many other researchers and authors that are connected only with a few other researchers [56]. Note that besides the above two message passing schemes, we also tried using GAT-like attention [51] in early experiments, without obtaining better results.

Multi-hop information.
Inspired by Jumping Knowledge Networks [57], instead of using only the final vertex representations $\mathbf{h}_v^{(T)}$ (i.e., obtained after T message passing steps), we also use the representations of the earlier steps $\mathbf{h}_v^{(1)},\ldots,\mathbf{h}_v^{(T-1)}$. Note that as one iterates, vertex representations capture more and more global information. However, retaining more local, intermediary information might be useful too. Thus, we concatenate the representations produced at the different steps, finally obtaining $\mathbf{h}_v = [\mathbf{h}_v^{(1)}||\mathbf{h}_v^{(2)}||\ldots||\mathbf{h}_v^{(T)}]$. These vertex representations are then passed on to one or more fully-connected layers to produce the output.

8.6 Experimental Evaluation

In this section, we first present the baselines against which we compared the proposed approach. We next give details about the employed experimental protocol. We last report on the performance of the different approaches and discuss the obtained results.

8.6.1 Baselines

We compare the proposed models against the following learning algorithms: (i) Lasso, a regression algorithm that performs both variable selection and regularization [49], (ii) SGDRegressor, a linear model which is trained by minimizing a regularized empirical loss with SGD, (iii) XGBoost, a scalable end-to-end tree boosting system [9], and a multi-layer perceptron (MLP). All the above models expect each input sample to be in the form of a vector. Given an author, the above vector is produced by the features extracted from the textual content of the author's papers and/or the features extracted from the co-authorship network.

8.6.2 Experimental Setup

To test the semi-supervised generalization capabilities of our model, we experimented with a 20/80 training/test split. For all algorithms, we standardize the input features by removing the mean and scaling to unit variance.

With regards to the hyperparameters of the proposed models, we use 2 message passing layers. The hidden-dimension size of the message passing layers is set to 32 and 64, respectively. In the case of the first model (i. e., GNN), batch normalization is applied to the output of every message passing layer. The representations produced by the second message passing layer are passed on to a multi-layer perceptron with one hidden layer of dimensionality 64. All dense layers use ReLU activation. The dropout rate is set equal to 0.1. To train all models, we use the Adam optimizer with learning rate of 0.01. We set the number of epochs to 300. The best model is chosen based on a validation experiment on a single 90% - 10% split of the training data and is stored into disk. At the end of training, the model is retrieved from the disk, and we use it to make predictions for the test instances. The MLP contains a hidden layer of dimensionality 64. All its remaining hyperparameters (e. g., dropout rate, activation function, etc.) take values identical to those of the two GNNs. For lasso, the weight of regularization is set to 1. For SGDRegressor, we use an l_2 regularizer with weight 0.0001. The initial learning rate is set to 0.01 and the number of epochs to 1000. For XGBoost, we use the default values of the hyperparameters.

8.6.3 Results and Discussion

The performance of the different models is illustrated in Table 8.1. We report the mean absolute error (MAE) and the mean squared error (MSE) of the different approaches. There are three different sets of features passed on to each algorithm (features extracted from graph, features extracted from text or from both). With regards to the performance of the different approaches, we first observe that neural network models outperform the other approaches in all settings and by considerable

Table 8.1 Performance of the different methods in the h-index prediction task

Method	Text Features		Graph Features		All Features	
	MAE	MSE	MAE	MSE	MAE	MSE
Lasso	4.99	66.91	3.28	29.51	3.28	29.51
SGDRegressor	8.48	112.20	6.20	78.38	8.01	120.91
XGBoost	4.22	64.43	3.04	34.83	2.91	**21.04**
MLP	4.10	59.77	**2.62**	**22.46**	2.59	21.44
GCN	**4.05**	**59.45**	2.68	24.32	**2.57**	21.29
GNN	4.07	60.00	2.66	23.82	2.58	21.85

margins, except for one case. We hypothesize that this stems from the inherent capability of neural networks to detect meaningful and useful patterns in large amounts of data. In this case, though semi-supervised, the training set still consists of more than 300,000 samples, which allows for effective use of the representational power of neural network models. On the other hand, we see that XGBoost performs better in terms of MSE in one setting since it optimizes the MSE criterion, while the neural architectures are trained by minimizing a MAE objective function, as also shown in Fig. 8.2. To train the proposed models, we chose to minimize MAE instead of MSE since MAE is more interpretable in the case of h-index, and provides a generic and even measure of how well our model is performing. Therefore, for very large differences between the h-index and the predicted h-index (e. g., $y = 120$ vs. $\hat{y} = 40$), the function does not magnify the error.

With regards to the different types of features, we can see that the graph metrics alone correspond to a much stronger predictor of the future h-index compared to the features extracted from the papers' textual content. Due to the rich information encoded into these features, MLP achieves similar performance to that of the two GNN models. This is not surprising since GNN and GCN consist of two message passing layers followed by a multi-layer perceptron which is identical in terms of architecture to MLP. The features extracted from the authors' papers seem not to capture the actual impact of the author. It is indeed hard to determine the impact of an author based solely on the textual content of a subset of the papers the author has published. Furthermore, these features have been produced from a limited number of an author's papers, and therefore, they might not be able to capture the author's relationship with similar authors because of the diversity of scientific abstracts and themes. GCN is in general the best-performing method and yields the lowest levels of MAE. Overall, given the MAE of 2.57, we can argue that GNNs can be useful for semi-supervised prediction of the authors' future h-index in real-world scenarios.

To qualitatively assess the effectiveness of the proposed models, we selected a number of well-known scholars in the field of data mining and present their h-index along with the predictions of our two models in Table 8.2. Keeping in mind that the overwhelming majority of authors have a relatively small h-index (as has been observed in Fig. 8.1), it is clear that these are some of the most extreme and hard cases which can pose a significant challenge to the proposed models. Even though

Table 8.2 The actual h-index of a number of authors and their predicted h-index

Author	h-index	Predicted h-index	
		GCN	GNN
Jiawei Han	131	94.75	115.09
Jie Tang	45	30.71	33.74
Yannis Manolopoulos	40	43.79	49.63
Lada Adamic	48	30.67	22.71
Zoubin Ghahramani	77	40.64	43.33
Michalis Vazirgiannis	36	25.76	27.89
Jure Leskovec	68	37.77	38.26
Philip S Yu	120	100.42	119.93

the objective function of the proposed models is MAE which does not place more weight on large errors (which can happen for authors that have high h-index values), still, as can be seen in Table 8.2, the models' predictions are relatively close to the actual h-index values of the authors in most cases.

8.7 Conclusion

In this paper, we developed two GNNs to deal with the problem of predicting the h-index of authors based on information extracted from co-authorship networks. The proposed models outperform standard approaches in terms of mean absolute error in a semi-supervised setting, i. e., when the training data is scant or not trustworthy, which is often the case in bibliographic data. We also underline the flexibility of our model which combines the graph structure with the textual content of the authors' papers and metrics extracted from the co-authorship network. In the future, we plan to investigate how multiple paper representations can be aggregated in the model, instead of computing the average of the top ones. Moreover, we will evaluate the performance of our models in other fields such as in physics and in chemistry.

References

1. Abbasi, A., Altmann, J., Hossain, L.: Identifying the effects of co-authorship networks on the performance of scholars: A correlation and regression analysis of performance measures and social network analysis measures. Journal of Informetrics 5(4), 594–607 (2011)
2. Abramo, G., D'Angelo, C.A., Solazzi, M.: The relationship between scientists' research performance and the degree of internationalization of their research. Scientometrics **86**(3), 629–643 (2011)
3. Acuna, D.E., Allesina, S., Kording, K.P.: Predicting scientific success. Nature **489**(7415), 201–202 (2012)

4. Adams, J.D., Black, G.C., Clemmons, J.R., Stephan, P.E.: Scientific teams and institutional collaborations: Evidence from US universities, 1981–1999. Research Policy **34**(3), 259–285 (2005)
5. Bird, S.B., Loper, E.: NLTK: The Natural Language Toolkit. In: Proceedings of the 42nd Annual Meeting of the Association for Computational Linguistics (ACL) - Poster and Demonstration (2004)
6. Bordons, M., Aparicio, J., González-Albo, B., Díaz-Faes, A.A.: The relationship between the research performance of scientists and their position in co-authorship networks in three fields. Journal of Informetrics **9**(1), 135–144 (2015)
7. Bornmann, L., Daniel, H.D.: Does the h-index for ranking of scientists really work? Scientometrics **65**(3), 391–392 (2005)
8. Bornmann, L., Mutz, R., Daniel, H.D.: Are there better indices for evaluation purposes than the h-index? A comparison of nine different variants of the h-index using data from biomedicine. Journal of the American Society for Information Science & Technology **59**(5), 830–837 (2008)
9. Chen, T., Guestrin, C.: Xgboost: A scalable tree boosting system. In: Proceedings of the 22nd ACM International Conference on Knowledge Discovery & Data Mining (KDD), pp. 785–794 (2016)
10. Dong, Y., Johnson, R.A., Chawla, N.V.: Will this paper increase your h-index? Scientific impact prediction. In: Proceedings of the 8th ACM International Conference on Web Search & Data Mining (WSDM), pp. 149–158 (2015)
11. Dong, Y., Johnson, R.A., Chawla, N.V.: Can scientific impact be predicted? IEEE Transactions on Big Data **2**(1), 18–30 (2016)
12. Ductor, L.: Does co-authorship lead to higher academic productivity? Oxford Bulletin of Economics & Statistics **77**(3), 385–407 (2015)
13. Eaton, J.P., Ward, J.C., Kumar, A., Reingen, P.H.: Structural analysis of co-author relationships and author productivity in selected outlets for consumer behavior research. Journal of Consumer Psychology **8**(1), 39–59 (1999)
14. Egghe, L.: Theory and practise of the g-index. Scientometrics **69**(1), 131–152 (2006)
15. Faloutsos, M., Faloutsos, P., Faloutsos, C.: On power-law relationships of the internet topology. ACM SIGCOMM Computer Communication Review **29**(4), 251–262 (1999)
16. Fan, W., Ma, Y., Li, Q., He, Y., Zhao, E., Tang, J., Yin, D.: Graph neural networks for social recommendation. In: Proceedings of the World Wide Web Conference, pp. 417–426 (2019)
17. Figg, W.D., Dunn, L., Liewchr, D.J., Steinberg, S.M., Thurman, P.W., Barrett, J.C., Birkinshaw, J.: Scientific collaboration results in higher citation rates of published articles. Pharmacotherapy: The Journal of Human Pharmacology & Drug Therapy **26**(6), 759–767 (2006)
18. Gilmer, J., Schoenholz, S.S., Riley, P.F., Vinyals, O., Dahl, G.E.: Neural message passing for quantum chemistry. In: Proceedings of the 34th International Conference on Machine Learning (ICML), pp. 1263–1272 (2017)
19. Goldberg, Y., Levy, O.: word2vec explained: deriving mikolov et al.'s negative-sampling word-embedding method. CoRR **abs/1402.3722** (2014)
20. Guns, R., Rousseau, R.: Recommending research collaborations using link prediction and random forest classifiers. Scientometrics **101**(2), 1461–1473 (2014)
21. Hébert-Dufresne, L., Grochow, J.A., Allard, A.: Multi-scale structure and topological anomaly detection via a new network statistic: The onion decomposition. Scientific Reports **6**, 31708 (2016)
22. Heiberger, R.H., Wieczorek, O.J.: Choosing collaboration partners. How scientific success in physics depends on network positions. CoRR **abs/1608.03251** (2016)
23. Hirsch, J.E.: An index to quantify an individual's scientific research output. Proceedings of the National Academy of Sciences **102**(46), 16569–16572 (2005)
24. Hsu, J.W., Huang, D.W.: Correlation between impact and collaboration. Scientometrics **86**(2), 317–324 (2011)
25. Hu, Z., Chen, C., Liu, Z.: How are collaboration and productivity correlated at various career stages of scientists? Scientometrics **101**(2), 1553–1564 (2014)
26. Hu, Z., Dong, Y., Wang, K., Sun, Y.: Heterogeneous graph transformer. In: Proceedings of The Web Conference, pp. 2704–2710 (2020)

27. Katz, J., Hicks, D.: How much is a collaboration worth? A calibrated bibliometric model. Scientometrics **40**(3), 541–554 (1997)
28. Kipf, T.N., Welling, M.: Semi-supervised classification with graph convolutional networks. In: Proceedings of the International Conference on Learning Representations (ICLR) (2017)
29. Kretschmer, H.: Author productivity and geodesic distance in bibliographic co-authorship networks, and visibility on the web. Scientometrics **60**(3), 409–420 (2004)
30. Lee, S., Bozeman, B.: The impact of research collaboration on scientific productivity. Social Studies of Science **35**(5), 673–702 (2005)
31. Malliaros, F.D., Giatsidis, C., Papadopoulos, A.N., Vazirgiannis, M.: The core decomposition of networks: Theory, algorithms and applications. The VLDB Journal **29**(1), 61–92 (2020)
32. Martín-Martín, A., Thelwall, M., Orduna-Malea, E., López-Cózar, E.D.: Google Scholar, Microsoft Academic, Scopus, Dimensions, Web of Science, and OpenCitations' COCI: A multidisciplinary comparison of coverage via citations. Scientometrics **126**(1), 871–906 (2021)
33. McCarty, C., Jawitz, J.W., Hopkins, A., Goldman, A.: Predicting author h-index using characteristics of the co-author network. Scientometrics **96**(2), 467–483 (2013)
34. Mikolov, T., Sutskever, I., Chen, K., Corrado, G.S., Dean, J.: Distributed representations of words and phrases and their compositionality. Advances in Neural Information Processing Systems **26**, 3111–3119 (2013)
35. Nikolentzos, G., Tixier, A., Vazirgiannis, M.: Message passing attention networks for document understanding. In: Proceedings of the 34th AAAI Conference on Artificial Intelligence, pp. 8544–8551 (2020)
36. Page, L., Brin, S., Motwani, R., Winograd, T.: The PageRank citation ranking: Bringing order to the web. Tech. rep., Stanford InfoLab (1999)
37. Panagopoulos, G., Malliaros, F.D., Vazirgiannis, M.: Influence maximization using influence and susceptibility embeddings. In: Proceedings of the 14th International AAAI Conference on Web & Social Media, pp. 511–521 (2020)
38. Panagopoulos, G., Tsatsaronis, G., Varlamis, I.: Detecting rising stars in dynamic collaborative networks. Journal of Informetrics **11**(1), 198–222 (2017)
39. Panagopoulos, G., Xypolopoulos, C., Skianis, K., Giatsidis, C., Tang, J., Vazirgiannis, M.: Scientometrics for success and influence in the Microsoft Academic Graph. In: Proceedings of the 8th International Conference on Complex Networks & their Applications (CNA), pp. 1007–1017 (2019)
40. Parish, A.J., Boyack, K.W., Ioannidis, J.P.: Dynamics of co-authorship and productivity across different fields of scientific research. PloS one **13**(1), e0189742 (2018)
41. Roediger III, H.L.: The h-index in science: A new measure of scholarly contribution. APS Observer **19**(4) (2006)
42. Sarigöl, E., Pfitzner, R., Scholtes, I., Garas, A., Schweitzer, F.: Predicting scientific success based on coauthorship networks. EPJ Data Science **3**(1), 9 (2014)
43. Scarselli, F., Gori, M., Tsoi, A.C., Hagenbuchner, M., Monfardini, G.: The graph neural network model. IEEE Transactions on Neural Networks **20**(1), 61–80 (2008)
44. Sen, P., Namata, G., Bilgic, M., Getoor, L., Galligher, B., Eliassi-Rad, T.: Collective classification in network data. AI magazine **29**(3), 93–93 (2008)
45. Servia-Rodríguez, S., Noulas, A., Mascolo, C., Fernández-Vilas, A., Díaz-Redondo, R.P.: The evolution of your success lies at the centre of your co-authorship network. PloS one **10**(3), e0114302 (2015)
46. Shen, Z., Ma, H., Wang, K.: A web-scale system for scientific knowledge exploration. In: Proceedings of Association for Computational Linguistics Conference (ACL) - System Demonstrations, pp. 87–92 (2018)
47. Sinha, A., Shen, Z., Song, Y., Ma, H., Eide, D., Hsu, B.J., Wang, K.: An overview of Microsoft Academic Service (MAS) and applications. In: Proceedings of the 24th International Conference on World Wide Web (WWW), pp. 243–246 (2015)
48. Sperduti, A., Starita, A.: Supervised neural networks for the classification of structures. IEEE Transactions on Neural Networks **8**(3), 714–735 (1997)
49. Tibshirani, R.: Regression shrinkage and selection via the lasso: a retrospective. Journal of the Royal Statistical Society: Series B (Statistical Methodology) **73**(3), 273–282 (2011)

50. Tulu, M.M., Hou, R., Younas, T.: Identifying influential nodes based on community structure to speed up the dissemination of information in complex network. IEEE Access **6**, 7390–7401 (2018)
51. Veličković, P., Cucurull, G., Casanova, A., Romero, A., Lio, P., Bengio, Y.: Graph attention networks. In: Proceedings of the International Conference on Learning Representations (ICLR) (2018)
52. Waltman, L., Van Eck, N.J.: The inconsistency of the h-index. Journal of the American Society for Information Science & Technology **63**(2), 406–415 (2012)
53. Weihs, L., Etzioni, O.: Learning to predict citation-based impact measures. In: Proceedings of the 17th ACM/IEEE Joint Conference on Digital Libraries (JCDL), pp. 1–10 (2017)
54. Wildgaard, L., Schneider, J.W., Larsen, B.: A review of the characteristics of 108 author-level bibliometric indicators. Scientometrics **101**(1), 125–158 (2014)
55. Wu, Z., Pan, S., Chen, F., Long, G., Zhang, C., Philip, S.Y.: A comprehensive survey on graph neural networks. IEEE Transactions on Neural Networks & Learning Systems **32**(1), 4–24 (2021)
56. Xu, K., Hu, W., Leskovec, J., Jegelka, S.: How powerful are graph neural networks? In: Proceedings of the International Conference on Learning Representations (ICLR) (2019)
57. Xu, K., Li, C., Tian, Y., Sonobe, T., Kawarabayashi, K.i., Jegelka, S.: Representation learning on graphs with jumping knowledge networks. In: Proceedings of the 37th International Conference on Machine Learning (ICML), pp. 5453–5462 (2018)
58. Yan, E., Ding, Y.: Applying centrality measures to impact analysis: A coauthorship network analysis. Journal of the American Society for Information Science & Technology **60**(10), 2107–2118 (2009)
59. Yan, E., Guns, R.: Predicting and recommending collaborations: An author-, institution-, and country-level analysis. Journal of Informetrics **8**(2), 295–309 (2014)
60. Zhang, C.T.: The e-index, complementing the h-index for excess citations. PLoS One **4**(5), e5429 (2009)
61. Zhu, X., Ghahramani, Z.: Learning from labeled and unlabeled data with label propagation. Tech. rep., Carnegie Mellon University (2002)
62. Zuckerman, H.: Nobel laureates in science: Patterns of productivity, collaboration, and authorship. American Sociological Review pp. 391–403 (1967)

Chapter 9
Identification of Promising Researchers through Fast-and-frugal Heuristics

Antônio de Abreu Batista-Jr, Fábio Castro Gouveia & Jesús P. Mena-Chalco

Abstract Predicting a researcher's future scientific achievements is essential for selection committees. Due to the unreflective use of metrics in evaluation processes, committees have to take controversial decisions while guiding exclusively by such metrics. This paper proposes a novel fast-and-frugal heuristics approach to identify prematurely prominent researchers. This study analyzes the careers of $1,631$ computer scientists from ACM data set used as applicants in our selection models. We compared the percentages of these applicants who had been successfully chosen by future promise-based heuristics and traditional-indicators-based heuristics. We founded that estimated future performance-based heuristics are more reliable than traditional-indicators-based heuristics.

9.1 Introduction

The fight for faculty positions has, in the recent decade, become intense, and, while when trying to assess the potential future accomplishments of individual faculty, committees have failed [9, 26]. Traditional metrics, such as Impact Factor and citation counts, continue to be used even though they have not been effective [4] in the selection process. Thus, a tool to foresee whether young scientists have a higher

Antônio de Abreu Batista-Jr
Department of Informatics, Federal University of Maranhão, São Luís, MA, Brazil,
e-mail: antonio.batista@ufma.br

Fábio Castro Gouveia
Oswaldo Cruz Foundation, House of Oswaldo Cruz, Museum of Life, Rio de Janeiro, RJ, Brazil,
e-mail: fgouveia@gmail.com

Jesús P. Mena-Chalco
Center for Mathematics, Computation and Cognition, Federal University of ABC, Santo André, SP, Brazil,
e-mail: jesus.mena@ufabc.edu.br

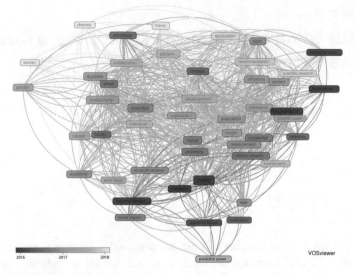

Fig. 9.1: VOSviewer overlay graph of the most frequent terms in titles and abstracts of articles citing Acuna [3]

probability of publishing successfully during their careers is of great value to selection committees. In recent years there has been growing interest in trying to predict the future accomplishments of individual faculty members. As a starting point to observe the big picture of studies that focus on predicting scientific success, we can look at the papers that cited the Acuna and collaborations predictive model [3]. Using VOSviewer v.1.6.15, a scientometrics software developed by the Leiden University [34], we can explore the co-occurrences of terms on their titles and abstracts. After removing the non-informative terminologies, we can see in an overlay graph the most frequent terms with a color scale of the mean year for of each. We found a total of 149 outputs on Scopus Database citing the Acuna and collaborators paper [3], and considering just the terms that occurred six or more times and the top 60% of them according to the relevance score of the software, we obtained the graph in Fig. 9.1 with 51 terms.

As can be observed, thematics related to gender, woman, and diversity are more recent than those focused on the idea of impact. Thus, by the graph, we can draw that equitable academic achievement assessments have gained importance.

Because the production and assessment of scholarship are so central to the faculty role [17], it has been incumbent upon decision-makers to strive to make assessments of scholarship fair and equitable. However, recently, due to a large number of candidates per available position, previous work, in particular, Frank has suggested hiring and promotion committees to solicit a few research products as the primary locus of evaluation and a research statement [12]. This latter intend to contextualize current work and outline long-term plans for evaluating the likely future impact of candidates. Nevertheless, this is a time-consuming strategy to implement across a large set of candidates [7].

On the other hand, a more practical approach [7] has recommended a decision strategy directed on indicators focusing on research excellence to select, in a first-round, the most promising candidates. Shortly after, a peer review panel assesses the previous results in detail.

Despite this interest, what becomes a prominent researcher is unclear. This paper proposes a novel fast-and-frugal heuristics approach to identify prominent researchers before becoming one. Fast-and-frugal heuristics are decision strategies that use part of the available information (only a few indicators) and ignore the rest.

Our main contribution is a framework to aid hiring committees in their task.

9.2 Related Work

Quantitative measures of scientific output and success in science already affect the evaluation of researchers and the funding of proposals [30], and consequently the future of science and the future career of researchers in science.

Nonetheless, the research community has moved from promoting measures such as the journal impact factor and h-index to critiquing them [32]. In a critical note about the Impact Factor, E. Garfield compared it to nuclear energy showing a cautious view on its use [14]. As in good hands, metrics could be constructively used and in wrong hands abused. For the research community, poorly constructed indicators can reify structural biases such as racism, sexism, and classism. H-index is an index that tries to represent the relative importance of a researcher, comparable in their given area of research activity, based on the n number of papers that have n or more citations [16].

In [15, 37] the authors found well-represented candidates for science, technology, engineering, and mathematics (STEM) faculty positions continue to dominate the hiring process. On the other hand, in [11] the authors concluded that, above a certain threshold, the benchmarks traditionally used to measure research success are ineffective while differentiating among applicants with and without job offers.

When guiding exclusively by the "Publish or Perish"[1] approach to success, committees have perpetuated the "Matthew effect" Merton[2] strengthening existing differences in productivity and recognition between scholars. The authors in [26] have advised focusing on multiple criteria and not just to look to past success as a chief indicator of future success, without considering systemic barriers.

Several studies, for instance [1, 3, 24, 36], have been conducted on the scholarly impact prediction of authors. A recent review of the literature on this topic [18] found that the distance of citing scholars in collaboration networks, the number of

---

[1] "Publish or Perish" describes the growing competition that pressures researchers to continuously produce publishable results.

[2] The Matthew effect is a social phenomenon describing the disproportionate rewards reaped by those in privileged positions.

papers in famous venues, the current h-index, the scholar's authority on a topic, the number of distinct conferences published in, the number of years since publishing the first paper, and the centrality in collaboration networks were the most used to predict scholar impact.

In [3] the authors proposed a linear regression model to predict neuroscientists' future h-index. However, as [13, 25] have highlighted, the model in [3] raises many doubts. The results do not seem to support the equations. Apart from this, the predictive power of these models depends heavily upon scientists' career time, producing inaccurate estimates for young researchers.

Even though models in [3, 5] predict future h-index better than does current h-index alone, in both, the accuracy of future h-index prediction decreases over time.

Unlike the previous two works, Lee focused on early career-related factors affecting scientists [24]. The factor that adds most to the future research performance (i.e. publication numbers) and future research impact (i.e. citation counts of publications) was the number of publications (both journal articles and conference papers) produced by the target scientists in their early career years. The authors in [40] state that a good publishing strategy for early career scholars is to publish some of their papers in the same journal [40].

On the other hand, as the authors' impact correlates to their publications' impact, many attempts have been made [6, 27, 28, 35] to predict how many citations a research article will receive. In [35] the authors developed a new method for predicting the total number of citations a paper will acquire based on an analytic model, while in [2] the authors developed a new method for that based on deep neural network learning techniques.

In [29], combining supervised machine learning and network analysis techniques, the authors found that the position of scientists in the collaboration network alone is - to a surprisingly large degree - indicative for the future citation success of their papers. Also, Laurance found that Pre-PhD publication success is indicative for long-term publication success (10 years post-PhD) [40].

It has now been hypothesized that the academic collaborator recommendation [8, 22] for junior researchers can be of great value in guiding hiring decisions.

According to [31], highly productive scientists possess an inherent talent or ability, called Q-factor. This latter quantifies an individual's ability to turn an idea into discovery with a given citation impact. In prior work [1], we discovered that the future Q of junior researchers is largely predictable using data of their five years after the beginning of their research career.

Despite the large amount of attention devoted to these issues, it remains unclear what factors contribute most to a researcher's future research impact. In this paper, we propose a novel fast-and-frugal heuristics approach to identify prominent researchers still flourishing. Based on the initial careers of $1,631$ scientists drawn from the Association for Computing Machinery (ACM) citation network data set Tang that constituted our test pool of applicants, we compared the percentages of these scientists who became prominent chosen by future and traditional indicators.

Table 9.1: The indicators used by heuristics to sort descendingly the candidates

Heuristic	Indicator	Description	Future	Current
H1	Q_i	Q value	x	
H2	$c_{r_i}^{\infty}$	Citation count	x	
H3	Q_i	Q value		x
H4	Q_{c_i}	Q value		x
H5	H_i	H-index*		x
H6	H_{c_i}	H-index		x

9.3 Method

We compared different heuristics intending to select competent researchers from a large set of candidates. We assessed the validity of one-cuc bibliometrics-based heuristics (Table 9.1) using top-k precision estimates.

A one-cue bibliometrics-based heuristic is composed of two steps. First, it orders the candidates using an indicator. Second, top-k ranked candidates are assessed again by a committee. Formally, it consists in classifying n candidates $1, 2, i, \ldots, n$ according to their performances, $P(1), P(2), P(i), \ldots, P(n)$ using solely an indicator-based either on past performance or on future promise estimated from past performance. If the rank position of $P(i) \leq x$, where x is the total number of applicants that can be selected, a hiring and promotion committee will assess $P(i)$; else applicant i is eliminated.

We use the index i to refer to the candidate as defined above and the variable c_i to refer to their highest h-index co-author. We only have considered candidates who have a c_i with at least 10 years of work (senior) at the 5th year of their career. If there is more than one eligible, the most productive one is selected.

The elimination process involves a single indicator (cue) that targets a relevant goal of the hirer: scientific competence. We measured it through the researcher's stable Q value. We used the following procedures to predict the stable Q and the future citation rate of representative papers. Note that we are using here just data from the five firsts year of publication of candidates.

Stable Q. We estimate the stable Q of candidate i through Eq. 9.1 proposed in [1]:

$$Q_i^{\text{Fut}} = 1.930 + 0.284 Q_i^{\text{Curr}} + 0.13 Q_{c_i}^{\text{Curr}} + 0.515 h_i^{\text{Curr}} \tag{9.1}$$

where Q_i^{Curr}, $Q_{c_i}^{\text{Curr}}$ and h_i^{Curr} are the current Q of the junior applicant, the current Q of their co-author c_i and the current h-index of the junior applicant, respectively.

Q is a unique parameter for scientist i which quantifies the ability of i to improve a project selected randomly with potential p_α and publish a paper of impact $c_{t,\alpha i} = Q_i p_\alpha$ after t years of their publication. Q_i^{Curr} was calculated using the Eq. 9.2 given in [31]:

$$Q_i^{\text{Curr}} = e^{(\frac{1}{N} \sum_{\alpha=1}^{N} \log_e c_{t,\alpha i}) - \mu_p}, \tag{9.2}$$

where $c_{t,\alpha i}$ is the citation count received by paper α after t years of their publication, and μ_p is the mean of p_α - a random variable that outlines the potential impact of project p_α selected - for all researchers belonging to the same specify-field.

Future citation count. We estimate the number of citations received by a research paper a over its lifetime c_a^∞ through Eq. 9.3 proposed by [35], defined in terms of paper "fitness", λ_a. The fitness parameter relates to the size of the citation peak following a paper's publication:

$$c_a^\infty = m(e^{\lambda_a} - 1) \tag{9.3}$$

where m represents the average number of references each new paper contains. Unlike [35], we determined $\lambda_a = \beta_1$ by taking the regression coefficients from the linear regression models (Eq. 9.4).

$$c_a^t = \beta_0 + \beta_1 t + \epsilon_t, \tag{9.4}$$

where c_a^t is the cumulative number of citations receiving by paper a and β_0, β_1 are the intercept and the regression coefficient for the number of years after publication t, respectively. ϵ_t is the error.

9.4 Evaluation

The goal of this section is to present the methodologies used in the evaluation and the data set.

9.4.1 Experimental Setup

We test with $8,150$ researchers who had at least 40 published articles, and, at their junior researcher phase, had authored at least one paper with a co-author with a career size of 10 or more years, derived from the ACM data set. Of this total, about 80% ($6,519$) belong to the train set used here to learn Eq. 9.1, and 20% ($1,631$) to the test set. We have considered, in our following discussions, only the researchers in the test data set. These latter constitute our pool of junior applicants (the average h-index value of them was 1.93).

We compared the rankings predicted by tested heuristics against the actual in the year 15. The goal of heuristics is to produce the latter effectively.

Data set. The data used for this paper are drawn from the ACM citation network data set Tang. The citation data was extracted from ACM by the AMiner[3], a free online service used to carry out web search and indexing, data mining tasks.

---

[3] http://www.aminer.cn/data, accessed on June 1, 2021.

Each paper is associated with abstract, authors, year, venue, and title. The database has close to 2.5 million computer science articles written by approximately 1, 600, 000 authors. The data set contains published articles in the period spanning from 1936 to 2017 and is publicly available at http://www.aminer.cn/citation.

We note that the data set has been widely used in the literature [19, 38, 39]. Also, it is relevant to highlight that we did not address problems of name disambiguation since we used a pre-processed database with unique author names.

Evaluation Metrics. There are several standard methods for comparing two permutations, such as Kendall's τ and Spearman's footrule [20]. In this work, we measure the correlation between the ranking of a heuristic and a ground truth ranking using Kendall's τ and the heuristic's ranking accuracy concerning the ground truth using top-k precision. These approaches have in fact been used in recent papers [10, 19].

Kendall's τ is computed based on the number of concordantly ordered pairs of items between the two lists. A coefficient of zero means "no association" between the rankings and a value of +1.0 or −1.0 means "perfect agreement" or "perfect inverse agreement", respectively.

Top-k precision calculates the percentage of shared items among the top-k ranked ones in each ranking. We select this metric because reducing the total of candidates to a manageable number is a necessary step.

9.4.2 Results

To show the degree of similarity between actual and predicted rankings, we test the null hypothesis that τ is zero (no agreement) against the alternative that it is non-zero. A p-value less than 0.01 means statistically significant and indicates strong evidence against the null hypothesis.

From Table 9.2 we can note a weak or moderate, positive monotonic correlation between the ranking predicted by a heuristic and the observed. On the other hand, as shown in Fig. 9.2, there is zero agreement ($\tau = 0$) between a random ranking and the observed.

Table 9.2: Kendall's τ correlation coefficient for each heuristic (p-value ≤ 0.01)

Heuristic	Kendall's τ	Correlation
H1	0.38*	weak
H2	0.47*	moderate
H3	0.36*	weak
H4	0.25*	weak
H5	0.34*	weak
H6	0.21*	weak

* [.00-.19] very weak, [.20-.39] weak, [.40-.59] moderate, [.60-.79] strong, [.80-1.0] very strong

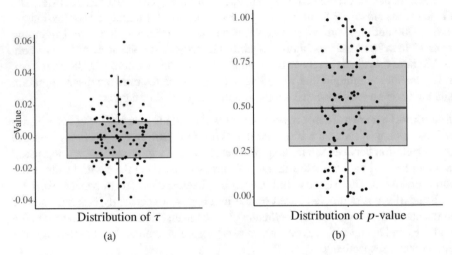

Fig. 9.2: Distribution of p-values and Kendall' τ coefficients for 100 comparations of rankings- a random and the observed (by Q)

We observe from Fig. 9.3 that the random choice of candidates resulted in a top 30% precision less than 40% (on average 30%-150 candidates). As expected, its precision is inferior to any one of our heuristics (Table 9.3 - rightmost column).

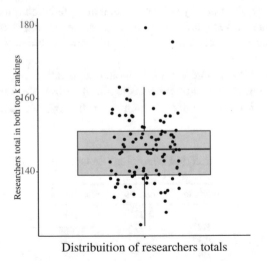

Fig. 9.3: Distribution of researchers totals among the top-30% ranked researchers in each ranking- the random and the observed

Table 9.3: Top-30% precision for each heuristic

Heuristic	Present in both rankings		Present only in year 1		Top-30% precision
	Group A*	Group B**	Group A	Group B	
H1	277	6	106	101	58
H2	287	29	96	78	64
H3	255	10	128	97	54
H4	207	39	176	68	50
H5	261	0	122	107	53
H6	195	15	188	92	43

* Researchers with h-index > 1.93 (mean) in year 5,
** Researchers with h-index ≤ 1.93 in year 5, Total of applicants 1601.

We note from Fig. 9.4 that the top-k (k = 30%) precision of the heuristics in group B is relatively much less than in group A. Group A includes researchers with h-index > *mean*, whereas Group B h-index ≤ *mean*. The disparity suggests that all indicators tested in the heuristics discriminate against researchers in group B to some degree.

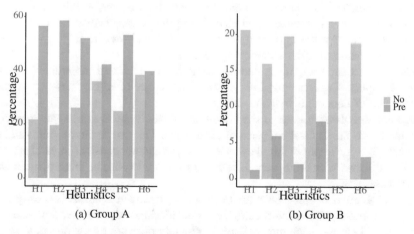

(a) Group A   (b) Group B

Fig. 9.4: Top-k precision rate per Group. In the dark green color is the top-k precision (Pre) and the light green color is its complement (No)

9.4.3 Discussion

The number of applicants vastly outnumbers the available academic faculty positions. How to identify a prominent researcher before becoming one is unclear. Traditionally, evaluators have looked to past success as a principal indicator of future success without considering systemic barriers. In this context, to foresee whether young scientists will publish successfully over their careers can assist hiring committees to perform their tasks.

In this study, we propose a novel fast-and-frugal heuristics approach to identify prominent researchers before becoming one. Our machine-predictions-based heuristics approach outperforms those based on traditional indicators, such as the h-index. Our experiments are consistent with previous results [1, 3, 23, 29, 33, 36].

However, it is perilous to make any causal interpretations of these results. Yet, we can speculate that what may be behind the higher performance of prediction-based heuristics is related to the potential of machines in predicting identifiable patterns. More in-depth and focused research would be needed to confirm that.

Also, we note that junior researchers with lower performance than average are much less likely than others to being noticed by any algorithms as prominent researchers in the future. This finding is consistent with previous studies [15, 21, 37] that have found unintentional bias in hiring introduced by algorithms and people. As a result, committees are indispensable due to the threats of unfairness and implicit discrimination.

On the other hand, some limitations may probably have influenced our results. The first limitation is due to the horizontal approach taken in the data set regarding the time dimension. Researchers in the 5th year of their career, working in different decades, were equally treated when considering the various heuristics. This choice ignores the fact that the current publishing behavior differs from the past. These differences could affect the accuracy of the heuristics.

One second shortcoming is that our sample only included researchers with at least 40 papers. Consequently, this may exclude cases of researchers that started strongly but did not confirm the tendency.

A third drawback relates to the condition of having at least one paper with a co-author who has at least ten years of experience. This constraint may also add to our analysis some implicit "Matthew effect" as it would not consider researchers that mainly published with their peers. However, this condition can be relaxed, but it was not the case here.

Lastly, whether or not a scientist becomes a prominent scientist is mostly predictable by their publication record, even considering only the first few years of publication. Our approach may not solely alleviate pressure on appraisers to make decisions quickly but mainly uncertainties arising from nonobjective decisions.

9.5 Conclusion

Assessing candidates for faculty positions requires an accurate prediction of their latent future performance. Through future promise-based heuristics, such as a faculty member's estimated future citation rate of a given paper, scientists and evaluators may reduce the efforts and time for decision making in recruitment processes. In this paper, we have tested the performance of these heuristics against past performance-based heuristics.

Our empirical evidence has led us to consider that, in recruitment, evaluating researchers based on future-promise is better than based solely on past performance. While the latter pressures researchers to continuously produce publishable results, the first decreases it. We also have found a novel heuristic for identifying talented researchers with lower than average h-indexes. However, further work needs to be done to improve accuracy.

On the other hand, predicting the future promise of good researchers is already hard. Nevertheless, predicting the progress of below-average researchers is more challenging than ever.

Future work should concentrate on enhancing the quality of heuristics and test other possible future-impact-based heuristics, e.g., future h-index, the future Q of the potential academic collaborator, and novel proposed impact prediction. We hope that our research will serve as a base for future studies focused on diversity, inclusion, and social justice, in the context of assessing the future promise of candidates for funding, hiring, and promotion.

Acknowledgements The authors would like to thank the anonymous reviewers for the critical comments and valuable suggestions. Antônio de Abreu Batista-Jr would like to thank Maranhão Research Foundation (FAPEMA) for the supporting grant #BD-08792/17. Fábio Castro Gouveia would like to thank the National Council for Scientific and Technological Development (http://www.cnpq.br) for the Scholarship in Research Productivity 2 (315521/2020-1) and the Universal Grant (430982/2018 6).

References

1. de Abreu Batista-Jr, A., Gouveia, F.C., Mena-Chalco, J.P.: Predicting the q of junior researchers using data from the first years of publication. Journal of Informetrics 15(2), 101130 (2021)
2. Abrishami, A., Aliakbary, S.: Predicting citation counts based on deep neural network learning techniques. Journal of Informetrics 13(2), 485–499 (2019)
3. Acuna, D.E., Allesina, S., Kording, K.P.: Predicting scientific success. Nature 489(201), 201–202 (2012)
4. Aroeira, R.I., A.R.B. Castanho, M.: Can citation metrics predict the true impact of scientific papers? The FEBS Journal 287(12), 2440–2448 (2020)
5. Ayaz, S., Masood, N., Islam, M.A.: Predicting scientific impact based on h-index. Scientometrics 114(3), 993–1010 (2018)
6. Bai, X., Zhang, F., Lee, I.: Predicting the citations of scholarly paper. Journal of Informetrics 13(1), 407–418 (2019)
7. Bornmann, L., Hug, S.: Bibliometrics-based heuristics: What is their definition and how can they be studied? - Research note. Profesional de la Información 29(4) (2020)
8. Chen, J., Wang, X., Zhao, S., Zhang, Y.: Content-enhanced network embedding for academic collaborator recommendation. Complexity 2021, 7035467 (2021)
9. Clauset, A., Larremore, D.B., Sinatra, R.: Data-driven predictions in the science of science. Science 355(6324), 477–480 (2017)
10. Fagin, R., Kumar, R., Sivakumar, D.: Comparing top k lists. SIAM Journal on Discrete Mathematics 17(1), 134–160 (2003)

11. Fernandes, J.D., Sarabipour, S., Smith, C.T., Niemi, N.M., Jadavji, N.M., Kozik, A.J., Hole-house, A.S., Pejaver, V., Symmons, O., Bisson Filho, A.W., Haage, A.: A survey-based analysis of the academic job market. eLife **9**, e54097 (2020)
12. Frank, M.C.: n-best evaluation for academic hiring and promotion. Trends in Cognitive Sciences **23**(12), 983–985 (2019)
13. García-Pérez, M.A.: Limited validity of equations to predict the future h index. Scientometrics **96**(3), 901–909 (2013)
14. Garfield, E.: Journal impact factor: A brief review. Canadian Medical Association Journal **161**(8), 979–980 (1999)
15. Gibbs Kenneth D, J., Basson, J., Xierali, I.M., Broniatowski, D.A.: Research: Decoupling of the minority PhD talent pool and assistant professor hiring in medical school basic science departments in the US. eLife **5**, e21393 (2016)
16. Hirsch, J.E.: An index to quantify an individual's scientific research output. Proceedings of the National Academy of Sciences **102**(46), 16569–16572 (2005)
17. Holden, G., Rosenberg, G., Barker, K.: Bibliometrics: A potential decision making aid in hiring, reappointment, tenure and promotion decisions. Social Work in Health Care **41**(3–4), 67–92 (2005)
18. Hou, J., Pan, H., Guo, T., Lee, I., Kong, X., Xia, F.: Prediction methods and applications in the science of science: A survey. Computer Science Review **34**, 100197 (2019)
19. Kanellos, I., Vergoulis, T., Sacharidis, D., Dalamagas, T., Vassiliou, Y.: Impact-based ranking of scientific publications: A survey and experimental evaluation. IEEE Transactions on Knowledge & Data Engineering **33**(4), 1567–1584 (2021)
20. Kendall, M., Gibbons, J.D.: Rank Correlation Methods, 5 edn. Charles Griffin (1990)
21. Köchling, A., Wehner, M.C.: Discriminated by an algorithm: A systematic review of discrimination and fairness by algorithmic decision-making in the context of HR recruitment and HR development. Business Research **13**(3), 795–848 (2020)
22. Kong, X., Wen, L., Ren, J., Hou, M., Zhang, M., Liu, K., Xia, F.: Many-to-many collaborator recommendation based on matching markets theory. In: Proceedings of the IEEE International Conferences on Dependable, Autonomic & Secure Computing, on Pervasive Intelligence & Computing, on Cloud & Big Data Computing, on Cyber Science & Technology Congress (DASC/PiCom/CBDCom/CyberSciTech), pp. 109–114 (2019)
23. Laurance, W.F., Useche, D.C., Laurance, S.G., Bradshaw, C.J.A.: Predicting publication success for biologists. BioScience **63**(10), 817–823 (2013)
24. Lee, D.H.: Predicting the research performance of early career scientists. Scientometrics **121**(3), 1481–1504 (2019)
25. Penner, O., Pan, R.K., Petersen, A.M., Kaski, K., Fortunato, S.: On the predictability of future impact in science. Scientific Reports **3**(1), 3052 (2013)
26. Peters, G.: Why not to use the Journal Impact Factor as a criterion for the selection of junior researchers: A comment on Bornmann and Williams (2017). Journal of Informetrics **11**(3), 888–891 (2017)
27. Revesz, P.Z.: A method for predicting citations to the scientific publications of individual researchers. In: Proceedings of the 18th International Database Engineering & Applications Symposium (IDEAS), pp. 9–18 (2014)
28. Ruan, X., Zhu, Y., Li, J., Cheng, Y.: Predicting the citation counts of individual papers via a BP neural network. Journal of Informetrics **14**(3), 101039 (2020)
29. Sarigöl, E., Pfitzner, R., Scholtes, I., Garas, A., Schweitzer, F.: Predicting scientific success based on coauthorship networks. EPJ Data Science **3**(1), 9 (2014)
30. Schweitzer, F.: Scientific networks and success in science. EPJ Data Science **3**(1), 35 (2014)
31. Sinatra, R., Wang, D., Deville, P., Song, C., Barabási, A.L.: Quantifying the evolution of individual scientific impact. Science **354**(6312) (2016)
32. Sugimoto, C.R.: Scientific success by numbers. Nature **593**(7857), 30–31 (2021)
33. van Dijk, D., Manor, O., Carey, L.B.: Publication metrics and success on the academic job market. Current Biology **24**(11), R516–R517 (2014)
34. Van Eck, N., Waltman, L.: Software survey: VOSviewer, a computer program for bibliometric mapping. Scientometrics **84**(2), 523–538 (2010)

35. Wang, D., Song, C., Barabási, A.L.: Quantifying long-term scientific impact. Science **342**(6154), 127–132 (2013)
36. Weihs, L., Etzioni, O.: Learning to predict citation-based impact measures. In: Proceedings of the 17th ACM/IEEE Joint Conference on Digital Libraries (JCDL), pp. 1–10 (2017)
37. Wright, C.B., Vanderford, N.L.: What faculty hiring committees want. Nature Biotechnology **35**(9), 885–887 (2017)
38. Yang, L., Zhang, Z., Cai, X., Guo, L.: Citation recommendation as edge prediction in heterogeneous bibliographic network: A network representation approach. IEEE Access **7**, 23232–23239 (2019)
39. Yudhoatmojo, S.B., Samuar, M.A.: Community detection on citation network of DBLP data sample set using LinkRank algorithm. Procedia Computer Science **124**, 29–37 (2017)
40. Zhang, Y., Yu, Q.: What is the best article publishing strategy for early career scientists? Scientometrics **122**(1), 397–408 (2020)

Chapter 10
Scientific Impact Vitality: The Citation Currency Ratio and Citation Currency Exergy Indicators

Gangan Prathap and Dimitrios Katsaros and Yannis Manolopoulos

Abstract Publication and impact measures for individual scientists are used worldwide for various purposes such as funding, promotion, and consequently, the development of such indicators comprises a fertile research area. In this article, we propose two simple citation-based indicators, a dimensionless *citation currency ratio* (CCR) and a size-dependent *citation currency exergy* (CCX) which are noncumulative indicators measuring current citation performance. These help to identify scientists who are at different stages of their career, with rising, steady, or fading visibility. Using a small-scale coherent sample of scientists from DLPB, Google Scholar and the recently published so-called Stanford[1] list [8], we demonstrate the applicability of these ideas and show that our methods provide substantial promise for future use.

10.1 Introduction

Like old soldiers, old scientists never die; their works' impact just fades away.
The quest for scientometric indicators which capture significant aspects of individual scientists performance keeps consistently growing the last fifteen years because of

Gangan Prathap [1,2]
[1] A. P. J. Abdul Kalam Technological University, Thiruvananthapuram, 695016 Kerala, India
[2] Institute of Computer Science & Digital Innovation, UCSI University, Kuala Lumpur, Malaysia
e-mail: gangan_prathap@hotmail.com

Dimitrios Katsaros
Department of Electrical & Computer Engineering University of Thessaly, 38221 Volos, Greece
e-mail: dkatsar@e-ce.uth.gr

Yannis Manolopoulos [1,2]
[1] School of Pure & Applied Sciences, Open University of Cyprus, 2210 Nicosia, Cyprus
[2] Department of Informatics, Aristotle University, 54124 Thessaloniki Greece
e-mail: yannis.manolopoulos@ouc.ac.cy, manolopo@csd.auth.gr

[1] It is for ease of reference only that this dataset is called "Stanford"; It is clear that it has not been endorsed by any means from Stanford authorities.

© The Author(s), under exclusive license to Springer Nature Switzerland AG 2021   209
Y. Manolopoulos, T. Vergoulis (eds.), *Predicting the Dynamics of Research Impact*,
https://doi.org/10.1007/978-3-030-86668-6_10

the availability of big scholarly data by bibliographic databases, and also because of their use in bureaucratic decision-making, despite criticism against such practises. Several hundreds of indicators for quantifying an individual scientist's performance have been proposed the last decade which use more and more detailed information concerning publication number, citation number citation networks, paper age, citation age, co-authorship networks, centrality measures, and so on. The great majority of these indicators are cumulative, i.e., they never decline; thus they favor those scientists of greater "scientific age", and/or they can not distinguish the scientists whose performance in terms of citation-based impact is getting significant in the present, and/or are rising stars.

Here, we seek to develop indicators that will capture the performance of a scientist whose citation-based impact per year is getting (gradually) bigger; to describe it in different words, we aim at developing indicators which can detect that the scientist's impact on an annual basis is not *fading away*. We term this characteristic of a scientist's impact evolution as the *vitality* of his/her research. This notion is different that the concept of rising star [4] or trendsetter [22]. Therefore, our goal is to develop scientometric indicators to capture the vitality of scientific impact, and in particular we are interested in developing indicators which will (practically) be noncumulative, and also they will be easy to calculate.

In this context, we propose two simple to calculate indicators, namely a dimensionless *citation currency ratio* (CCR), and a size-dependent *citation currency exergy* (CCX) to capture these trends quantitatively. We study a coherent cohort of scientists to see how this happens across the board. We use data from DPLB and Google Scholar as well as the recently published Stanford list [8], to demonstrate the applicability of these ideas. The obtained results are encouraging showing promise that these indicators can be used in the future to capture the aforementioned aspect of citation-based performance of a scientist.

The rest of this article is organized as follows: Sect. 10.2 surveys the related work. Sect. 10.3 reviews some basic definitions from past literature and develops the new scientometric indicators. Sect. 10.4 presents two case-studies to assert the usefulness of the new indicators, and finally Sect. 10.5 concludes the article.

10.2 Related Work

During the last fifteen years the research efforts pertaining to scientometric indicators have been growing steadily, especially after the development of large bibliographic databases, such as Google Scholar, Microsoft Academic, Elsevier Scopus, etc. At that time, the introduction of the Hirsch h-index [7] was a path-breaking idea that among other scientific consequences, it popularized the concept of scientometric indicators. The work in [13] offers solid proof that the research pertaining to the h-index has been growing after 2005 (besides some small slowing down in 2014 and in 2017), and this is the general trend with other indicators as well. Surveys of such indicators can be found in [24, 25]. Since the focus of the present article is about the indicators capturing the aggregated citations' dynamics (i.e., *evolution in time*) of scientists, in the rest of this section, we will survey only such indicators.

The goal of identifying young (in terms of career) scientists who currently have relatively low profiles, but may eventually emerge as prominent scientists – the so-called *rising star scientists* – is a recent, challenging, hot topic, and it is extended to other social media, e.g., in geo-social networks [14]. *ScholarRank* was proposed in [27] which considers the citation counts of authors, the mutual influence among coauthors and the mutual reinforce process among different entities in academic networks to infer who are expected to increase their performance dramatically. In [16] several performance indicators are compiled for each scholar and monitored over time; then these indicators are used in a clustering framework which groups the scholars into categories. *StarRank* was proposed in [15] which is an improved *PageRank* [12] method to calculate the initial "promise" of rising stars, and then this "promise" is diffused in an artificial social network constructed via explicit and implicit links, and finally a prediction of a scholars' ranking in the future is conducted. Departing from existing methods for rising star detection, which explore the co-author networks or citation networks, the work reported in [3] uses an article's textual content, and [6] uses a form of similarity clustering of scholars. Finally, [4] presents a survey on rising star identification by classifying existing methods into ranking-, prediction-, clustering-, and analysis-based methods, and also discusses the pros and cons of these methods.

Another group of indicators that is related to the present work includes those indicators which apply an aging mechanism to citations and articles to identify the scientists whose performance degrades in time. Two popular indices of this group are the *trend h-index* and *contemporary h-index* [10, 22] which however have been described in the context of the *h*-index mechanism, capturing both productivity and impact. Nevertheless, the trend/contemporary *h*-index's idea of citation/paper aging can be applied in isolation to a scientist's citation set. Along, the same line but slightly different are the *Optimum-time* and *Speed of withdrawal* [2]. Finally, the idea of measuring by approximation the citation acceleration [26] considers also temporal aspects, and thus it is partially relevant to the present work.

Moreover, our work is only remotely related to identification of sleeping beauties in science [11] and the subsequent work on the topic concerning the rediscovering paper of a sleeping beauty [23], how sleeping beauties get cited in patents [21], the derivative analysis of sleeping beauties' citation curves [5], and case studies [1] because those works deal with individual article's temporal citation performance.

Almost all aforementioned works are based on extensive citation network analysis, and/or exploitation of the timestamp of each citation. Our present work departs from these practices, and it is based only on citation information received at a specific year and on total citations received up to this year, thus making the calculation of the proposed indicators far more less computationally expensive.

10.3 The New Scientometric Indicators

We will start by presenting some scientometric indicators from earlier works which are necessary for defining the new ones. So firstly, we will provide the definitions of energy and exergy, and in the next subsection, we will introduce the *citation currency ratio* and *citation currency exergy*. To facilitate our presentation, we will make use of the symbols reported in Table 10.1.

In search of a single number indicator able to adequately describe the whole performance of a scientist, the concepts of energy, exergy and entropy were introduced in [19] borrowing those terms from thermodynamics. When fractional citation counting [9] is necessary, these concepts can be extended appropriately [17]. So, the energy of a single paper which has received c_i citations is equal – by definition – to c_i^2. We recall the following definitions:

Definition 1 (from [19]) The Energy of a set of publications $P = \{p_1, p_2, \ldots, p_n\}$ (i.e., $||P|| = n$), where the article p_i has received c_i citations is defined as follows:

$$E \equiv \sum_{i=1}^{||P||} c_i^2. \tag{10.1}$$

Definition 2 (from [19]) The Exergy of a set of publications $P = \{p_1, p_2, \ldots, p_n\}$ (i.e., $||P|| = n$), where the article p_i has received c_i citations is defined as follows:

$$X = \frac{1}{||P||} \left(\sum_{i=1}^{||P||} c_i \right)^2 = \frac{C^2}{||P||}. \tag{10.2}$$

Table 10.1 The set of symbols used throughout the article. The above quantities are assumed to have been measured during a specific time window, e.g., from 1996 to 2019

Definition	Symbol
Set of articles published by a scientist	P
Cardinality of the set of articles published by a scientist	$||P||$
Citations received by the i-th article	c_i
Total number of citations received by all P articles	C
Citations received only during year yr (by articles published until year yr)	^{yr}c e.g., ^{2015}c
Total (cumulative) citations received up until year yr (by articles published until year yr)	^{yr}C e.g., ^{2019}C
Energy of a set of publications (e.g., of P)	E
Exergy of a set of publications (e.g., of P)	X
Entropy of a set of publications (e.g., of P)	S

Apparently, it holds that $E > X$. The difference is called *entropy (S)* in [19], i.e., $S = E - X$.

The exergy X is an excellent single, scalar measure of a scientist's performance, especially when only aggregate information, i.e., C and $||P||$ is available. The entropy on the other hand measures how much "uneven" (unordered) is the scientist's publication portfolio; it measures whether the published articles of a scientist continue to attract citations in the course of time or if his/her impact is "fading away".

10.3.1 The Citation Currency Ratio and Citation Currency Exergy Indicators

The astute reader will have realized by now that in our scientometric world, exergy corresponds to the portion of "impact flow", and thus it is a measure of how the impact of a scientist degrades (decays) over the years. However, exergy is defined using cumulative citation performance, and just like the raw measures ^{yr}c and ^{yr}C tends to be a proxy for cumulative impact. Both ^{yr}c and ^{yr}C vary by orders of magnitude depending on the citation intensity of the field or subfield, and on the size of collaboration. Over a lifetime, there are scientists who have collaborated with large numbers of co-workers, whereas others have worked in much smaller teams.

To deal with these issues and introduce some notion of fairness, we introduce the *Citation Currency Ratio* defined as follows:

Definition 3 The *Citation Currency Ratio (CCR)* of a scientist in a specific year yr is the ratio of the citations ^{yr}c received during that year to the cumulative number ^{yr}C of citations until that year, i.e.,:

$$^{yr}CCR = \frac{^{yr}c}{^{yr}C},\qquad(10.3)$$

or dropping year information when it is clear at which year it is being calculated:

$$CCR = \frac{c}{C}.\qquad(10.4)$$

CCR is a dimensionless parameter; it starts from a value of 1 and never exceeds this number, diminishing to zero as the portfolio fades away. This is a welcome feature in that at the very beginning of a scientific career one would start with a CCR value of 1, whereas at the very end, when the portfolio of work ceases to gather citations, it becomes zero. Note that at each stage, CCR is like a size-independent quality term; using the quality-quasity-quantity terminology from [18], c is a quasity term, and C is a quantity term.

Starting from this new indicator, we propose a second-order composite exergy term [20] defined as follows:

Definition 4 The *Citation Currency Exergy (CCX)* of a scientist in a specific year yr is the product of the CCR times the number of citations ^{yr}c received during that year, i.e.,:

$$^{yr}CCX = {}^{yr}CCR \times {}^{yr}c \implies {}^{yr}CCX = \frac{{}^{yr}c^2}{{}^{yr}C}, \tag{10.5}$$

or dropping year information when it is clear at which year it is being calculated:

$$CCX = CCR \times c \implies CCX = \frac{c^2}{C}. \tag{10.6}$$

This is a size-dependent term and gives a relative measure of current activity in comparison with cumulative activity.

Along the same lines of methodology as previously, we can also compute currency indicators for the h-index and for $i10$. We use the 5-year values of these as the "recent" metrics. In each case we get the ratios, hCR and $i10R$, and the respective exergies hCX, and $i10CX$.

10.3.1.1 Example of CCR and CCX

To get a glimpse of these indicators' behaviour we selected four scientists from one of our datasets, which will be presented in detail in Sect. 10.4.1. Table 10.2 displays the names of these scientists from the list of 219 Greek prolific authors. Georgios Giannakis (GG) is the most highly cited; Christos Davatzikos (CD) occupies a middle position, whereas Iakovos Venieris (IV) and Agathoniki Trigoni (NT) have a modest position of all scientists with more than 150 publications in this list.

Table 10.3 displays year-wise citations c of these four scientists. For each scientist, at each instant, c is the proxy for current citation impact. For each scientist, cumulative citations C can be computed as shown in Table 10.3, for the window up to this instant. Table 10.3 as well as Figs. 10.1-10.2 show how CCR and CCX indicators evolve for the four scientists, who are at noticeably different stages of their scientific career. A younger promising scientist, a rising star, tends to have a higher CCR. An older scientist who is a "fading" star will tend to lower CCR. Both c and C differ considerably, and this is compounded in the composite indicator CCX. These values can rise as for CD, or fall, rise and fall, as for GG and IV, as seen in Figs. 10.1 and 10.2. IV's CCR peaked in 2000 and CCX in 2002. In the case of GG, CCR peaked in 2005 and CCX in 2006.

Table 10.2 Four scientists from the list of prolific Greek scientists in the DPLB at different positions in the performance spectrum taken from Google Scholar Citations. Undefined variables will be clarified in Sect. 10.4.1

Author name	$\|\|P\|\|$	C	h	$i10$	C_5	h_5	$i10\_5$
Georgios Giannakis	972	77789	145	718	24863	76	412
Christos Davatzikos	282	45420	106	387	24293	72	311
Agathoniki Trigoni	166	5734	40	97	4063	33	78
Iakovos Venieris	166	3132	26	84	713	13	21

Table 10.3 Year-wise citations c since 1996 of the four scientists, computation of the cumulative citations C, the citation currency ratio CCR, and the citation currency exergy CCX

Year	CDavatzikos				IVenieris				GGiannakis				ATrigoni			
	c	C	CCR	CCX	c	C	CCR	CCX	c	C	CCR	CCX	c	C	CCR	CCX
1996					42	42	1.000	42.0	386	386	1.000	386.0				
1997					38	80	0.475	18.1	558	944	0.591	329.8				
1998	148	148	1	148	52	132	0.394	20.5	473	1417	0.334	157.9				
1999	186	334	0.55	103.58	37	169	0.219	8.1	566	1983	0.285	161.6				
2000	244	578	0.42	103.00	74	243	0.305	22.5	694	2677	0.259	179.9				
2001	281	859	0.32	91.92	96	339	0.283	27.2	877	3554	0.247	216.4				
2002	383	1242	0.30	118.10	128	467	0.274	35.1	1160	4714	0.246	285.4				
2003	446	1688	0.26	117.84	134	601	0.223	29.9	1522	6236	0.244	371.5				
2004	594	2282	0.25	154.61	160	761	0.210	33.6	2245	8481	0.265	594.3				
2005	762	3044	0.21	190.75	156	917	0.170	26.5	3108	11589	0.268	833.5	33	33	1	33.0
2006	852	3896	0.21	186.32	150	1076	0.141	21.1	3831	15420	0.248	951.8	70	103	0.679	47.5
2007	1048	4944	0.19	222.15	183	1250	0.146	26.8	4222	19642	0.215	907.5	80	183	0.437	34.9
2008	1174	6118	0.18	225.28	145	1395	0.104	15.1	4543	24185	0.188	853.4	84	267	0.314	26.4
2009	1401	7519	0.16	261.04	168	1563	0.107	18.1	4391	28576	0.154	674.7	129	396	0.325	42.0
2010	1463	8982	0.17	238.29	148	1711	0.086	12.8	4447	33023	0.135	598.8	166	562	0.295	49.0
2011	1868	10850	0.15	321.60	143	1854	0.077	11.0	4165	37188	0.112	466.5	158	720	0.219	34.6
2012	2008	12858	0.14	313.58	134	1988	0.067	9.0	4321	41509	0.104	449.8	202	922	0.219	44.2
2013	2133	14991	0.14	303.49	159	2147	0.074	11.8	4549	46058	0.099	449.3	207	1129	0.183	37.9
2014	2553	17544	0.13	371.51	137	2284	0.060	8.2	4738	50796	0.093	441.9	261	1390	0.187	49.0
2015	2852	20396	0.14	398.78	128	2412	0.053	6.8	4106	54902	0.075	307.1	399	1789	0.223	88.9
2016	3522	23918	0.14	518.62	107	2519	0.042	4.5	4374	59276	0.074	322.8	335	2124	0.157	52.8
2017	4035	27953	0.14	582.45	126	2645	0.048	6.0	4255	63531	0.067	285.0	452	2576	0.175	79.3
2018	4658	32611	0.14	665.32	127	2772	0.046	5.8	4219	67750	0.062	262.7	720	3296	0.218	157.2
2019	5476	38087	0.14	787.31	130	2902	0.045	5.8	4305	72055	0.060	257.2	1100	4396	0.250	275.2

10.4 Case Studies of Greek and Indian Scientists

In this section we will first present in detail the datasets we have collected over which we will evaluate the appropriateness of the proposed indicators, and then we will show the actual experimental results.

10.4.1 Datasets Collection and Measured Indicators

To showcase the usefulness of the proposed indicators, we created two datasets; one with scientists with Greek origin and the second with Indian scientists. The first dataset is curated by identifying the prolific authors from DBLP[2] which have Greek last names. They can be citizens of Greece, of Cyprus, or of any other country such as USA, UK, Australia, or elsewhere, affiliated with any academic or research institution in Greece, Cyprus and so on. They all have a record of at least 150 publications according to DBLP. In passing, publications according to DBLP can be journal or conference papers, as well as books edited or authored, technical reports

---

[2] https://dblp.org/statistics/prolific1.html

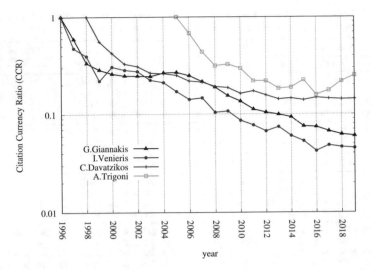

Fig. 10.1 Evolution of *CCR*s of four sampled Greek scientists

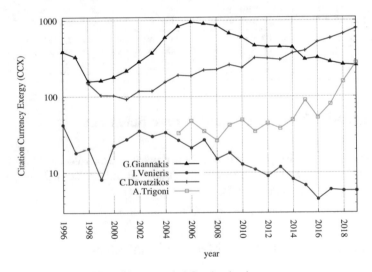

Fig. 10.2 Evolution of *CCX*s of four sampled Greek scientists

etc[3]. For each author of this set, we checked their profile in Google Scholar (GS). Those without GS profiles were removed from further consideration, and, thus, a

---

[3] We mention that these "publications" contain many CoRR articles, which comprise unreferred works uploaded to arxiv.org. In recent years this duplication of articles in DBLP is quite significant, accounting for some authors even to 20% of their total number of publications.

set of 219 prolific Greek researchers has been gathered (curation date: the 17th of December 2020)[4].

For each GS profile, we can retrieve the citations in all years (C) to all publications (P), and also a so-called "recent" value of this metric (C_5), which is the number of citations to all publications in the past five years. Also, we can get the well-known h-index [7], the so-called h_5, which is the h-index which is the largest number h such that h publications in the past five years have at least h citations. The $i10$-index is the number of publications which have at least 10 citations, whereas $i10\_5$ is the corresponding value for the last 5 years.

The second dataset uses the Indian academic space as an example. The latest Stanford list [8], gives two separate lists which cover the cumulative and "recent" windows. The citation data from Scopus is frozen as of May 6, 2020. The Single Year list (Table-S7-singleyr-2019) gives the citation impact, nc1919(ns) during the single calendar year 2019 for more than 160000 scientists from a ranking list of more than 6 million scientists. This corresponds to the "recent" citation value c for our calculations. The Career list (Table-S6-career-2019) assesses scientists for career-long citation impact, nc9619(ns) from 1996 until the end of 2019 for nearly 160000 scientists from a ranking list of nearly 3 million scientists. This corresponds to our cumulative citations C from 1996 to 2019. The qualifier (ns) indicates that self-citations have been excluded. From this, the citation currency ratio CCR and the citation currency exergy CCX can be computed for every scientist who appears simultaneously in both lists. The Single Year list gives many rising stars (i.e., they are here but absent in the Career list). Similarly, from the Career list, we can also see the fading (sinking) stars, where they are absent from the Single Year list. Many names remain common to both lists, and it is only for this cohort that we can compute CCR and CCX. All updated databases and code are made freely available in Mendeley[5].

We will focus our attention only on the names from India's premier research intensive Higher Educational Institution, namely the Indian Institute of Science at Bengaluru, India which appear in the Stanford list [8]. There are 97 scientists in the Single Year list, 94 in the Career list and only 65 in the common list. It is this last cohort which we shall use to continue our demonstration. Table 10.4 is an extract showing the top-20 scientists from the Indian Institute of Science, Bengaluru, India from the so-called Stanford list [8]. The actual names are shown as they offer a reality check and there are no surprises. The Stanford list also makes available for each scientist the respective year of first publication (firstyr) and year of most recent publication (lastyr). We shall use this to show how the citation indicators evolve with year of first publication, which we take as the date of launch of the scientific career.

---

[4] The dataset is publicly available upon request.

[5] https://dx.doi.org/10.17632/btchxktzyw

Table 10.4 The top 20 scientists from the Indian Institute of Science, Bengaluru, India from the so-called Stanford list [8]

Author name	firstyr	lastyr	c	C	CCR	CCX
Desiraju, Gautam R.	1977	2020	3098	34086	0.091	281.57
Sood, A. K.	1979	2020	2035	15555	0.131	266.23
Ramaswamy, Sriram	1979	2020	1097	7133	0.154	168.71
Sarma, D. D.	1980	2020	1420	12662	0.112	159.25
Munichandraiah, N.	1981	2019	1118	7432	0.150	168.18
Raj, Baldev	1982	2018	1591	10199	0.156	248.19
Gopakumar, K.	1984	2020	1183	7163	0.165	195.38
Sitharam, T. G.	1987	2020	612	2390	0.256	156.71
Ghose, Debasish	1988	2019	826	4506	0.183	151.42
Shivakumara, C.	1991	2020	757	3719	0.204	154.09
Ramachandra, T. V.	1992	2020	667	2744	0.243	162.13
Madras, Giridhar	1993	2020	2435	14247	0.171	416.17
Nanda, Karuna Kar	1994	2020	819	4000	0.205	167.69
Kumar, Jyant	1994	2020	561	1916	0.293	164.26
Suwas, Satyam	1996	2020	1173	3820	0.307	360.19
Somasundaram, Kumaravel	1996	2019	1053	5995	0.176	184.9 6
Basu, Bikramjit	1998	2020	1437	6398	0.225	322.7
Mukherjee, Partha Sarathi	2000	2020	1334	8657	0.154	205.56
Biju, Akkattu T.	2003	2020	970	5042	0.192	186.61
Barpanda, Prabeer	2004	2020	717	2971	0.241	173.04

10.4.2 Case Study 1: Application of the New Indicators to Prolific Greek Scientists

Using the dataset with the prolific Greek scientists, we aim to show the dynamics of their profile by means of currency indicators. In particular, we will calculate the dimensionless *citation currency ratio (CCR)* and a size-dependent citation currency exergy (CCX) to capture these trends quantitatively.

Figs. 10.3-10.5 show the dispersion of currency ratios CCR, hCR and $i10R$ and the respective values of C, h, and $i10$ for the 219 Greek prolific authors. Logarithmic scales are used and in each case the power trendlines are shown. We clearly see that in the case of citations and the h-index, at higher levels of C and h, indicating mature scientists, the CCR goes down perceptibly. However, such a signal is not evident in the case of the $i10$-index for reasons still under examination. We also see that in all cases, there is little correlation.

10.4.3 Case Study 2: Application of the New Indicators to Prolific Indian Scientists

Using the second dataset, namely with the Indian scientists we explore the usefulness of our proposed indicators. Fig. 10.6 shows the evolution of the citation currency ratio CCR for the 65 scientists from the Indian Institute of Science, Bengaluru, India from the Stanford list [8]. The exponential law trendline shows that younger scientists (by scientific age) have higher CCR. The egregious outlier seen in the figure is that of G.N. Ramachandran, whose scientific career spanned from 1942 to 1994 with

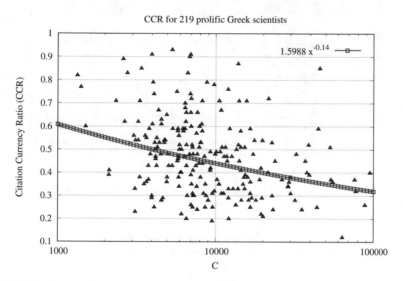

Fig. 10.3 The dispersion of CCR and the total citations C for the 219 prolific Greek scientists

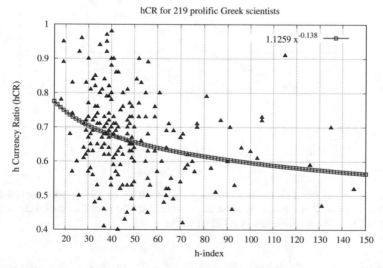

Fig. 10.4 The dispersion of Currency Ratio hCR and the h-index for the 219 prolific Greek scientists

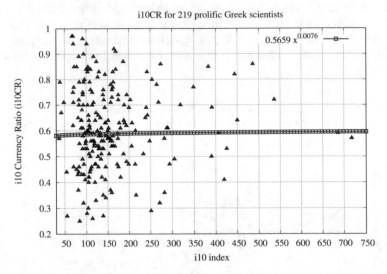

Fig. 10.5 The dispersion of Currency Ratio $i10CR$ and the $i10$-index for the 219 prolific Greek scientists

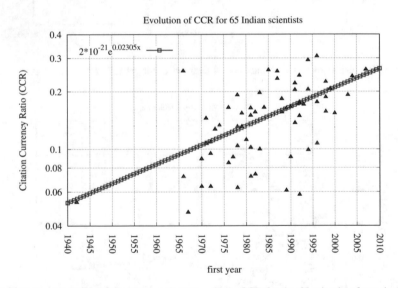

Fig. 10.6 The evolution of the Citation Currency Ratio CCR for the 65 scientists from the Indian Institute of Science, Bengaluru, India from the Stanford list. The powerlaw trendline shows that younger scientists (by scientific age) have higher CCRs

$c = 330, C = 6233, CCR = 0.053$, and $CCX = 17.47$. Twenty-five years on, his scientific legacy continues to attract citations.[6]

---

[6] Many in India believe that he was very unfortunately overlooked for a Nobel Prize.

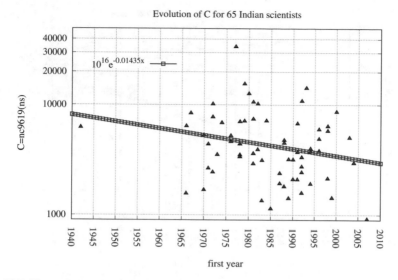

Fig. 10.7 The evolution of the size-dependent cumulative citations C for the 65 scientists from the Indian Institute of Science, Bengaluru, India from the Stanford list. The exponential law trendline shows that younger scientists (by scientific age) have lower cumulative citations C

Fig. 10.7 shows the evolution of the size-dependent cumulative citations, C for the 65 scientists. The exponential law trendline shows that younger scientists (by scientific age) have lower cumulative citations C. The highest performer here is Gautam R. Desiraju, with a career spanning from 1977 to 2020 and with $c = 3098, C = 34086, CCR = 0.091$, and $CCX = 281.57$.

Fig. 10.8 shows the evolution of the composite size-dependent cumulative citation currency exergy CCX for the 65 scientists. The exponential law trendline shows that younger scientists (by scientific age) have higher citation currency exergies CCX. The highest-ranking performer by this criterion is Giridhar Madras, with career spanning from 1993 to 2020, and $c = 2435, C = 14247, CCR = 0.171$, and $CCX = 416.17$. This is an excellent illustration of how current and cumulative activity go strongly together at this prime stage of a career.

Fig. 10.9 shows the dispersion of CCR with $C = nc9619(ns)$ for the cohort of 65 scientists. The power law trendline shows that there is a negative slope and correlation with C. This is to be expected from what we have seen earlier. Younger scientists have lower C and higher CCR but higher citation currency exergies CCX. The older scientists slowly fade away. G. N. Ramachandran remains an exception while many doyens of science from India of his era have now vanished from the Single Year lists.

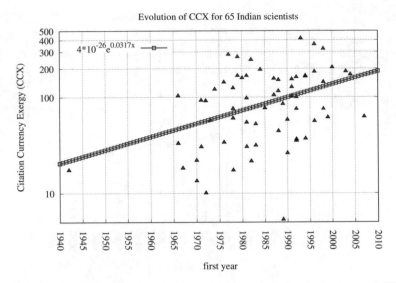

Fig. 10.8 The evolution of the composite size-dependent cumulative citation currency CCX for the 65 scientists from the Indian Institute of Science, Bengaluru, India from the Stanford list. The exponential law trendline shows that younger scientists (by scientific age) have higher citation currency exergies CCX

10.5 Concluding Remarks

Most scientists are active over long periods. Some are younger and some are at the end of their careers. The old truism about soldiers is valid: old scientists never die; their impact in terms of citations just fade away. In this paper we introduced two new indicators, a dimensionless *Citation Currency Ratio* (CCR) and a size-dependent *Citation Currency Exergy* (CCX) to capture these stages quantitatively. A welcome feature of CCR is that it starts from a value of 1 and never exceeds this and diminishes to zero as the portfolio fades away. Thus, at the very beginning of a scientific career one would start with a CCR value of 1 and at the very end, only when the portfolio of work ceases to gather citations, and not when the agent stops work, it becomes zero. We saw this in the case of G. N. Ramachandran, whose scientific career spanned from 1942 to 1994, that twenty-five years on, his scientific legacy continues to attract citations. We study cohorts of scientists to see how this happens across the board. We use data from DPLB and Google Scholar as well as another recently published dataset, the Stanford list [8], demonstrate the applicability of these ideas. These help to identify scientists who are at different stages of their career, with rising, steady, or fading visibility.

Acknowledgements The authors are grateful to suggestions and criticism from professors G. Madras and S. Gopalakrishnan of the Indian Institute of Science at Bengaluru for the reality check on the application to their institute.

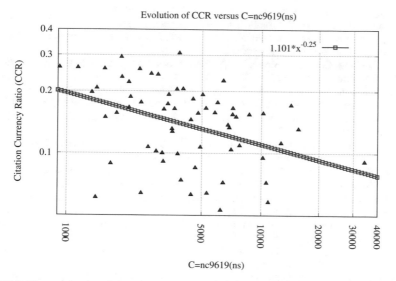

Fig. 10.9 The scatter plot of the dispersion of CCR with $C = nc9619(ns)$ for the 65 scientists from the Indian Institute of Science, Bengaluru, India from the Stanford list. The power law trendline shows that there is a negative slope and correlation with C. This is to be expected from what we have seen earlier. Younger scientists (by scientific age) have lower C and higher CCR, but higher citation currency exergies CCX

References

1. Bornmann, L., Ye, A., Ye, F.: Identifying "hot papers" and papers with "delayed recognition" in large-scale datasets by using dynamically normalized citations impact scores. Scientometrics **116**(2), 655–674 (2018)
2. Bouabid, H.: Revisiting citation aging: A model for citation distribution and life-cycle prediction. Scientometrics **88**(1), 199–211 (2011)
3. Daud, A., Abbas, F., Amjad, T., Alshdadi, A.A., Alowibdi, J.S.: Finding rising stars through hot topics detection. Future Generation Computer Systems **115**, 798–813 (2021)
4. Daud, A., Song, M., Hayat, M.K., Amjad, T., Abbasi, R.A., Dawood, H., Ghani, A.: Finding rising stars in bibliometric networks. Scientometrics **124**(1), 633–661 (2020)
5. Fang, H.: Analysing the variation tendencies of the numbers of yearly citations for sleeping beauties in science by using derivative analysis. Scientometrics **115**(2), 1051–1070 (2018)
6. Gogoglou, A., Sidiropoulos, A., Katsaros, D., Manolopoulos, Y.: A scientist's impact over time: The predictive power of clustering with peers. In: Proceedings of the International Database Engineering & Applications Symposium (IDEAS), pp. 334–339 (2016)
7. Hirsch, J.E.: An index to quantify an individual's scientific research output. Proceedings of the National Academy of Sciences **102**(46), 16569–16572 (2005)
8. Ioannidis, J.P.A., Boyack, K.W., Baas, J.: Updated science-wide author databases of standardized citation indicators. PLoS Biology **18**(10), e3000918 (2020)
9. Katsaros, D., Akritidis, L., Bozanis, P.: The f index: Quantifying the impact of coterminal citations on scientists' ranking. Journal of the American Society for Information Science & Technology **60**(5), 1051–1056 (2009)
10. Katsaros, D., Sidiropoulos, A., Manolopoulos, Y.: Age-decaying h-index for social networks of citations. In: Proceedings of the Workshop on Social Aspects of the Web (SAW), pp. 27–43 (2007)

11. Ke, Q., Ferrara, E., Radicchi, F., Flammini, A.: Defining and identifying sleeping beauties in science. Proceedings of the National Academy of Sciences **112**(24), 7426–7431 (2015)
12. Langville, A.N., Meyer, C.D.: Google's PageRank and Beyond: The Science of Search Engine Rankings. Princeton University Press (2006)
13. Lin, D., Gong, T., Liu, W., Meyer, M.: An entropy-based measure for the evolution of h index research. Scientometrics **125**(2), 2283–2298 (2020)
14. Ma, Y., Yuan, Y., Wang, G., Bi, X., Wang, Z., Wang, Y.: Rising star evaluation based on extreme learning machine in geo-social networks. Cognitive Computation **12**, 296–308 (2020)
15. Ning, Z., Liu, Y., Zhang, J., Wang, X.: Rising star forecasting based on social network analysis. IEEE Access **5**, 24229–24238 (2017)
16. Panagopoulos, G., Tsatsaronis, G., Varlamis, I.: Detecting rising stars in dynamic collaborative networks. Journal of Informetrics **11**(1), 198–222 (2017)
17. Prathap, G.: Fractionalized exergy for evaluating research performance. Journal of the American Society for Information Science & Technology **62**(11), 2294–2295 (2011)
18. Prathap, G.: Quasity, when quantity has a quality all of its own - toward a theory of performance. Scientometrics **88**(2), 555–562 (2011)
19. Prathap, G.: The Energy-Exergy-Entropy (or EEE) sequences in bibliometric assessment. Scientometrics **87**(3), 515–524 (2011)
20. Prathap, G.: Eugene Garfield: From the metrics of science to the science of metrics. Scientometrics **114**(2), 637–650 (2018)
21. van Raan, A., Winnick, J.: Do younger beauties prefer a technological prince? Scientometrics **114**(2), 701–717 (2018)
22. Sidiropoulos, A., Katsaros, D., Manolopoulos, Y.: Generalized Hirsch h-index for disclosing latent facts in citation networks. Scientometrics **72**(2), 253–280 (2007)
23. Song, Y., Situ, F., Zhu, H., Lei, J.: To be the prince to wake up sleeping beauty - The discovery of the delayed recognition studies. Scientometrics **117**(1), 9–24 (2018)
24. Todeschini, R., Baccini, A.: Handbook of Bibliometric Indicators: Quantitative Tools for Studying and Evaluating Research. Wiley (2016)
25. Vitanov, N.K.: Science Dynamics and Research Production: Indicators, Indexes, Statistical Laws and Mathematical Models. Springer (2016)
26. Wilson, M.C., Tang, Z.: Noncumulative measures of researcher citation impact. Quantitative Science Studies **1**(3), 1309–1320 (2020)
27. Zhang, J., Ning, Z., Bai, X., Wang, W., Yu, S., Xia, F.: Who are the rising stars in academia? In: Proceedings of the 16th ACM/IEEE Joint Conference on Digital Libraries (JCDL), pp. 211–212 (2016)

Chapter 11
Detection, Analysis, and Prediction of Research Topics with Scientific Knowledge Graphs

Angelo A. Salatino, Andrea Mannocci, and Francesco Osborne

Abstract Analysing research trends and predicting their impact on academia and industry is crucial to gain a deeper understanding of the advances in a research field and to inform critical decisions about research funding and technology adoption. In the last years, we saw the emergence of several publicly-available and large-scale Scientific Knowledge Graphs fostering the development of many data-driven approaches for performing quantitative analyses of research trends. This chapter presents an innovative framework for detecting, analysing, and forecasting research topics based on a large-scale knowledge graph characterising research articles according to the research topics from the Computer Science Ontology. We discuss the advantages of a solution based on a formal representation of topics and describe how it was applied to produce bibliometric studies and innovative tools for analysing and predicting research dynamics.

11.1 Introduction

Analysing research trends and predicting their impact on academia and industry is key to gain a deeper understanding of the research advancements in a field and to inform critical decisions about research funding and technology adoption. These

Angelo A. Salatino
Knowledge Media Institute - The Open University, Milton Keynes, United Kingdom
e-mail: angelo.salatino@open.ac.uk

Andrea Mannocci
Istituto di Scienza e Tecnologie dell'Informazione "A. Faedo", Italian National Research Council, Pisa, Italy
e-mail: andrea.mannocci@isti.cnr.it

Francesco Osborne
Knowledge Media Institute - The Open University, Milton Keynes, United Kingdom
e-mail: francesco.osborne@open.ac.uk

© The Author(s), under exclusive license to Springer Nature Switzerland AG 2021
Y. Manolopoulos, T. Vergoulis (eds.), *Predicting the Dynamics of Research Impact*,
https://doi.org/10.1007/978-3-030-86668-6_11

225

analyses were initially performed through qualitative approaches in a top-down fashion, relying on experts' knowledge and manual inspection of the state of the art [34, 76]. However, the ever-increasing number of research publications [11] has made this manual approach less feasible [51].

In the last years, we witnessed the emergence of several publicly-available, large-scale Scientific Knowledge Graphs (SKGs) [28], describing research articles and their metadata, e.g., authors, organisations, keywords, and citations. These resources fostered the development of many bottom-up, data-driven approaches to perform quantitative analyses of research trends. These methods usually cover one or more of three tasks: (i) detection of research topics, (ii) scientometric analyses, and (iii) prediction of research trends. In the first task, the articles are classified according to a set of research topics, typically using probabilistic topics models or different types of classifiers [8, 9, 12, 31, 78]. In the second task, the topics are analysed according to different bibliometrics over time [20, 27, 30]. For instance, each topic may be associated with the number of relevant publications or citations across the years to determine if their popularity in the research community is growing or decreasing. In the third task, regression techniques or supervised forecasters are typically used for predicting the future performance of research topics [35, 61, 75].

This chapter presents an innovative framework for detecting, analysing, and forecasting research topics based on a large-scale knowledge graph characterising research articles according to the research topics from the Computer Science Ontology (CSO) [63, 64]. In describing this framework we leverage and reorganise our previous research works. Specifically, we discuss the advantage of a solution based on a formal representation of research topics and describe how it was applied to produce bibliometrics studies [32, 45] and novel approaches for analysing [2, 51, 58] and predicting [48, 58, 61] research dynamics.

The chapter is articulated in six sections. In Sect. 11.2, we review state-of-the-art approaches for topic detection, analysis, and forecasting. In Sect. 11.3, we formally define a scientific knowledge graph, then we describe CSO [63] and the CSO Classifier [62], which is a tool for automatically annotating research papers with topics drawn from CSO. We then discuss how we generate the Academia/Industry DynAmics (AIDA) Knowledge Graph [3], a new SKG that describes research articles and patents according to the relevant research topics.

Sect. 11.4 describes three recent methods for analysing research trends that adopt this framework:

- *the EDAM methodology* [51] for supporting systematic reviews (Sect. 11.4.1);
- *the ResearchFlow framework* [58] for analysing trends across academia and industry (Sect. 11.4.2);
- *the AIDA Dashboard* [2] for supporting trend analysis in scientific conferences (Sect. 11.4.3).

In Sect. 11.5, we discuss three approaches that use the same framework for forecasting trends:

- *Augur* [61] (Sect. 11.5.1), a method for predicting the emergence of novel research topics;

- *the ResearchFlow forecaster* [58] (Sect. 11.5.2), a supervised deep learning classifier for predicting the impact of research topics on the industrial sector;
- *the Technology-Topic Framework (TTF)* [48] (Sect. 11.5.3), an approach for predicting the technologies that will be adopted by a research field.

Finally, in Sect. 11.6, we outline the conclusions and future work.

11.2 Literature Review

Detecting, analysing and predicting topics in the scientific literature has attracted considerable attention in the last two decades [9, 18, 22, 42, 59, 69]. In this section, we review the state of the art according to the main themes of this chapter. In particular, we initially review existing approaches for detecting topics in scholarly corpora. Then, we show approaches performing the analysis of research trends and finally, we describe the approaches for predicting research trends.

11.2.1 Topic Detection

Topic detection in the scholarly communication field aims at identifying relevant subjects within a set of scientific documents. This operation can support several subsequent tasks, such as suggesting relevant publications [68], assisting the annotation of articles [52] and video lessons [10], assessing academic impact [15], recommending items [14, 38], and building knowledge graphs of research concepts (e.g., AI-KG [19], ORKG [29], TKG [57]).

In the body of literature, we can find several approaches that can be characterised according to four main categories: (i) topic modelling [8, 20, 27], (ii) supervised machine learning approaches [25, 31, 43], (iii) approaches based on citation networks [12], and (iv) approaches based on natural language processing [18, 30].

For what concerns the topic modelling category, we can find the Latent Dirichlet Analysis (LDA) developed by Blei et al. [8]. LDA is a three-level hierarchical Bayesian model to retrieve latent – or hidden – patterns in texts. The basic idea is that each document is modelled as a mixture of topics, where a topic is a multinomial distribution over words, which is a discrete probability distribution defining the likelihood that each word will appear in a given topic. In brief, LDA aims to discover the latent structure, which connects words to topics and topics to documents. Over the years, LDA influenced many other approaches, such as Griffiths et al. [26] and Bolelli et al. [9] in which they designed generative models for document collections. Their models simultaneously modelled the content of documents and the interests of authors. Besides, Bolelli et al. [9] exploit citations information to evaluate the main terms in documents. However, since topic modelling aims at representing topics as a distribution of words, it is often tricky to map them to research subjects.

For supervised machine learning approaches, Kandimalla et al. [31] propose a deep attentive neural network for classifying papers according to 104 Web of Science[1] (WoS) subject categories. Their classifier was trained on 9 million abstracts from WoS, and it can be directly applied to abstract without the need for additional information, e.g., citations. However, the fact that they can map research papers to a limited set of only 104 subject categories is due to the intensive human labelling effort behind the generation of a gold standard that foresees all possible research topics, and that is also balanced with regard to the number of papers labelled per topic. Indeed, broad areas tend to have many published papers and thus are highly represented, while very specific areas tend to have fewer papers.

Among the citation network approaches, there is Boyack and Klavans [12] who developed a map of science using 20 million research articles over 16 years using co-citation techniques. Through this map, it is possible to observe the disciplinary structure of science in which papers of the same area tend to group together. The main drawback of citation-based approaches is that they are able to assign each document to one topic only, while a document is seldom monothematic.

For the category of natural language processing approaches, we find Decker [18] who introduced an unsupervised approach that generates paper-topic relations by exploiting keywords and words extracted from the abstracts to analyse the trends of topics on different timescales. Additionally, Duvvuru et al. [21] relied on keywords to build their co-occurring network and subsequently perform statistical analysis by calculating degree, strength, clustering coefficient, and end-point degree to identify clusters and associate them to research topics. Some recent approaches use word embeddings aiming to quantify semantic similarities between words. For instance, Zhang et al. [78] applied clustering techniques on a set of words represented as embeddings. However, all these approaches to topic detection need to generate the topics from scratch rather than exploiting a domain vocabulary, taxonomy, or ontology, resulting in noisier and less interpretable results [49].

In brief, state-of-the-art approaches for detecting topics either use keywords as proxies for topics, or match terms to manually curated taxonomies, or apply statistical techniques to associate topics to bags of words. Unfortunately, most of these solutions are suboptimal as they (i) fail to manage polysemies and synonyms, respectively, when a keyword may denote different topics depending on the context and when multiple labels exist for the same research area (i.e., productive ambiguity vs maximal determinacy); and (ii) fail to model and take advantage of the semantic relations that may hold between research areas, treating them as lists of unstructured keywords.

11.2.2 Research Trends Analysis

Research trends analysis deals with the dynamic component of topics, as it aims at observing their development over time. In the literature, we can find a variety

---

[1] Web of Science - https://clarivate.com/webofsciencegroup/solutions/web-of-science

of approaches based on citation patterns between documents [27, 30], or co-word analysis [20]. Jo et al. [30] developed an approach that combines the distribution of terms (i.e., n-grams) with the distribution of the citation graph related to publications containing that term. They assume that if a term is relevant for a topic, documents containing that term will have a stronger connection (i.e., citations) than randomly selected ones. Then, the algorithm identifies the set of terms having citation patterns that exhibit synergy. Similarly, He et al. [27] combined the citation network with Latent Dirichlet Allocation (LDA) [8]. They detect topics in independent subsets of a corpus and leverage citations to connect topics in different time frames. However, these approaches suffer from time lag, as newly published papers usually require some time, if not several years, before getting noticed and being cited [70].

Di Caro et al. [20] designed an approach for observing how topics evolve over time. After splitting the collection of documents according to different time windows, their approach selects two consecutive slices of the corpus, extracts topics using LDA, and analyses how such topics change from one time window to the other. The central assumption is that by comparing the topics generated in two adjacent time windows, it is possible to observe how topics evolve as well as capture their birth and death.

Other methods for analysing trends in research include overlay mapping techniques, which create maps of science and enable users to assess trends [13, 37]. Although these approaches provide a global perspective, the interpretation of such maps heavily relies on visual inspection by human experts.

11.2.3 Research Trends Prediction

Predicting research trends deals with anticipating the emergence of new topics. This task is particularly complex, mostly because of the limited availability of gold standards, i.e., extensive datasets of publication records manually and reliably labelled with research topics, that can be used to train a machine learning model to forecast trends. Indeed, regarding this aspect, we can find a very limited number of approaches in the literature.

Widodo et al. [75] performed a time series analysis to make predictions on PubMed data, and their experiments show that such time series are more suitable for machine learning techniques rather than statistical approaches. Instead, Krampen et al. [35] also rely on time series computing the Single Moving Average method to predict research trends in psychology. Watatani et al. [74] developed a bibliometric approach based on co-citation analysis, which provides insights for technology forecasting and strategic research, as well as development planning and policy-making.

In brief, in the literature, we can find many approaches capable of tracking topics over time. However, they focus on recognised topics, which are already associated with a good number of publications. Forecasting new research topics is yet an open challenge.

11.3 Topic Detection with Scientific Knowledge Graphs

In this section, we define the Scholarly Knowledge Graph (Sect. 11.3.1), and in particular, we introduce the main one used to fuel our research: Microsoft Academic Graph (MAG). In Sect. 11.3.2, we describe the Computer Science Ontology (CSO) and the CSO Classifier, which are at the foundation of our ontology-based topic detection method. Finally, in Sect. 11.3.3, we showcase the Academia/Industry DynAmics (AIDA) Knowledge Graph, an SKG derived by enriching MAG with the topics from CSO.

11.3.1 KG of Scholarly Data

Knowledge graphs are large collections of entities and relations describing real-world objects and events, or abstract concepts. Scientific Knowledge Graphs focus on the scholarly domain and describe the actors (e.g., authors, organisations), the documents (e.g., publications, patents), the research knowledge (e.g., research topics, tasks, technologies), and any other contextual information (e.g., project, funding) in an interlinked manner. Such descriptions have formal semantics allowing both computers and people to process them efficiently and unambiguously. Being interconnected within a network, each entity contributes to the description of the entities related to it, providing a wider context for their interpretation. SKGs provide substantial benefits to researchers, companies, and policymakers by powering several data-driven services for navigating, analysing, and making sense of research dynamics. Some examples include Microsoft Academic Graph (MAG)[2] [73], AMiner [67], ScholarlyData[3] [47], PID Graph[4] [23], SciGraph[5], Open Research Knowledge Graph[6] [29], OpenCitations[7] [53], and OpenAIRE research graph[8] [44].

More formally, given a set of entities E, and a set of relations R, a scientific knowledge graph is a directed multi-relational graph G that comprises triples (subject, predicate, object) and is a subset of the cross product $G \subseteq E \times R \times E$. Fig. 11.1 depicts an excerpt of a scientific knowledge graph representing some metadata about this current chapter.

In this chapter, we will focus on *Microsoft Academic Graph* (MAG) [66, 73], a pan-publisher, longitudinal scholarly knowledge graph produced and actively maintained by Microsoft, which contains scientific publication records, citation relations,

---

[2] Microsoft Academic Graph - `https://www.microsoft.com/en-us/research/project/academic`

[3] ScholarlyData - `http://www.scholarlydata.org`

[4] PID Graph - `https://www.project-freya.eu/en/pid-graph/the-pid-graph`

[5] SciGraph datasets - `https://sn-scigraph.figshare.com`

[6] Open Research Knowledge Graph - `https://www.orkg.org/orkg`

[7] OpenCitations - `https://opencitations.net`

[8] OpenAIRE research graph - `https://graph.openaire.eu`

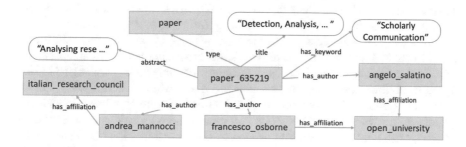

Fig. 11.1: Visualisation of a scientific knowledge graph. Entities are represented as rectangles and relations as arrows

authors, institutions, journals, conferences, and fields of study. We decided to adopt MAG because it is the most extensive datasets of scholarly data publicly available [72], containing more than 250 million publications as of January 2021. In addition, it is widely adopted by the scientometrics community, making our results easier to replicate.

At the time of writing, MAG is released at no cost (*per se*) for research purposes, while a small fee is required to cover the costs and maintenance of the provisioning infrastructure, which is based on Azure. MAG can be either processed on the cloud, leveraging Azure cloud computing solutions (whose usage and costs fall outside the scope of the present chapter) or downloaded as a whole. In the latter case, the current due costs are around 100$ a go for bandwidth (likely to increase slightly over time as the dataset grows and is regenerated fortnightly), plus a few dollars a month for data parking in case the dataset is kept on the cloud as a backup, rather than deleted right away. In this form, MAG is provided as a collection of TSV files via Microsoft Azure Storage and features an ODC-BY[9] licence, an aspect that is essential to ensure the transparency and reproducibility of any resulting analysis in compliance with Open Science best practices. The information contained in this dataset consists of six types of main entities in the scholarly domain — publications, authors, affiliations, venues (journals and conferences), fields of study, and events (specific conference instances). These entities are connected through relations like citations, authorship, and others. The relations between such entities are described in details in [66]. The dataset contains publication metadata, such as DOI, title, year of publication, and venue of publication; moreover, MAG provides, whenever possible, key data often not available in other publicly accessible datasets (e.g., Crossref[10], DBLP[11]), including authors' and affiliations' identifiers, which are often required to address some compelling research questions (e.g., collaboration networks). Although paper

---

[9] Open Data Commons Attribution Licence (ODC-BY) v1.0 - https://opendatacommons.org/licenses/by/1.0

[10] Crossref API - https://github.com/CrossRef/rest-api-doc

[11] DBLP - https://dblp.uni-trier.de

abstracts are available for many records in MAG, this, however, does not redistribute the publications' full-texts, somehow limiting full-scale NLP applications.

For the sake of completeness, in the case current costs and technical skills associated with MAG provisioning are not viable, other alternative SKGs repackaging MAG are available for free. *The Open Academic Graph*[12] (OAG) is a large SKG unifying two billion-scale academic graphs: MAG and AMiner. In mid-2017, the first version of OAG, which contains 166,192,182 papers from MAG and 154,771,162 papers from AMiner, and 64,639,608 linking (matching) relations between the two graphs, was released. The current version of OAG combines the MAG snapshot as of November 2018 and AMiner snapshots of July 2018 and January 2019. In this new release, authors, venues, newer publication data, and the corresponding matchings are available as well. Another feasible alternative is the *OpenAIRE dataset DOIboost* [36] which provides an "enhanced" version of Crossref that integrates information from Unpaywall, ORCID and MAG, such as author identifiers, affiliations, organisation identifiers, and abstracts. DOIboost is periodically released on Zenodo[13].

11.3.2 Ontology-based Topic Detection

In the following section, we describe the two main technologies enabling our methodology for ontology-based topic detection: the Computer Science Ontology (Sect. 11.3.2) and the CSO Classifier (Sect. 11.3.2).

The Computer Science Ontology
The Computer Science Ontology is a large-scale ontology of research areas in the field of Computer Science. It was automatically generated using the Klink-2 algorithm [49] on a dataset of 16 million publications, mainly in the field of Computer Science [50]. Compared to other solutions available in the state of the art (e.g., the ACM Computing Classification System), the Computer Science Ontology includes a much higher number of research topics, which can support a more granular representation of the content of research papers, and it can be easily updated by rerunning Klink-2 on more recent corpora of publications.

The current version of CSO[14] includes 14 thousand semantic topics and 159 thousand relations. The main root is Computer Science; however, the ontology includes also a few additional roots, such as Linguistics, Geometry, Semantics, and others. The CSO data model[15] is an extension of SKOS[16] and it includes four main semantic relations:

---

[12] Open Academic Graph - `https://www.microsoft.com/en-us/research/project/open-academic-graph`

[13] DOIboost laster release - `https://zenodo.org/record/3559699`

[14] CSO is available for download at `https://w3id.org/cso/downloads`

[15] CSO Data Model - `https://cso.kmi.open.ac.uk/schema/cso`

[16] SKOS Simple Knowledge Organisation System - `http://www.w3.org/2004/02/skos`

- superTopicOf, which indicates that a topic is a super-area of another one (e.g., Semantic Web is a super-area of Linked Data).
- relatedEquivalent, which indicates that two given topics can be treated as equivalent for exploring research data (e.g., Ontology Matching and Ontology Mapping).
- contributesTo, which indicates that the research output of one topic contributes to another.
- owl:sameAs, this relation indicates that a research concept is identical to an external resource, i.e., DBpedia.

The Computer Science Ontology is available through the CSO Portal[17], a web application that enables users to download, explore, and visualise sections of the ontology. Moreover, users can use the portal to provide granular feedback at different levels, such as rating topics and relations, and suggesting missing relations.

In the last years CSO was adopted by the wider community and supported the creation of many innovative approaches and applications, including ontology-driven topic models (e.g., CoCoNoW [7]), systems for predicting the academic impact (e.g., ArtSim [15]), recommender systems for articles (e.g., SBR [68]) and video lessons [10], visualisation frameworks (e.g., ScholarLensViz [40], ConceptScope [77]), temporal knowledge graphs (e.g., TGK [57]), and tools for identify domain experts (e.g., VeTo [71]). It was also adopted for several large-scale analyses of the literature (e.g., Cloud Computing [41], Software Engineering [16], Ecuadorian publications [16]).

The CSO Classifier

The CSO Classifier [62] is a tool for automatically classifying research papers according to the Computer Science Ontology. This application takes in input the metadata associated with a research document (title, abstract, and keywords) and returns a selection of research concepts drawn from CSO, as shown in Fig. 11.2.

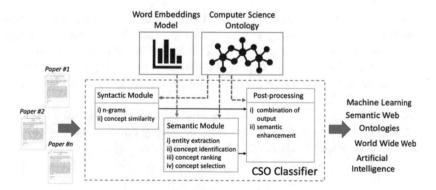

Fig. 11.2: Workflow of the CSO Classifier

---

[17] Computer Science Ontology Portal - https://cso.kmi.open.ac.uk

The CSO Classifier works in three steps. First, it finds all topics in the ontology that are explicitly mentioned in the paper (*syntactic module*). Then it identifies further semantically related topics by means of part-of-speech tagging and word embeddings (*semantic module*). Finally, it enriches this set by including the super-areas of these topics according to CSO (*post-processing module*).

In particular, the syntactic module splits the text into unigrams, bigrams, and trigrams. Each *n*-gram is then compared with concept labels in CSO using the Levenshtein similarity. As a result, this module returns all matched topics having similarities greater than or equal to the pre-defined threshold.

The semantic module instead leverages a pre-trained Word2Vec word embedding model that captures the semantic properties of words [46]. We trained this model[18] using titles and abstracts of over 4.6 million English publications in the field of Computer Science from MAG. We preprocessed this data by replacing spaces with underscores in all *n*-grams matching the CSO topic labels (e.g., "semantic web" turned into "semantic_web"). We also performed a collocation analysis to identify frequent bigrams and trigrams (e.g., "highest_accuracies", "highly_cited_journals") [46]. This solution allows the classifier to disambiguate concepts better and treat terms such as "deep_learning" and "e-learning" as entirely different words.

More precisely, to compute the semantic similarity between the terms in the document and the CSO concepts, the semantic module uses part-of-speech tagging to identify candidate terms composed by a combination of nouns and adjectives and decomposes them into unigrams, bigrams, and trigrams. For each *n*-gram, it retrieves its most similar words from the Word2Vec model. The *n*-gram tokens are initially glued with an underscore, creating one single word, e.g., "semantic_web". If this word is not available within the model vocabulary, the classifier uses the average of the embedding vectors of all its tokens. Then, it computes the relevance score for each topic in the ontology as the product between the number of times it was identified in those *n*-grams (frequency) and the number of unique *n*-grams that led to it (diversity). Finally, it uses the elbow method [65] for selecting the set of most relevant topics.

The resulting set of topics returned by the two modules is enriched in the post-processing module by including all their super-topics in CSO. For instance, a paper tagged as *neural network* is also tagged with *machine learning* and *artificial intelligence*. This solution yields an improved characterisation of high-level topics that are not directly referred to in the documents. More details about the CSO Classifier are available in [62].

The latest release of the CSO Classifier can be installed via *pip* from PyPI with `pip install cso-classifier`; or it can be simply downloaded from `https://github.com/angelosalatino/cso-classifier`.

---

[18] The model parameters are: *method* = skipgram, *embedding-size* = 128, *window-size* = 10, *min-count-cutoff* = 10, *max-iterations* = 5.

11.3.3 The Academia/Industry DynAmics Knowledge Graph

To support new approaches for monitoring and predicting research trends across academia and industry, we build on CSO and the CSO Classifier to generate the Academia/Industry DynAmics (AIDA) Knowledge Graph [3]. This novel resource describes 21 million publications and 8 million patents according to the research topics drawn from CSO. 5.1 million publications and 5.6 million patents are further characterised according to the type of the author's affiliation (e.g., academia, industry) and 66 industrial sectors (e.g., automotive, financial, energy, electronics) organised in a two-level taxonomy.

AIDA was generated using an automatic pipeline that integrates several knowledge graphs and bibliographic corpora, including Microsoft Academic Graph, Dimensions, English DBpedia, the Computer Science Ontology [64], and the Global Research Identifier Database (GRID). Its pipeline consists of three phases: (i) topics extraction, (ii) extraction of affiliation types, and (iii) industrial sector classification.

To extract topics from publications and patents, we run the CSO Classifier. Instead, to extract the affiliation types of a given document, we queried GRID[19], an open database identifying over 90 thousand organisations involved in research. GRID describes research institutions with a multitude of attributes, including their type, e.g., education, company, government, and others. A document is classified as "academic" if all its authors have an educational affiliation and as "industrial" if all its authors have an industrial affiliation. Documents whose authors are from both academia and industry are classified as "collaborative"

Finally, we characterised documents from industry according to the Industrial Sectors Ontology (INDUSO)[20]. In particular, from GRID, we retrieved the Wikipedia entry of companies, and we queried DBpedia, the knowledge graph of Wikipedia. Then, we extracted the objects of the properties "About:Purpose" and "About:Industry", and mapped them to INDUSO sectors. All industrial documents were associated with the industrial sectors based on their affiliations.

Table 11.1 reports the number of publications and patents from academy, industry, and collaborative efforts. Most scientific publications (69.8%) are written by academic institutions, but industry is also a strong contributor (15.3%). Conversely, 84% of the patents are from industry and only 2.3% from academia. Collaborative efforts appear limited, including only 2.6% of the publications and 0.2% of patents.

The data model of AIDA is available at http://aida.kmi.open.ac.uk/ontology and it builds on SKOS[21] and CSO[22]. It focuses on four types of entities: *publications, patents, topics,* and *industrial sectors*.

The main information about publications and patents are given by means of the following semantic relations:

---

[19] Global Research Identifier Database (GRID) - https://grid.ac
[20] INDUSO - http://aida.kmi.open.ac.uk/downloads/induso.ttl
[21] SKOS Simple Knowledge Organisation System - http://www.w3.org/2004/02/skos
[22] CSO Data Model - https://cso.kmi.open.ac.uk/schema/cso

Table 11.1: Distribution of publications and patents classified as Academia, Industry and Collaboration

	Publications	Patents
Total documents	20,850,710	7,940,034
Documents with GRID IDs	5,133,171	5,639,252
Academia	3,585,060	133,604
Industry	787,151	4,741,695
Collaborative	133,781	16,335
Additional categories with GRID ID	627,179	747,618

- `hasTopic`, which associates to the documents all their relevant topics drawn from CSO;
- `hasAffiliationType` and `hasAssigneeType`, which associates to the documents the three categories (academia, industry, or collaborative) describing the affiliations of their authors (for publications) or assignees (for patents);
- `hasIndustrialSector`, which associates to documents and affiliations the relevant industrial sectors drawn from the Industrial Sectors Ontology (INDUSO) we describe in the next sub-section.

A dump of AIDA in Terse RDF Triple Language (Turtle) is available at `http://aida.kmi.open.ac.uk/downloads`.

11.4 Research Trends Analysis

In this section, we discuss different methodologies that use the framework described in the previous section for analysing the evolution of research topics in time. In particular, we first introduce the EDAM methodology for supporting systematic reviews (Sect. 11.4.1) and present two exemplary bibliometric analyses, then we describe the ResearchFlow approach for analysing research trends across academia and industry (Sect. 11.4.2), and finally, we introduce the AIDA Dashboard, a web application for examining research trends in scientific conferences (Sect. 11.4.3).

11.4.1 The EDAM Methodology for Systematic Reviews

A systematic review is *"a means of evaluating and interpreting all available research relevant to a particular research or topic area or phenomenon of interest"* [33]. Systematic reviews are used by researchers to create a picture of the state of the art in a research field, often including the most important topics and their trends. Given a set of research questions, and by following a systematically defined and reproducible process, a systematic review can assist the selection of a set of research articles (primary studies) that contribute to answering them [54].

In this section, we describe the EDAM (Expert-Driven Automatic Methodology) [51], which is a recent methodology for creating systematic reviews, thus limiting the number of tedious tasks that have to be performed by human experts. Specifically, EDAM can produce a comprehensive analysis of topic trends in a field by (i) characterising the area of interest using an ontology of topics, (ii) asking domain experts to refine this ontology, and (iii) exploiting this knowledge base for classifying relevant papers and producing useful analytics. The main advantage of EDAM is that the authors of the systematic review are still in full control of the ontological schema used to represent the topics.

Fig. 11.3 illustrates the workflow of EDAM. First, the researchers define the research questions that will be answered by the produced analytics and select the data sources. Then, they reuse or create the ontology that will support the classification process. The creation can also be supported by automatic methodologies for ontology learning such as (i) statistical methods for deriving taxonomies from keywords [39]; (ii) natural language processing approaches (e.g., FRED [24], LOD-ifier [6], Text2Onto [17]); (iii) approaches based on deep learning (e.g., recurrent neural networks [55]); (iv) specific approaches for generating research topic ontologies (e.g., Klink-2 [49]). The resulting ontology is corrected and refined by domain experts. When the ontology is ready, the authors define the primary studies' inclusion criteria according to the domain ontology and other metadata of the papers (e.g., year, venue, language). The inclusion criteria are typically expressed as a search string, which uses simple logic constructs, such as AND, OR, and NOT [4]. The simplest way for mapping categories to papers is to associate to each category each paper that contains the label of the category or any of its subcategories. However, this step can also take advantage of a more advanced classification approach, such as the CSO Classifier described in Sect. 11.3.2. The last step is the computation of analytics to answer the research questions.

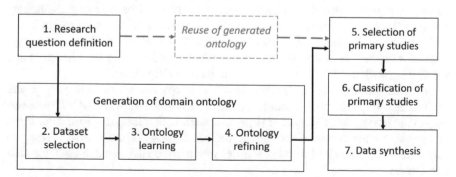

Fig. 11.3: Steps of a systematic mappings adopting the EDAM methodology. The gray-shaded elements refer to the alternative step of reusing the previously generated ontology

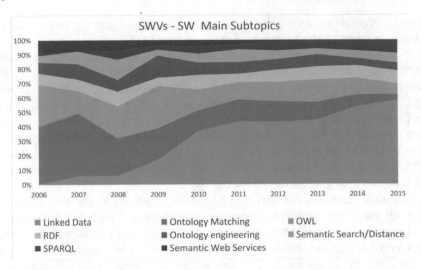

Fig. 11.4: Exemplary analysis in Semantic Web: Percentage of articles associated with the eight main subtopics of Semantic Web

EDAM has also been used in a recent review that analyses the main research trends in the field of Semantic Web [32]. This study compares two datasets covering respectively the main Semantic Web venues (e.g., ISWC, ESWC, SEMANTiCS, SWJ, and JWS) and 32,431 publications associated with the Semantic Web from a dump of Scopus. The authors annotated the articles with CSO and produced several analyses based on the number of papers and citations in each research topic over the years. Specifically, articles were associated with a given topic if their title, abstract, or keywords field contained: (i) the label of the topic (e.g., "semantic web"), (ii) a relevantEquivalent of the topic (e.g., "semantic web tecnologies"), (iii) a skos:broaderGeneric of the topic (e.g., "ontology matching"), or (iv) a relevantEquivalent of any skos:broaderGeneric of the topic (e.g., "ontology mapping"). Fig. 11.4 shows, for example, the popularity of the main topics in the main venue of the Semantic Web community. The analysis highlights two main dynamics confirmed by domain experts: the fading of Semantic Web Services and the rapid growth of Linked Data after the release of DBpedia in 2007 [5].

A further instantiation of the EDAM methodology can be found in [45], where Mannocci et al. performed a circumstantial, narrow-aimed study for the 50th anniversary of the International Journal of Human-Computer Studies (IJHCS) and delivered an analysis on the evolution throughout the last 50 years of the journal making an extensive comparison with another top-tier venue in the HCI research field: the CHI conference.

Among several other analyses (e.g., bibliometrics, spatial scientometrics), one aspect of the study focuses on topic analysis in both venues. To this end, it instantiates EDAM using CSO and the CSO classifier as reference ontology and classifier, respectively.

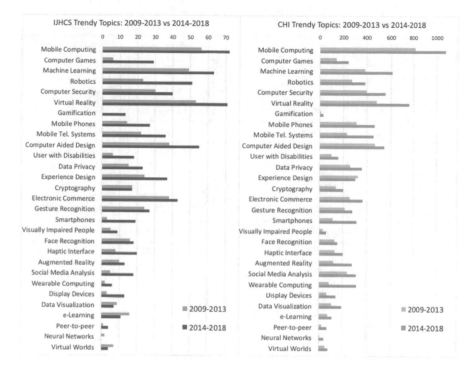

Fig. 11.5: Exemplary analysis in Human Computer Interaction: Growing Topics in IJHCS and CHI

Fig. 11.5 shows a comparison of the most growing topics in the two venues according to their number of articles in 2009-2013 and 2014-2018. These plots highlight very clearly which topics are gaining momentum in a community.

Fig. 11.6 shows the geographical distribution of some of the most important topics in the two venues under analysis. This kind of analytics allows to easily assess the contribution of different countries for each topic. For instance, Japan is ranked seventh when considering the full CHI dataset, but in Robotics is third. Similarly, South Korea, ranked 12th and 10th in IJHCS and CHI, is particularly active in Computer Security, ranking in fourth and sixth place when considering this topic.

11.4.2 ResearchFlow: Analysing Trends Across Academia and Industry

ResearchFlow [58] is a recent methodology for quantifying the diachronic behaviour of research topics in academia and industry that builds on the Academia/Industry DynAmics (AIDA) Knowledge Graph [3], described in Sect. 11.3.3.

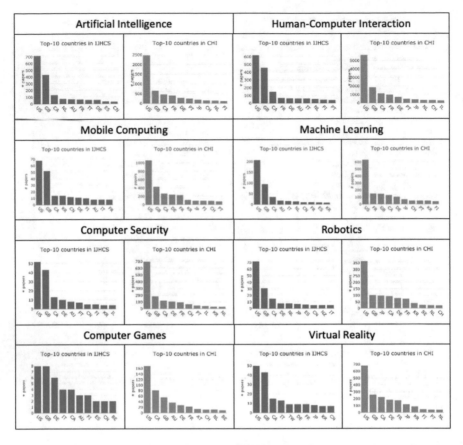

Fig. 11.6: Overall country contributions in IJHCS and CHI across the top-8 topics

Specifically, ResearchFlow represents the evolution of research topics in CSO according to four time series reporting the frequency of (i) papers from academia, (ii) papers from industry, (iii) patents from academia, and (iv) patents from industry. Such characterisation allows us to (i) study the diachronic behaviour of topics to characterise their trajectory across academia and industry, (ii) compare each pair of signals to understand which one typically precedes the other and in which order they usually tackle a research topic, and (iii) assess how signals influence each other by identifying pairs of signals that are highly correlated, after compensating for a time delay.

Fig. 11.7 shows the distribution of the most frequent 5,000 topics from CSO in a bi-dimensional diagram according to two indexes: academia-industry (x-axis) and papers-patents (y-axis). The academia-industry index for a given topic t is the difference between the documents in academia and industry related to t, over the whole set of documents. A topic is predominantly academic if this index is positive;

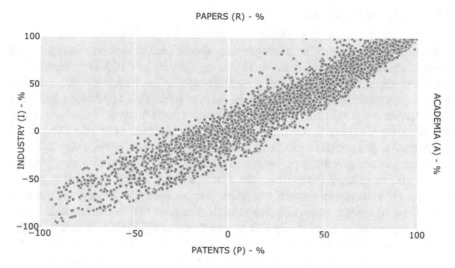

Fig. 11.7: Distribution of the most frequent 5,000 topics according to their academia-industry and publications-papers indexes

otherwise, it is mostly populated by industry. On the other hand, the papers-patents index of a given topic t is the difference between the number of research papers and patents related to t, over the whole set of documents. If this index is positive, a topic tends to be associated with a higher number of publications; otherwise, it has a higher number of patents.

From the figure, we can observe that topics are tightly distributed around the bisector: the ones which attract more interest from academia are prevalently associated with publications (top-right quadrant). In contrast, the ones in industry are mostly associated with patents (bottom-left quadrant).

We also analysed which time series typically precedes another in first addressing a research topics. In particular, we determined when a topic emerges in all associated signals, and then we compared the time elapsed between each couple of signals. To avoid false positives, we considered a topic as "emerged" when associated with at least ten documents. Our results showed that 89.8% of the topics first emerged in academic publications, 3.0% in industrial publications, 7.2% in industrial patents, and none in academic patents. On average, publications from academia preceded publications from industry by 5.6±5.6 years, and in turn, the latter preceded patents from industry by 1.0±5.8 years. Publications from academia also preceded patents from industry by 6.7±7.4 years.

11.4.3 The AIDA Dashboard

The framework discussed in this chapter can also be used to support tools that allow users to analyse research dynamics. This is the case of the AIDA Dashboard [2], a novel web application for analysing scientific conferences which integrates statistical analysis, semantic technologies, and visual analytics. This application has been developed to support the editorial team at Springer Nature in assessing and evaluating the evolution of conferences through a number of bibliometric analytics. The AIDA Dashboard builds on AIDA SKG and, compared to other state-of-the-art solutions, it introduces three novel features. First, it associates conferences with a very granular representation of their topics from the CSO [64] and uses it to produce several analyses about its research trends over time. Second, it enables an easy comparison and raking of conferences according to several metrics within specific fields (e.g., Digital Libraries) and timeframes (e.g., last five years). Finally, it offers several parameters for assessing the involvement of industry in a conference. This includes the ability to focus on commercial organisations and their performance, to report the ratio of publications and citations from academia, industry, and collaborative efforts, and to distinguish industrial contributions according to the 66 industrial sectors from INDUSO.

AIDA Dashboard is highly scalable and allows users to browse the different facets of a conference according to eight tabs: *Overview*, *Citation Analysis*, *Organisations*, *Countries*, *Authors*, *Topics*, *Related Conferences*, and *Industry*.

Fig. 11.8 shows the *Overview* tab of the International Semantic Web Conference (ISWC). This is the main view of a conference that provides introductory information about its performance, the main authors and organisation, and the conference rank in its main fields in terms of average citations for papers during the last five years. More detailed and refined analytics are available in the other tabs.

Fig. 11.8: The Overview of the International Semantic Web Conference (ISWC) according to the AIDA Dashboard

11.5 Research Trends Prediction

In this section, we explore two approaches that use our framework for predicting research trends. In Sect. 11.5.1, we present Augur, a method for forecasting the emergence of new research trends which takes advantage of the collaborative networks of topics. In Sect. 11.5.2, we present an approach that was developed within the ResearchFlow framework and aims at predicting the impact of research topics on the industrial landscape. In Sect. 11.5.3, we introduce the Technology-Topic Framework (TTF) [48], a novel methodology that takes advantage of a semantically enhanced technology-topic model for predicting the technologies that will be adopted by a research community.

11.5.1 Predicting the Emergence of Research Topics

Augur [61] is an approach that aims to effectively detect the emergence of new research topics by analysing topic networks and identifying research areas exhibiting an overall increase in the pace of collaboration between already existing topics. This dynamic has been shown to be strongly associated with the emergence of new topics in a previous study by Salatino et al. [60].

A topic network is a fully weighted graph in which nodes represent topics, and the edges identify whether the connected pair of topics co-occurred together in at least one research paper. Additionally, the node weights count the number of papers in which a topic appears, and edge weights count the number of papers the two topics co-appear together. The original implementation of Augur extracted these networks using CSO on the Rexplore dataset [50].

Augur takes as input five topic networks characterising the collaborations between topics over five subsequent years (e.g., 2017-2021), as shown in the left side of Fig. 11.9. Over these networks, it then applies the Advanced Clique Percolation Method, a clustering algorithm for identifying groups of topics that over time intensify their pace of collaboration (centre of Fig. 11.9). The resulting clusters outline the portions of the network that may be nurturing research topics that will emerge in the following years (right side of Fig. 11.9).

To evaluate this approach, we built a gold standard of 1408 emerging trends in the 2000-2011 timeframe. Each emerging topic has been associated with a list of related topics considered as "ancestors". The latter is crucial because Augur returns clusters of topics, which can also be considered ancestors of the future emerging trends. To this end, the evaluation consisted of checking whether the clusters of topics returned by Augur could be mapped with the ancestors of the emerging trends in the gold standard.

We evaluated Augur against other four state-of-the-art algorithms: Fast Greedy (FG), Leading Eigenvector (LE), Fuzzy C-Means (FCM), and Clique Percolation Method (CPM). The evaluation has been performed across time, and as shown in Table 11.2, Augur outperformed the alternative approaches in terms of both precision

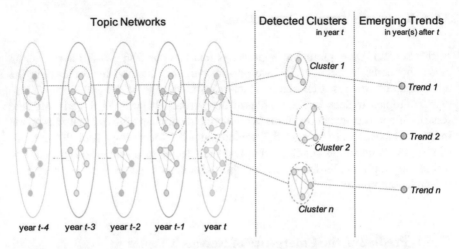

Fig. 11.9: Augur approach. On the left, the topic networks, whereas on the right there are emerging trends. The dashed circles represent groups of research topics nurturing new research trends

Table 11.2: Values of Precision and Recall for the five approaches along time. In bold the best results. The table and the experiments were previously reported in [61]

| | FG | | LE | | FCM | | CPM | | ACPM | |
Year	Precis.	Recall	Precis.	Recall	Precis.	Recall	Precis.	Recall	Precis.	Recall
1999	0.27	0.11	0.00	0.00	0.00	0.00	0.06	0.01	**0.86**	**0.76**
2000	0.21	0.07	0.14	0.02	0.96	0.01	0.05	0.00	**0.78**	**0.70**
2001	0.13	0.04	0.11	0.01	0.00	0.00	0.17	0.00	**0.77**	**0.72**
2002	0.14	0.04	0.11	0.01	0.00	0.00	0.29	0.01	**0.82**	**0.80**
2003	0.09	0.02	0.20	0.02	0.00	0.00	0.08	0.02	**0.83**	**0.79**
2004	0.11	0.05	0.06	0.00	0.00	0.00	0.00	0.00	**0.84**	**0.68**
2005	0.07	0.11	0.06	0.01	0.00	0.00	0.00	0.00	**0.71**	**0.66**
2006	0.01	0.01	0.07	0.01	0.00	0.00	0.00	0.00	**0.43**	**0.51**
2007	0.01	0.08	0.00	0.00	0.00	0.00	0.00	0.00	**0.28**	**0.44**
2008	0.01	0.04	0.00	0.00	0.00	0.00	0.00	0.00	**0.15**	**0.33**
2009	0.00	0.00	0.00	0.00	0.00	0.00	0.00	0.00	**0.09**	**0.76**

and recall in all years. Further details of Augur, the gold standard and the evaluation are available in Salatino et al. [61].

11.5.2 The ResearchFlow Forecaster

In this section, we show how the ResearchFlow approach introduced in Sect. 11.4.2 can also support a very accurate forecaster to predict the impact of research topics in industry.

A good measure to assess the impact of research trends on commercial organisations is the number of relevant patents granted to companies. The literature proposes a range of approaches for this task [1, 56]. However, most of these methods focus only on patents since they are limited by current datasets that do not typically integrate research articles nor can they distinguish between documents produced by academia or industry. We thus hypothesised that a scientific knowledge graph like AIDA, which integrates all the information about publications and patents and their origin, should offer a richer set of features, ultimately yielding a better performance in comparison to approaches that rely solely on the number of publications or patents.

To train a forecaster, we created a gold standard, in which, for each topic in CSO, we selected all the timeframes of five years in which the topic had not yet emerged (less than ten patents). We then labelled each of these samples as *True* if the topic produced more than 50 industrial patents in the following ten years and *False* otherwise. The resulting dataset includes 9, 776 labelled samples, each composed of four time series: (i) papers from academia **RA**, (ii) papers from industry **RI**, (iii) patents from academia **PA**, (iv) patents from industry **PI**. We then trained five machine learning classifiers on the gold standard: Logistic Regression (**LR**), Random Forest (**RF**), AdaBoost (**AB**), Convoluted Neural Network (**CNN**), and Long Short-term Memory Neural Network (**LSTM**). We ran each of them on 17 possible combinations of the four time series to assess which set of features would yield the best results. **RA-RI** concatenates the time series of research papers in academia (**RA**) and industry (**RI**) together. Instead, R and P sum the contribution of all papers and patents, respectively. Finally, **RA-RI-PA-PI** concatenates the four time series. performed a 10-fold cross-validation of the data and measured the performance of the classifiers by computing the average precision (**P**), recall (**R**), and F1 (**F**).

Table 11.3 reports the results of our experiment. LSTM outperforms all the other solutions, yielding the highest F1 for 12 of the 17 feature combinations and the highest average F1 (73.7%). CNN (72.8%) and AB (72.3%) also produce competitive results.

As hypothesised, taking advantage of the full set of features available in AIDA – broken into its times series – significantly outperforms (with F1 = 84.6%) the version which uses only the number of patents by companies (75.2%). Considering the origin (academia and industry) of the publications and the patents also increases performance: RA-RI (80.7%) significantly outperforms R (68.2%) and PA-PI (75.2%) is marginally better than P (74.8%).

In conclusion, the experiments confirm that the granular representation of research topics in AIDA can effectively support deep learning approaches for forecasting the impact of research topics on the industrial sector. It also validates the intuition

Table 11.3: Performance of the five classifiers on 17 combinations of time series. In bold the best F1 (F) for each combination. The table and the experiments were previously reported in [58].

	LR			RF			AB			CNN			LSTM		
	P%	R%	F%	P%	R%	F%	P%	R%	F%	P%	R%	F%	P%	R%	F%
RA	70.8	45.2	55.2	63.3	55.8	59.2	66.0	58.4	61.9	64.1	66.3	**65.0**	65.2	64.2	64.6
RI	83.5	67.1	74.4	78.9	69.8	74.0	80.0	73.1	76.4	79.2	75.1	**77.0**	79.1	74.8	76.9
PA	58.3	15.3	24.2	60.4	15.4	24.5	59.3	16.0	**25.2**	60.5	15.7	24.9	60.8	15.6	24.8
PI	76.5	69.0	72.5	73.9	68.4	71.0	75.6	71.8	73.6	73.7	76.6	75.0	74.1	76.6	**75.2**
R	73.7	48.8	58.7	65.5	59.7	62.5	68.6	63.1	65.6	67.6	69.2	**68.3**	67.2	69.4	68.2
P	76.5	68.6	72.3	72.8	67.6	70.0	74.4	71.6	73.0	73.2	76.1	74.6	73.1	76.6	**74.8**
RA, RI	85.7	70.9	77.6	80.5	76.0	78.2	82.6	76.6	79.5	78.9	75.1	76.8	82.2	79.3	**80.7**
RA, PA	70.3	47.0	56.3	63.1	55.5	59.0	66.5	59.3	62.6	64.5	65.1	64.5	65.4	64.2	**64.6**
RA, PI	79.6	73.7	76.5	77.2	74.3	75.7	79.1	76.5	77.7	75.2	76.3	75.7	77.4	81.9	**79.5**
RI, PA	83.3	67.0	74.3	77.9	70.8	74.1	79.6	73.0	76.1	78.6	75.6	77.0	79.1	75.2	**77.1**
RI, PI	83.4	77.3	80.2	81.0	77.3	79.1	82.7	78.6	80.6	82.0	78.6	80.2	81.7	81.2	**81.4**
PA, PI	76.7	68.6	72.4	74.2	69.0	71.5	75.9	71.5	73.6	71.1	70.8	70.9	73.8	76.7	**75.2**
RA, RI, PA	85.2	71.4	77.7	80.8	75.4	78.0	82.5	77.0	79.6	82.6	78.1	**80.3**	82.6	78.2	**80.3**
RA, RI, PI	85.4	79.8	82.5	84.5	80.5	82.4	84.6	81.2	82.9	83.8	84.7	84.2	84.1	85.4	**84.7**
RA, PA, PI	79.6	73.9	76.6	77.5	74.4	75.9	79.2	76.5	77.8	78.9	78.6	78.6	77.4	81.4	**79.2**
RI, PA, PI	83.6	77.5	80.4	81.1	78.0	79.5	82.7	78.6	80.6	82.2	80.9	**81.5**	81.1	81.0	81.1
RA, RI, PA, PI	85.4	79.8	82.5	83.8	80.0	81.8	84.6	81.2	82.9	84.7	81.3	82.9	83.2	86.1	**84.6**

that including features from research articles can be very useful when predicting industrial trends.

11.5.3 Predicting the Spreading of Technologies

Every new piece of research, no matter how ground-breaking, adopts previous knowledge and reuses tools and methodologies. Typically, a technology is expected to originate in the context of a research area and then spread and contribute to several other fields. Unfortunately, given the tremendous amount of knowledge produced yearly, it is difficult to digest it and pick up on an interesting piece of research from a different field. Thus, the transfer of a technology from one research area (e.g., Semantic Web) to a different, and possibly conceptually distant, one (e.g., Digital Humanities) may take several years, slowing down the whole research progress. It is, therefore, crucial to be able to anticipate technology spreading across the different research community.

The Technology-Topic Framework (TTF) [48] is a recent approach that uses a semantically enhanced technology-topic model to predict the technologies that will be adopted by a research field. It aims at suggesting promising technologies to scholars from a field, thus helping to accelerate the knowledge flow and the pace of technology propagation. It is based on the hypothesis that *technologies that exhibit similar spreading patterns across multiple research topics will tend to be adopted by similar topics*. It thus characterises technologies in terms of a set of topics drawn from a large-scale ontology of research areas over a given time period and applies machine learning on these data so to forecast technology spreading.

The prototype of TTF takes as input three knowledge bases: (i) a dataset of 16 million research papers from 1990 to 2013 mainly in the field of Computer Science, described by means of their titles, abstracts, and keywords; (ii) a list of 1,118 input technologies extracted automatically from paper abstracts and from generic KBs (e.g., DBpedia [5]) and then manually curated, which are associated to at least 10 relevant publications in the research paper dataset; (iii) 173 relevant research topics and their relationships drawn from the Computer Science Ontology [63, 64]. TTF generates a 3-dimensional co-occurrence matrix that tracks the number of papers in which a given technology is associated with a given topic in a given year. It then uses this matrix to train a classifier on a technology recent history to predict its future course. The Random Forest classifier performed best on this data, obtaining a precision of 74.4% and a recall of 47.7%, outperforming alternative solutions such as Logistic Regression, Decision Tree, Gradient Boosting, SVM, and a simple Neural Network (see [48] for the full details).

Table 11.4 reports the results of the version of TTF using Random Forest on the most frequent 24 topics in the experiment. We can see that TTF obtains the best performance on topics associated with a large set of publications and technologies, such as Information Retrieval and Databases. However, it is also able to obtain satisfactory results in many other fields, especially the ones that are usually associated with a coherent set of technologies, such as Computer Networks, Sensors, and Peer-to-peer Systems. This confirms the hypothesis that is indeed possible to learn from historical spreading patterns and forecast technology propagation, at least for the set of topics that are more involved in technology propagation events.

Table 11.4: Performance of Random Forest on the first 24 topics, with at least 50 positive labels, ordered by F1 score. The table and the experiments were previously reported in [48]

Topics	Prec%	Rec%	F1%	Topics	Prec%	Rec%	F1%
information retrieval	92.6	66.8	77.6	wireless networks	64.7	47.8	55.0
database systems	82.6	65.9	73.3	sensor networks	71.9	43.6	54.3
world wide web	88.6	56.1	68.7	software engineering	70.6	44.0	54.2
artificial intelligence	83.6	55.2	66.5	distributed comp. sys.	67.5	45.0	54.0
computer architecture	68.3	63.3	65.7	quality of service	59.6	48.6	53.5
computer networks	82.1	54.0	65.2	imaging systems	100.0	35.8	52.8
image coding	96.8	46.9	63.2	data mining	60.8	45.3	52.0
P2P networks	78.9	50.8	61.9	computer vision	92.3	36.0	51.8
telecom. traffic	70.8	48.1	57.3	programming languages	65.3	42.0	51.2
wireless telecom. sys.	74.4	46.4	57.1	problem solving	69.0	39.7	50.4
sensors	78.8	43.7	56.2	semantic web	77.8	37.1	50.2
web services	83.3	42.2	56.0	image quality	74.2	37.7	50.0

11.6 Conclusion

In this chapter, we presented a framework for detecting, analysing, and predicting research topics based on a large-scale knowledge graph characterising research articles according to the research topics from the Computer Science Ontology (CSO) [63, 64]. We first illustrated how to annotate a scientific knowledge graph describing research articles and their metadata with a set of research topics from a domain ontology. We then discussed several methods that build on this knowledge graph for analysing research from different perspectives. Finally, we presented two approaches for predicting research trends based on this framework.

References

1. Altuntas, S., Dereli, T., Kusiak, A.: Analysis of patent documents with weighted association rules. Technological Forecasting & Social Change **92**, 249–262 (2015)
2. Angioni, S., Salatino, A., Osborne, F., Recupero, D.R., Motta, E.: The AIDA Dashboard: Analysing conferences with semantic technologies. In: Proceedings of the 19th International Semantic Web Conference (ISWC), pp. 271–276 (2020)
3. Angioni, S., Salatino, A.A., Osborne, F., Recupero, D.R., Motta, E.: Integrating knowledge graphs for analysing academia and industry dynamics. In: Proceedings of the ADBIS, TPDL and EDA 2020 Common Workshops and Doctoral Consortium, pp. 219–225 (2020)
4. Aromataris, E., Riitano, D.: Constructing a search strategy and searching for evidence. American Journal of Nursing **114**(5), 49–56 (2014)
5. Auer, S., Bizer, C., Kobilarov, G., Lehmann, J., Cyganiak, R., Ives, Z.: DBpedia: A nucleus for a web of open data. In: Proceedings of the 6th International Semantic Web Conference (ISWC), 2nd Asian Semantic Web Conference (ASWC), pp. 722–735 (2007)
6. Augenstein, I., Padó, S., Rudolph, S.: Lodifier: Generating linked data from unstructured text. In: Proceedings of the 9th Extended Semantic Web Conference (ESWC), pp. 210–224 (2012)
7. Beck, M., Rizvi, S.T.R., Dengel, A., Ahmed, S.: From automatic keyword detection to ontology-based topic modeling. In: Proceedings of the 14th IAPR International Workshop on Document Analysis Systems (DAS), pp. 451–465 (2020)
8. Blei, D.M., Ng, A.Y., Jordan, M.I.: Latent Dirichlet Allocation. Journal of Machine Learning Research **3**, 993–1022 (2003)
9. Bolelli, L., Ertekin, Ş., Giles, C.L.: Topic and trend detection in text collections using Latent Dirichlet Allocation. In: Proceedings of the 31st European Conference on IR Research (ECIR), pp. 776–780 (2009)
10. Borges, M.V.M., dos Reis, J.C.: Semantic-enhanced recommendation of video lectures. In: Proceedings of the 19th IEEE International Conference on Advanced Learning Technologies (ICALT), vol. 2161, pp. 42–46 (2019)
11. Bornmann, L., Mutz, R.: Growth rates of modern science: A bibliometric analysis based on the number of publications and cited references. Journal of the Association for Information Science & Technology **66**(11), 2215–2222 (2015)
12. Boyack, K.W., Klavans, R.: Creation of a highly detailed, dynamic, global model and map of science. Journal of the Association for Information Science & Technology **65**(4), 670–685 (2014)
13. Boyack, K.W., Klavans, R., Börner, K.: Mapping the backbone of science. Scientometrics **64**(3), 351–374 (2005)
14. Cena, F., Likavec, S., Osborne, F.: Anisotropic propagation of user interests in ontology-based user models. Information Sciences **250**, 40–60 (2013)

15. Chatzopoulos, S., Vergoulis, T., Kanellos, I., Dalamagas, T., Tryfonopoulos, C.: ArtSim: Improved estimation of current impact for recent articles. In: Proceedings of the ADBIS, TPDL and EDA 2020 Common Workshops and Doctoral Consortium, pp. 323–334 (2020)
16. Chicaiza, J., Reátegui, R.: Using domain ontologies for text classification. a use case to classify computer science papers. In: Proceedings of the 2nd Iberoamerican Conference and 1st Indo-American Conference on Knowledge Graphs & Semantic Web (KGSWC), pp. 166–180 (2020)
17. Cimiano, P., Völker, J.: Text2onto. In: Proceedings of the 10th International Conference on Application of Natural Language & Information Systems (NLDB), pp. 227–238 (2005)
18. Decker, S.L.: Detection of bursty and emerging trends towards identification of researchers at the early stage of trends. University of Georgia (2007)
19. Dessì, D., Osborne, F., Recupero, D.R., Buscaldi, D., Motta, E., Sack, H.: AI-KG: An automatically generated knowledge graph of artificial intelligence. In: Proceedings of the 19th International Semantic Web Conference (ISWC), vol. 3, pp. 127–143 (2020)
20. Di Caro, L., Guerzoni, M., Nuccio, M., Siragusa, G.: A bimodal network approach to model topic dynamics. CoRR **abs/1709.09373** (2017)
21. Duvvuru, A., Radhakrishnan, S., More, D., Kamarthi, S., Sultornsanee, S.: Analyzing structural and temporal characteristics of keyword system in academic research articles. Procedia Computer Science **20**, 439–445 (2013)
22. Erten, C., Harding, P.J., Kobourov, S.G., Wampler, K., Yee, G.: Exploring the computing literature using temporal graph visualization. In: SPIE Proceedings of the International Conference on Visualization & Data Analysis, vol. 5295, pp. 45–56 (2004)
23. Fenner, M., Aryani, A.: Introducing the PID Graph (2019). URL https://blog.datacite. org/introducing-the-pid-graph
24. Gangemi, A., Presutti, V., Reforgiato Recupero, D., Nuzzolese, A.G., Draicchio, F., Mongiovì, M.: Semantic web machine reading with FRED. Semantic Web **8**(6), 873–893 (2017)
25. García-Silva, A., Gómez-Pérez, J.M.: Classifying scientific publications with BERT - Is self-attention a feature selection method? In: Proceedings of the 43rd European Conference on IR Research (ECIR), vol. 1, pp. 161–175 (2021)
26. Griffiths, T.L., Steyvers, M.: Finding scientific topics. Proceedings of the National Academy of Sciences **101**(suppl 1), 5228–5235 (2004)
27. He, Q., Chen, B., Pei, J., Qiu, B., Mitra, P., Giles, C.L.: Detecting topic evolution in scientific literature: How can citations help? In: Proceedings of the 18th ACM Conference on Information & Knowledge Management (CIKM), pp. 957–966 (2009)
28. Hogan, A., Blomqvist, E., Cochez, M., d'Amato, C., de Melo, G., Gutiérrez, C., Gayo, J.E.L., Kirrane, S., Neumaier, S., Polleres, A., Navigli, R., Ngomo, A.N., Rashid, S.M., Rula, A., Schmelzeisen, L., Sequeda, J.F., Staab, S., Zimmermann, A.: Knowledge graphs. CoRR **abs/2003.02320** (2020)
29. Jaradeh, M.Y., Oelen, A., Farfar, K.E., Prinz, M., D'Souza, J., Kismihók, G., Stocker, M., Auer, S.: Open research knowledge graph: Next generation infrastructure for semantic scholarly knowledge. In: Proceedings of the 10th International Conference on Knowledge Capture (K-CAP), pp. 243–246 (2019)
30. Jo, Y., Lagoze, C., Giles, C.L.: Detecting research topics via the correlation between graphs and texts. In: Proceedings of the 13th ACM International Conference on Knowledge Discovery & Data Mining (KDD), pp. 370–379 (2007)
31. Kandimalla, B., Rohatgi, S., Wu, J., Giles, C.L.: Large scale subject category classification of scholarly papers with deep attentive neural networks. Frontiers in Research Metrics & Analytics **5**, 31 (2021)
32. Kirrane, S., Sabou, M., Fernández, J.D., Osborne, F., Robin, C., Buitelaar, P., Motta, E., Polleres, A.: A decade of semantic web research through the lenses of a mixed methods approach. Semantic Web **11**(6), 979–1005 (2020)
33. Kitchenham, B.: Procedures for performing systematic reviews. Tech. rep., Keele University (2004)
34. Kitchenham, B.A., Charters, S.: Guidelines for performing systematic literature reviews in Software Engineering. Tech. rep., Keele University and Durham University Joint Report (2007)

35. Krampen, G., von Eye, A., Schui, G.: Forecasting trends of development of psychology from a bibliometric perspective. Scientometrics **87**, 687–694 (2011)
36. La Bruzzo, S., Manghi, P., Mannocci, A.: OpenAIRE's DOIBoost - Boosting Crossref for research. In: Proceedings of the 15th Italian Research Conference on Digital Libraries (IRCDL), pp. 133–143 (2019)
37. Leydesdorff, L., Rafols, I., Chen, C.: Interactive overlays of journals and the measurement of interdisciplinarity on the basis of aggregated journal-journal citations. Journal of the American Society for Information Science & Technology **64**(12), 2573–2586 (2013)
38. Likavec, S., Osborne, F., Cena, F.: Property-based semantic similarity and relatedness for improving recommendation accuracy and diversity. International Journal on Semantic Web & Information Systems **11**(4), 1–40 (2015)
39. Liu, X., Song, Y., Liu, S., Wang, H.: Automatic taxonomy construction from keywords. In: Proceedings of the 18th ACM International Conference on Knowledge Discovery & Data Mining (KDD), pp. 1433–1441 (2012)
40. Löffler, F., Wesp, V., Babalou, S., Kahn, P., Lachmann, R., Sateli, B., Witte, R., König-Ries, B.: ScholarLensViz: A visualization framework for transparency in semantic user profiles. In: Proceedings of the Demos and Industry Tracks: From Novel Ideas to Industrial Practice, co-located with 19th International Semantic Web Conference (ISWC), pp. 20–25 (2020)
41. Lula, P., Dospinescu, O., Homocianu, D., Sireteanu, N.A.: An advanced analysis of cloud computing concepts based on the computer science ontology. Computers, Materials & Continua **66**(3), 2425–2443 (2021)
42. Lv, P.H., Wang, G.F., Wan, Y., Liu, J., Liu, Q., Ma, F.C.: Bibliometric trend analysis on global graphene research. Scientometrics **88**, 399–419 (2011)
43. Mai, F., Galke, L., Scherp, A.: Using deep learning for title-based semantic subject indexing to reach competitive performance to full-text. In: Proceedings of the 18th ACM/IEEE Joint Conference on Digital Libraries (JCDL), p. 169–178 (2018)
44. Manghi, P., Atzori, C., Bardi, A., Baglioni, M., Schirrwagen, J., Dimitropoulos, H., La Bruzzo, S., Foufoulas, I., Löhden, A., Bäcker, A., Mannocci, A., Horst, M., Jacewicz, P., Czerniak, A., Kiatropoulou, K., Kokogiannaki, A., De Bonis, M., Artini, M., Ottonello, E., Lempesis, A., Ioannidis, A., Manola, N., Principe, P.: Openaire research graph dump (2020). URL `https://zenodo.org/record/4279381#.YK5jZagzZaQ`
45. Mannocci, A., Osborne, F., Motta, E.: The evolution of IJHCS and CHI: A quantitative analysis. International Journal of Human-Computer Studies **131**, 23–40 (2019)
46. Mikolov, T., Chen, K., Corrado, G., Dean, J.: Efficient estimation of word representations in vector space. In: Workshop Proceedings of the 1st International Conference on Learning Representations (ICLR) (2013)
47. Nuzzolese, A.G., Gentile, A.L., Presutti, V., Gangemi, A.: Conference linked data: The ScholarlyData project. In: Proceedings of the 15th International Semantic Web Conference (ISWC), vol. 2, pp. 150–158 (2016)
48. Osborne, F., Mannocci, A., Motta, E.: Forecasting the spreading of technologies in research communities. In: Proceedings of the Knowledge Capture Conference (K-CAP), pp. 1:1–1:8 (2017)
49. Osborne, F., Motta, E.: Klink-2: Integrating multiple web sources to generate semantic topic networks. In: Proceedings of the 14th International Semantic Web Conference (ISWC), vol. 1, pp. 408–424 (2015)
50. Osborne, F., Motta, E., Mulholland, P.: Exploring scholarly data with Rexplore. In: Proceedings of the 12th International Semantic Web Conference (ISWC), vol. 1, pp. 460–477 (2013)
51. Osborne, F., Muccini, H., Lago, P., Motta, E.: Reducing the effort for systematic reviews in software engineering. Data Science **2**(1–2), 311–340 (2019)
52. Peroni, S., Osborne, F., Di Iorio, A., Nuzzolese, A.G., Poggi, F., Vitali, F., Motta, E.: Research articles in simplified HTML: A web-first format for HTML-based scholarly articles. PeerJ Computer Science **3**, e132 (2017)
53. Peroni, S., Shotton, D.: OpenCitations, an infrastructure organization for open scholarship. Quantitative Science Studies **1**(1), 428–444 (2020)

54. Petersen, K., Vakkalanka, S., Kuzniarz, L.: Guidelines for conducting systematic mapping studies in software engineering: An update. Information & Software Technology **64**, 1–18 (2015)
55. Petrucci, G., Ghidini, C., Rospocher, M.: Ontology learning in the deep. In: Proceedings of the 20th International Conference on Knowledge Engineering & Knowledge Management (EKAW), pp. 480–495 (2016)
56. Ramadhan, M.H., Malik, V.I., Sjafrizal, T.: Artificial neural network approach for technology life cycle construction on patent data. In: Proceedings of the 5th International Conference on Industrial Engineering & Applications (ICIEA), pp. 499–503 (2018)
57. Rossanez, A., dos Reis, J.C., da Silva Torres, R.: Representing scientific literature evolution via temporal knowledge graphs. In: Proceedings of the 6th Workshop on Managing the Evolution & Preservation of the Data Web (MEPDaW), co-located with the 19th International Semantic Web Conference (ISWC), pp. 33–42 (2020)
58. Salatino, A., Osborne, F., Motta, E.: Researchflow: Understanding the knowledge flow between academia and industry. In: Proceedings of the 22nd International Conference on Knowledge Engineering & Knowledge Management (EKAW), pp. 219–236 (2020)
59. Salatino, A., Thanapalasingam, T., Mannocci, A., Osborne, F., Motta, E.: Classifying research papers with the computer science ontology. In: Proceedings of the ISWC Posters & Demonstrations, Industry and Blue Sky Ideas Tracks co-located with 17th International Semantic Web Conference (ISWC) (2018)
60. Salatino, A.A., Osborne, F., Motta, E.: How are topics born? Understanding the research dynamics preceding the emergence of new areas. PeerJ Computer Science **3**, e119 (2017)
61. Salatino, A.A., Osborne, F., Motta, E.: Augur: Forecasting the emergence of new research topics. In: Proceedings of the 18th ACM/IEEE on Joint Conference on Digital Libraries (JCDL), p. 303–312 (2018)
62. Salatino, A.A., Osborne, F., Thanapalasingam, T., Motta, E.: The CSO classifier: Ontology-driven detection of research topics in scholarly articles. In: Proceedings of the 23rd International Conference on Theory & Practice of Digital Libraries (TPDL), pp. 296–311 (2019)
63. Salatino, A.A., Thanapalasingam, T., Mannocci, A., Birukou, A., Osborne, F., Motta, E.: The computer science ontology: A comprehensive automatically-generated taxonomy of research areas. Data Intelligence **2**(3), 379–416 (2020)
64. Salatino, A.A., Thanapalasingam, T., Mannocci, A., Osborne, F., Motta, E.: The Computer Science ontology: A large-scale taxonomy of research areas. In: Proceedings of the 17th International Conference on The Semantic Web (ISWC), vol. 2, pp. 187–205 (2018)
65. Satopaa, V., Albrecht, J., Irwin, D., Raghavan, B.: Finding a "kneedle" in a haystack: Detecting knee points in system behavior. In: Proceedings of the 31st IEEE International Conference on Distributed Computing Systems Workshops (ICDCS), pp. 166–171 (2011)
66. Sinha, A., Shen, Z., Song, Y., Ma, H., Eide, D., Hsu, B.J.P., Wang, K.: An overview of Microsoft Academic Service (MAS) and applications. In: Proceedings of the 24th International Conference on World Wide Web (WWW Companion), p. 243–246 (2015)
67. Tang, J., Zhang, J., Yao, L., Li, J., Zhang, L., Su, Z.: Arnetminer: Extraction and mining of academic social networks. In: Proceedings of the ACM International Conference on Knowledge Discovery & Data Mining (KDD), pp. 990–998 (2008)
68. Thanapalasingam, T., Osborne, F., Birukou, A., Motta, E.: Ontology-based recommendation of editorial products. In: Proceedings of the 17th International Semantic Web Conference (ISWC), vol. II, pp. 341–358 (2018)
69. Tseng, Y.H., Lin, Y.I., Lee, Y.Y., Hung, W.C., Lee, C.H.: A comparison of methods for detecting hot topics. Scientometrics **81**(1), 73–90 (2009)
70. Van Raan, A.F.: Sleeping Beauties in science. Scientometrics **59**(3), 467–472 (2004)
71. Vergoulis, T., Chatzopoulos, S., Dalamagas, T., Tryfonopoulos, C.: VeTo: Expert set expansion in academia. In: Proceedings of the 24th International Conference on Theory & Practice of Digital Libraries (TPDL), pp. 48–61 (2020)
72. Visser, M., van Eck, N.J., Waltman, L.: Large-scale comparison of bibliographic data sources: Scopus, Web of Science, Dimensions, Crossref, and Microsoft Academic. Quantitative Science Studies **2**(1), 20–41 (2021)

73. Wang, K., Shen, Z., Huang, C., Wu, C.H., Dong, Y., Kanakia, A.: Microsoft Academic Graph: When experts are not enough. Quantitative Science Studies **1**(1), 396–413 (2020)
74. Watatani, K., Xie, Z., Nakatsuji, N., Sengoku, S.: Global competencies of regional stem cell research: Bibliometrics for investigating and forecasting research trends. Regenerative Medicine **8**(5), 659–68 (2013)
75. Widodo, A., Fanany, M.I., Budi, I.: Technology forecasting in the field of Apnea from online publications: Time series analysis on Latent Semantic. In: Proceedings of the 6th International Conference on Digital Information Management (ICDIM), pp. 127–132 (2011)
76. Wohlin, C., Prikladnicki, R.: Systematic literature reviews in software engineering. Information & Software Technology **55**(6), 919–920 (2013)
77. Zhang, X., Chandrasegaran, S., Ma, K.L.: Conceptscope: Organizing and visualizing knowledge in documents based on domain ontology. In: Proceedings of the CHI Conference on Human Factors in Computing Systems, pp. 1–13 (2021)
78. Zhang, Y., Lu, J., Liu, F., Liu, Q., Porter, A., Chen, H., Zhang, G.: Does deep learning help topic extraction? A kernel k-means clustering method with word embedding. Journal of Informetrics **12**(4), 1099–1117 (2018)

Chapter 12
Early Detection of Emerging Technologies using Temporal Features

Xiaoli Chen and Tao Han

Abstract This paper proposes a framework for detecting emerging technologies at an early stage based on temporal features. The framework included modeling of the technology citation network as well as multisource social media streaming data. In each section, we use statistical analysis to demonstrate the significance of temporal features, laying a solid foundation for data representation and prediction. We employ feature engineering and graph embedding to represent temporal data. To identify emerging technologies, the supervised machine learning and outlier detection methods were used. We also put both methods to test with labeled data. In the multisource social media streaming model and the citation network growth model, we demonstrate the pipeline of predicting emerging technology. However, one shoe doesn't fit all. Deploying the framework in a real-world prediction scenario still requires careful design.

12.1 Introduction

In recent years, the Fortune Global 500[1] has seen an increase in the number of new innovative companies. The success of these innovative companies is partly due to their success in capitalizing the emerging technologies. Sensing and responding to technological changes is a critical capability for individuals, organizations, and nations in today's world. This ability is also referred to as technological opportunism [50], and it has an increasing competitive value. Hence, technology forecasting has been

Xiaoli Chen[1,2] and Tao Han[1,2]
[1] National Science Library, Chinese Academy of Sciences
[2] Department of Library, Information and Archives Management, School of Economics and Management, University of Chinese Academic of Sciences
e-mail: chenxl@mail.las.ac.cn

[1] It is an annual list released by Fortune Media IP Limited, the latest list can be found at https://fortune.com/global500/. Lasted retrieved on Feb 28, 2021.

© The Author(s), under exclusive license to Springer Nature Switzerland AG 2021
Y. Manolopoulos, T. Vergoulis (eds.), *Predicting the Dynamics of Research Impact*,
https://doi.org/10.1007/978-3-030-86668-6_12

gaining researchers' and entrepreneurs' attention, to minimize the uncertainty of the future, and better prepare for future disruption.

Several concepts like emerging technologies, transformative technologies, and disruptive technologies [12] are difficult to distinguish. In this paper, we treat them as interchangeable terms, representing technologies with relatively rapid growth, radical novelty, and the potential to have a significant impact. Characterizing and forecasting these technologies at an early stage can help us gain a significant competitive advantage. Please see for more elaborate and detailed surveys [6, 43, 45, 57].

In terms of identifying emerging technologies, To what extent or quantity can we be certain that the technology is promising? And how can we quantify the size of the emerging market? From the pioneering work of forecasting emerging technologies by citation network [49], more recent research [4, 52] has confirmed that the emerging arcs in the citation network exhibit patterns of gaining more connections in a short period, indicating a type of growth. Having more connections or being cited also indicates that their radical novelty and impact have been recognized. Lexical-based approaches [2, 21, 24, 56], on the other hand, are intended to be more accurate in calibrating novelty. Scientometric indicators [17, 31, 39, 54, 57] and social media indicators [25, 43] are used to assess human resources, capital resources, and social awareness and devotion to technology, which can be another source of prominence and novelty.

This paper proposes a framework for fully demystifying temporal features, which includes a forecasting model based on temporal network growth and a forecasting model based on temporal trend features in multiple social media source monitoring. In the first model, we create an A/B test to identify the distinguishing features of emerging technologies in the temporal citation network. To identify emerging technologies at an early stage, we also employ a temporal embedding-based outlier detection method. We collect stream data from multiple social media data sources in the second model. These two models can help policymakers make decisions about emerging technologies.

12.2 Related Work

Technology forecasting is a hot topic for research policy studies and innovation management studies. We will first overview the data source, qualifying methods, and quantifying methods in Sect. 12.2.1. As our method belongs to quantifying methods, the two key challenges in quantifying methods of data representation and forecasting methods are overviewed in Sect. 12.2.2 and Sect. 12.2.3.

12.2.1 Qualifying Methods, Quantifying Methods and Data Source

Since the early days of technology forecasting by the government, business, and researchers, the way we forecast has changed in three ways: data source, qualifying methods, and quantifying methods.

In the primitive stage, the way we do technology assessment and technology road mapping [27, 33, 55] were usually made by expert consultation. The evaluation criteria include technology maturity, product, industry, and external environment. All of these metrics are purely evaluated by experts. Business models are also widely used to evaluate technology [13]. These qualifying methods also lay a solid foundation for quantifying methods.

Quantity models aim to target different perspectives of emerging technology. Citation network [4, 49, 52] based works confirmed the emerging arcs in citation network show patterns of getting more connection in a short time. Getting more connections or cited also means their radical novelty and have an impact. However,lexical based approaches [2, 21, 24, 51, 56] are meant to be more accurate in calibrating novelty. Scientometric indicators [14, 17, 29, 31, 39, 54, 57] and social media indicators [25, 37, 43] are leveraged to measure the human resources, capital resources, and social awareness and devotion of the technology, which can be another perspective of prominent impact and novelty.

Various data source has been used in quantity models, like scientific papers [4], patents [39], R&D statistics [13], questionnaires, citations [52], social media or web data [10, 43],and business data. The digestion and representation of these data become a great challenge for quantity models. Data representation is one of the challenges for quantity models.

12.2.2 Data Representation Methods

Feature engineering is mainly used to represent data and further make quantity predictions. Features used in forecasting models include patent bibliometrics [39], publication bibliometrics [4], citation metrics [4], knowledge spillover, and co-keyword [16, 26] based semantic metrics. These hand-crafted features can represent the characteristics of emerging technologies in some aspects, particularly the novelty and prominent impact. While there are some window sliced methods for calibrating growth, few studies demonstrate what the effective growth metric is and how to describe temporal growth.

Recently, embedding-based methods are adopted, like word2vec [36], BERT [15]. These models are more accurate in representing the semantic meaning of the unstructured text in scientific literature. They are the more widely used methods for assessing radical novelty. To identify emerging technology nodes, graph embedding methods can be used to represent each publication node by extracting features from citation networks. The static graph embedding cannot account for the evolution of emerging technology. There are two alternatives: first, extract temporal statistic features from each node in the time series data and combine them with static graph embedding. Second, employ dynamic temporal embedding methods [3, 18, 42]. In the paper, we use the former method, which can be easily incorporated with hand-crafted features.

12.2.3 Forecasting Methods

To determine a technology to be emerging or not, mathematically it is a binary classification or regression problem. most of the previous studies are these two types, they choose regression function or machine learning methods to determine an emerging technology.

Mariani proposes an age-normalized measure of patent centrality as a key feature to distinguish important and non-important publications [34, 35]. Porter uses lexical change measures to predict emerging technology on four attributes: novelty, persistence, community, and growth [39]. Kyebambe proposes a novel algorithm to automatically label data and then use the labeled data to train learners to forecast emerging technologies [29].

Another type of method is clustering or topic modeling [1], where they carry the presumption that emerging technologies will cluster as a significant subgroup in the data. Small proposes an emergence potential function as a post-processing step of co-citation clustering [48]. Salatino showed the pace of collaboration in topic graphs, which is also a representation of connecting with more diverse existing technologies [44, 45]. Chen develops the destabilization index and the consolidation index using patent citation networks. It explores the metrics' ability to measure technologies' radicalness [8].

Our method aims to codify temporal feature engineering from multisource data, as well as temporal feature engineering in citation networks via embedding. The other contribution of this paper is that we demonstrate a forecasting method based on outlier detection that is more reasonable because the two categories of emerging and not emerging are highly unbalanced.

12.3 Framework and Method

This paper proposes a systematic framework based on data science for persistently monitoring and forecasting emerging technology. We fully describe the framework pipeline in Sect. 12.3.1. Two key methods used in this framework are introduced in Sect. 12.3.2 and Sect. 12.3.3.

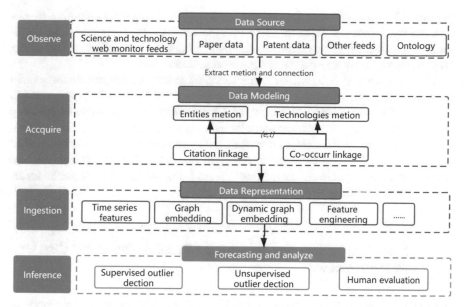

Fig. 12.1 Data science-driven technology monitoring and forecasting framework. The framework includes four steps: observe, acquire, ingestion and inference

12.3.1 Framework

In today's complex and ever-changing world, the data about technology is ubiquitous. This necessitates the development of a persistent system capable of automatically harvesting data from multiple sources, learning to acquire a model, ingesting information, and inferring the evolution of technology on a large scale. These four steps are depicted in Fig. 12.1.

The framework aims to model the human, capital, or intellectual resources to imagine every possible emerging scenario, capture every signal, or have access to all critical data. At the stage of development, we have included science and technology web monitor feeds. The paper data and patent data are purchased from a commercial supplier. Other types of data include company databases, expert databases, and product databases. The goal of ontology data is to reduce the difficulty of ingesting domain terms and domain technology [9].

The data model we use to fuse the multiple data sources aims to provide (wh, t) tuples and the (wh_1, wh_2, t) triples. The wh means "where","who","what", "technology". We use natural language processing methods to extract these entities and their timestamp. (wh_1, wh_2, t) describes the temporal network of the wh. the relations between wh can be either citation linkage or co-occurrence.

To cope with various scenarios, we need different data representation methods [11]. Time series features to aid in identifying emerging weak signals. The citation network and co-occur networks can be better represented using graph embeddings. The goal of dynamic graph embeddings is to detect weak emerging signals in a changing network. Because some vivid features known by humans can be directly coped with machine features, feature engineering is an effective and direct way to improve the performance of data representation.

The final section is based on machine intelligence and human evaluation. In contrast to previous research, emerging technology detection may be better suited to the outlier detection method. In Section 12.3.2, We will show a case study of detecting emerging technology in the semiconductor industry using a dynamic citation network model, as well as experiment with various data representation methods and outlier detection to achieve the best performance. In Section 12.3.3, to achieve the best performance, we experiment with digesting and fusing multiple social media feeds, and we use ensemble machine learning methods.

12.3.2 Emerging Technology Forecasting Based on Temporal Citation Network Growth

In this model, we create an A/B test to identify the distinguishing features of emerging technologies in their temporal citation network. To identify emerging technologies at an early stage, we also employ a temporal embedding-based outlier detection method.

Data collection and A/B testing

We collect "Semiconductor" emerging technologies from Computer History Museum[2]. We gather emerging "Semiconductor" technologies from the Computer History Museum[3]. The references of each emerging technologies' introduction web page arc the emerging papers[4]. We design an A/B testing experiment. The controlled data (which are non-emerging papers, denoted as $NM1$) and the emerging papers (denoted as $M1$) share the common reference. Both groups published in the same date range. We collect the backward citations of emerging papers, denoted as P0. The citations of $P0$ can be divided into two groups: emerging papers $M1$ and non-emerging papers $NM1$. Both groups have citations. These citations (denoted as $C1$) may overlap and can be connected as a temporal citation network. The data collection and citation network constructing process is depicted in Fig. 12.2. We want to examine the difference between the two groups in their temporal citation network.

With the emphasis on the growth feature, we use the steepest slope of the citations time series to describe the pace of growth $gmaxrate$ (this scheme will be denoted as $ctss\text{-}gr$). Given the time instant T, the maximum growth rate is the difference between the peak value and bottom value in the time series of citation counts, divide

---

[2] https://www.computerhistory.org/siliconengine/timeline/

[3] https://www.computerhistory.org/siliconengine/timeline/

[4] Since we use the paper to represent the birth of technology, emerging papers will be phrased as emerging technologies

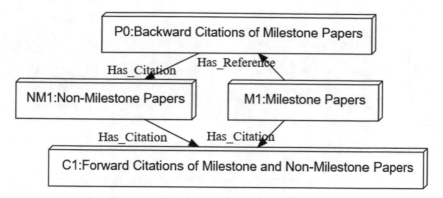

Fig. 12.2 Data collect and process in citation network growth model

the period of the two values as Eq. 12.1. This value could be negative; for some time series, the maximum change could be the downward trend. This growth feature will be added to the *cts* representation scheme, to validate if the effectiveness of the proposed growth feature. The *ctss-gr* scheme has the dimension of $(N_V, 6, \tau)$ at the observing time τ.

$$gmaxrate = \frac{\max(c_0, c_1, \ldots, c_T) - \min(c_0, c_1, \ldots, c_T)}{index(\max(c_0, c_1, \ldots, c_T)) - index(\min(c_0, c_1, \ldots, c_T))} \quad (12.1)$$

Fig. 12.3 illustrates data distribution under three temporal features: *range, sum* and *gmaxrate. range* is the date range between the first cited year and last cited year. *gmaxrate* measures how fast the citation times reached the peak. These three features are better performed in distinguishing the emerging technology and non-emerging technology at the embryo stage than other statistic features in our analysis as shown in Figs. 12.4-12.7, which are the Correlation between *gmaxrate* and *prange* feature at 5, 10, 15, 20 years after debutant, respectively.

Temporal embeddings and outlier detection

In this model, we will explore how a dynamic temporal citation network can help in predicting emerging technology. We experiment with different representation methods. In Base feature representations, we use *count, mean, std, max, min, sum, range* of each technology's citation time series data. In temporal embedding, we use the Python package tsfresh[5] to automatically extract features from citation time series data. Graph embeddings we choose over deepWalk [38], node2vec [20], and LINE [53] method. In temporal graph embedding, we use the ensemble of temporal embedding and graph embedding. *gmaxrate* feature is added to each representation scheme.

We use an outlier detection algorithm to detect emerging technologies. The final results are the average ensemble of ten different algorithms: Angle-based Outlier

---

[5] https://tsfresh.readthedocs.io/en/latest/text/list_of_features.html

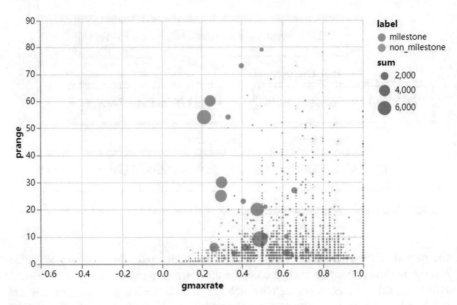

Fig. 12.3 Correlation between *gmaxrate* and *prange* feature

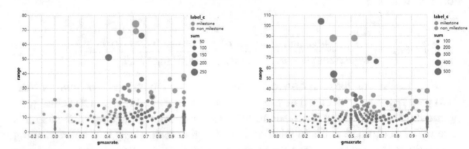

Fig. 12.4 Correlation between *gmaxrate* and *prange* feature of 5 year after debutant

Fig. 12.5 Correlation between *gmaxrate* and *prange* feature of 10 year after debutant

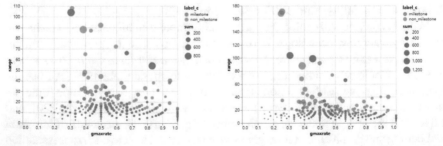

Fig. 12.6 Correlation between *gmaxrate* and *prange* feature of 15 year after debutant

Fig. 12.7 Correlation between *gmaxrate* and *prange* feature of 20 year after debutant

Detector (ABOD) [28], Cluster-based Local Outlier Factor (CBLOF) [22], Feature Bagging [30], Histogram-base Outlier Detection (HBOS) [19], Isolation Forest [32], K Nearest Neighbors (KNN), Average KNN [40], Local Outlier Factor (LOF) [5], One-class SVM (OCSVM) [46], Principal Component Analysis (PCA) [47]. The main aim of our work is not to compare different outlier detection algorithm. The Pyod python package [58] also encourages the use of an ensemble way since no single algorithm can guarantee the best performance in every scenario. In this case, we use ensemble methods to get a robust result.

Results and analysis

The Average of Maximum results of different feature patterns and different embryo stages (5 to 20 years, respectively) are shown in Table 12.1, with $gmaxrate$ feature the performance is better at the different embryo stage. The only exception is the minor superior of Temporal embedding over $gmaxrate$, which may mean temporal embedding already captured growth well. However, we have shown that we can achieve a reasonable accuracy of emerging technology forecasting at the embryo stage.

Table 12.1 Results of different representation methods and different embryo stages

Representations	Embryo@5	Embryo@10	Embryo@15	Embryo@20
Base feature	0.6344	0.6345	0.6559	0.664
Base feature + $gmaxrate$	0.6523	0.6578	**0.6707**	0.672
Temporal embedding	**0.747**	0.6744	0.5615	0.6757
Temporal embedding + $gmaxrate$	0.7344	0.658	0.6139	**0.7183**
Graph embeddings	0.5020	0.5037	0.5056	0.4778
Graph embeddings + $gmaxrate$	0.5086	0.5114	0.5127	0.4696
Temporal graph embedding (TGE)	0.5786	0.668	0.6481	0.593
TGE + $gmaxrate$	0.5656	**0.6867**	0.6255	0.5758

The main contribution of this model is that we use A/B testing to investigate temporal features of emerging technology. We have used citation network growth, which measures the pace of acceptance by the scientific community as a dynamic feature of emerging technology. If technology is mentioned at a rapid pace, it is more likely to be an emerging technology. We use temporal embedding-based methods to capture the characteristics of the temporal citation network based on observations of temporal features. The results demonstrated the efficacy of temporal embedding methods, particularly for emerging technology in its early stages.

12.3.3 Emerging Technology Forecasting Based on Monitoring Multiple Social Media Source

We collect stream data from multiple social media data sources in the second model. The social media data sources in this model are used to reflect the social acceptance and influence of the technology. We use features of time-series data from multiple social media data sources. We used data from Gartner Hype Cycle, Google Trends, Google Book Ngram viewer, Scopus scientific papers, and US Patents, respectively.

Social media source and feature engineering

Gartner Hype Cycle

To convert the Gartner Hype Cycle data into computable features, we transform each data point in two dimensional Gartner Hype Cycle to three dimensional representation as in Fig. 12.8, which has *time, phase* and *plateau will be reached in* dimensions, denoted as (x_t), (x_{ph}) and (x_{pl}) respectively. With data been mapped to three dimensions, we can obtain following features of candidate Hype Cycle technology:

1. the number of occurrences: N,
2. the last phase of Phase: PH,
3. Phase change: $PH = PH_{te} - PH_{ts}$, where te and ts represent the last year candidate technology showed up and the first year candidate technology showed up respectively.,
4. Slope of Phase change: $k_{PH} = PH/(te - ts)$,
5. the last expectation time PL,
6. the amount of change in the expectation time: $PL = PL_{te} - PL_{ts}$,
7. Slope of the expectation time: $k_{PL} = PL/(te - ts)$

In the consideration of calculating above seven feature, the intermediate state is ignored, and only the first and last states are considered. Another import issue here is that (x_{ph}) and (x_{pl}) must be on same scale, which means after encoding, the technology with greater value should be more matured.

Google Trends index[6]

Google Trends can be used as a measure of people's attention towards the candidate technology: We use the Python version of Google Trend API[7] to get trend data of each candidate technology. Google Trend API will return two data frames of interest over time and interest over the region. The former is time-series data, in which we need to extract key features from time series. We use the Python library tsfresh, which will return the most relevant features from 65 statistic values, which we denote as $GTTS$.

We use the latter data frame to compute the influential metric measured by region $GTR = NR/NC$, where NR is the number of regions the candidate technology has been recognized, NC is the total number of regions. Another trick of Google Trend API is that it provides us suggested term, which is a similar term to candidate technology. We use candidate technology and the suggested term to get trend data in this part and the following part.

---

[6] https://trends.google.com

[7] https://github.com/GeneralMills/pytrends

Google book Ngram viewer
Google Books Ngram[8] provides Ngram data for a portion of Google Books scanned and digitized (4 of human published books) and can be used to query the frequency curve of a term in all publications from 1800 to the present. Through this platform, we can measure the attention and growth of technology. We use the Python version of Google Books Ngram API[9], which also returns time series data of candidate technology. We also use the tsfresh library to extract key features of Google Books Ngram data, denoted as $GBTS$.

Scientific literature trend from Scopus
Scientific literature data is mainly used to measure the growth potential of candidate technology. And we assume that if a technology has potential, then more organizations and authors will study this technology. We use Scopus API[10] to get articles, citations, and authors' time-series data. And we also extract key features of articles time series, citations time series, and authors time series using tsfresh, denoted by $SNTS$, $SCTS$, and $SATS$ respectively.

Patents trend from US Patents
Patent data is primarily used to measure growth potential in the evolution of candidate technologies. And we also assume that if a technology has potential, then more organizations and authors will study this technology. We use USPTO patent search API[11] to get patent numbers over the years and extract key features of PNTS using

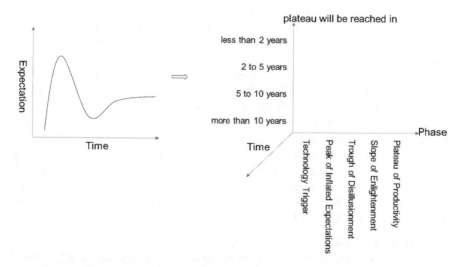

Fig. 12.8 Gartner Hype Cycle three-dimensional representation

---

[8] https://books.google.com/ngrams

[9] https://github.com/econpy/google-ngrams

[10] https://github.com/scopus-api/scopus

[11] https://github.com/USPTO/PatentPublicData

tsfresh. With the above features been computed, some of them are single values, some of them are lists. We concatenate them as features of candidate technology for our machine learning models.

Supervised classification model based on temporal features

Technology is constantly evolving, and even well-known emerging technologies have the potential to be disrupted again, so when collecting data, you must pay close attention to the selection of appropriate time windows. Whether or not the main criterion for emerging technology is generally accepted by the public at a given time is not judged by whether or not it is still considered an emerging technology. In the case of RFID, RFID was once an emerging technology, but it is now being overturned by lower-cost QR codes.

If we annotate which technologies are emerging based on nowhere, it would be a great challenge. However, Gartner Hype Cycle makes this challenge possible since we can make our decision based on Gartner's choice. Among Gartner's technology predictions, some create a splash, some remain to stand the test of time, which we think is the emerging technology we are looking for.

We ask several annotators to label each technology to be emerging to minimize biases. However, some of the technology is vague to judge, like "Business Process Fusion" in 2003 and 2004 Gartner Hype Cycle, which are hard to evaluate. In our work, we ask several graduate students to label candidate technology to be emerging or not based on the three criteria introduced in Sect. 12.3. We average their labeled data as our training data.

However, the goal of this research is to give a framework for forecasting emerging technology based on historical data in a supervised machine learning way. We can add more experts' wisdom in giving labels to historical data to improve the model's performance.

We manually label 387 of technology based on the Gartner Hype Cycle from 1995 to 2018. We use Google Open Refine[12] to clean the technical terms since the same term sometimes uses a different name in different years. Using the feature engineering method, we get $(387, 3981)$ dimension feature matrix. We split the data into 2/3 and 1/3 as training data and test data for our machine learning method.

We selected four supervised machine learning algorithms for data analysis, namely Logistic Regression [23], Random Forest [41], RBF kernel SVM [7], and Ensemble methods [7] (which is the combination of Logistic Regression, Random Forest, RBF kernel SVM on the proportion of 2:1:1). The classification boundary of the four algorithms is shown in Fig.12.9.

Results and Analysis

In this section, we discussed how to use machine learning methods to identify potential emerging technologies. The main contribution of our work is to forecast the future based on historical data using machine learning. Allow data to decide whether technology is emerging or not. We collect various time-series data from scientific literature platforms (Scopus papers and US patents) and social media

---

[12] http://openrefine.org/

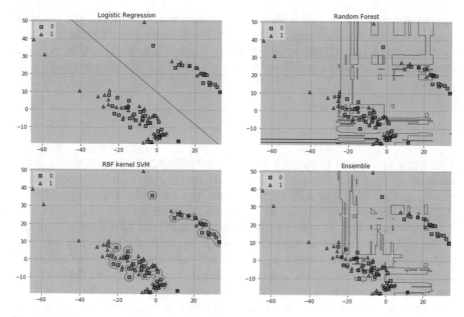

Fig. 12.9 Classification boundary of the four algorithms

platforms to determine the technological and social impact of technology (Google Trends and Google Book Ngram). To reduce the self-defined metric's bias and try to keep as much information of time series data as possible, we extract 65 time-series features for each time series data using Python Package 'tsfresh'. In the four machine learning algorithms, Logistic Regression performs the best in predicting emerging technology, with the precision, recall, and f1-score of 0.71, 0.65, 0.63, as shown in Table 12.2.

Table 12.2 Evaluation of different supervised technology forecasting algorithms

Method	precision	recall	f1-score
Logistic Regression	0.71	0.65	0.63
Random Forest	0.54	0.54	0.53
RBF Kernel SVM	0.6	0.59	0.59
ENSEMBLE	0.65	0.64	0.63

This model demonstrates how to forecast technology in a data-driven manner. However, there is still room for advancement. To begin, keeping a close eye on technological advancements will help us improve the performance of our model. A forecasting system should monitor not only the rate of change of the parameters of interest but also the second derivative of change—that is, the rate of change. Accelerating rates of change may be an indicator of impending disruption, as evidenced by the features. Second, because the candidate technology is only 30 years old, we may get more accurate labels with a longer development time, or use more experts' wisdom to improve the accuracy of the labels. Finally, emerging technology, like the Gartner Hype Cycle, should be divided into stages. Then, to predict the potential emerging stage, we could extend our model to a multi-classification model.

12.4 Conclusion

This paper proposes a framework for fully demystifying temporal features, which includes an emerging technology forecasting model based on temporal network growth and an emerging technology forecasting model based on temporal trend features based on multiple social media source monitoring.

In the first model, we create an A/B test to identify the distinguishing features of emerging technologies in their temporal citation network. To identify emerging technologies at an early stage, we employ a temporal embedding-based outlier detection method at different embryo stages after the technology's first debutant. Our result showed the average performance of the model with $gmaxrate$ metric performs better compared to models without the metric. That is if the technology is cited at an accelerating speed, it is more likely to be an emerging technology. The temporal embedding method with $gmaxrate$ features performs the best.

We collect stream data from multiple social media data sources in the second model. We use this model to reflect the social acceptance and influence of the technology. We use features of time-series data from multiple social media data sources. The test on Gartner Hype Cycle data demonstrated the feasibility of the method.

Overall, no single source or method can guarantee robust emerging technology forecasting. data-driven technology forecasting should consider every possibility of signals the emerging technology may reflect. In this paper, we designed a framework to cover two such scenarios, in which we may detect signals of emerging. In both scenarios, we proposed models of data representation and the data discrimination method. We also use data to validate the models. However, the limitation of both models doesn't consider the novelty and potential social benefits of the technology, this remains to be further studied and incorporated with current models for more accurate forecasting.

Acknowledgements The authors thank anonymous reviewers for their helpful comments that improved the manuscript. This work is supported in part by the National Natural Science Foundation of China under contracts No 71950003, and the Programs for the Young Talents of National Science Library, Chinese Academy of Sciences (Grant No. 2019QNGR003).

References

1. Abdolreza, M., Rost, K.: Identification and monitoring of possible disruptive technologies by patent-development paths and topic modeling. Technological Forecasting & Social Change **104**, 16–29 (2016)
2. Arora, S.K., Youtie, J., Shapira, P., Gao, L., Ma, T.: Entry strategies in an emerging technology: A pilot web-based study of graphene firms. Scientometrics **95**(3), 1189–1207 (2013)
3. Asatani, K., Mori, J., Ochi, M., Sakata, I.: Detecting trends in academic research from a citation network using network representation learning. PLOS ONE **13**(5), e0197260 (2018)
4. Boyack, K.W., Klavans, R., Small, H., Ungar, L.: Characterizing the emergence of two nanotechnology topics using a contemporaneous global micro-model of science. Journal of Engineering & Technology Management **32**, 147–159 (2014)
5. Breunig, M., Kriegel, H., Ng, R., Sander, J.: LOF: Identifying density-based local outliers. In: Proceedings of the ACM International Conference on Management of Data (SIGMOD), pp. 93–103 (2000)
6. Burmaoglu, S., Sartenaer, O., Porter, A., Li, M.: Analysing the theoretical roots of technology emergence: An evolutionary perspective. Scientometrics **119**(1), 97–118 (2019)
7. Cao, H., Naito, T., Ninomiya, Y.: Approximate RBF kernel SVM and its applications in pedestrian classification. In: Proceedings of the 1st International Workshop on Machine Learning for Vision-based Motion Analysis (MLVMA) (2008)
8. Chen, J., Shao, D., Fan, S.: Destabilization and consolidation: Conceptualizing, measuring, and validating the dual characteristics of technology. Research Policy **50**(1), 104115 (2021)
9. Chen, X.: Evolving taxonomy based on graph neural networks. In: Proceedings of the 20th ACM/IEEE Joint Conference on Digital Libraries (JCDL), pp. 465–466 (2020)
10. Chen, X., Han, T.: Disruptive technology forecasting based on Gartner hype cycle. In: Proceedings of the IEEE Technology & Engineering Management Conference (TEMSCON), pp. 1–6 (2019)
11. Chen, X., Zhang, Z.: Representing and reconstructing PhySH: Which embedding competent? In: Proceedings of the 8th International Workshop on Mining Scientific Publications, pp. 48–53 (2020)
12. Christensen, C.: The Innovator's Dilemma: When New Technologies Cause Great Firms to Fail. MA: Harvard Business School Press, Boston (1997)
13. Collins, R., Hevner, A., Linger, R.: Evaluating a disruptive innovation: Function extraction technology in software development. In: Proceedings of the 44th Hawaii International International Conference on Systems Science (HICSS), pp. 1–8 (2011)
14. Daim, T.U., Rueda, G., Martin, H., Gerdsri, P.: Forecasting emerging technologies: Use of bibliometrics and patent analysis. Technological Forecasting & Social Change **73**(8), 981–1012 (2006)
15. Devlin, J., Chang, M.W., Lee, K., Toutanova, K.: BERT: Pre-training of deep bidirectional transformers for language understanding. In: Proceedings of the Conference of the North American Chapter of the Association for Computational Linguistics: Human Language Technologies (NAACL-HLT), vol. 1, pp. 4171–4186 (2019)
16. Dotsika, F., Watkins, A.: Identifying potentially disruptive trends by means of keyword network analysis. Technological Forecasting & Social Change **119**, 114–127 (2017)
17. Ganguly, A., Nilchiani, R., Farr, J.: Defining a set of metrics to evaluate the potential disruptiveness of a technology. Technology Engineering Management Journal **22**(1), 34–44 (2010)
18. Ghosh, R., Kuo, T.T., Hsu, C.N., Lin, S.D., Lerman, K.: Time-aware ranking in dynamic citation networks. In: Proceedings of the 11th IEEE International Conference on Data Mining Workshops (ICDMW), pp. 373–380 (2011)
19. Goldstein, M., Dengel, A.: Histogram-based outlier score (HBOS): A fast unsupervised anomaly detection algorithm. In: Proceedings of the Posters & Demo Track of the 35th German Conference on Artificial Intelligence (KI), pp. 59–63 (2012)
20. Grover, A., Leskovec, J.: node2vec: Scalable feature learning for networks. Proceedings of the 22nd ACM International Conference on Knowledge Discovery & Data Mining (KDD) pp. 855–864 (2016)

21. Guo, H., Weingart, S., Börner, K.: Mixed-indicators model for identifying emerging research areas. Scientometrics **89**(1), 421–435 (2011)
22. He, Z., Xu, X., Deng, S.: Discovering cluster-based local outliers. Pattern Recognition Letters **24**(9–10), 1641–1650 (2003)
23. Hosmer, D.W., Lemeshow, S.: Applied Logistic Regression, 2 edn. Wiley (2000)
24. Joung, J., Kim, K.: Monitoring emerging technologies for technology planning using technical keyword based analysis from patent data. Technological Forecasting & Social Change **114**, 281–292 (2017)
25. Ke, Q.: Technological impact of biomedical research: The role of basicness and novelty. Research Policy **49**(7), 104071 (2020)
26. Kim, J., Park, Y., Lee, Y.: A visual scanning of potential disruptive signals for technology roadmapping: Investigating keyword cluster, intensity, and relationship in futuristic data. Technology Analysis & Strategic Management pp. 1–22 (2016)
27. Kostoff, R., Boylan, R., Simons, G.: Disruptive technology roadmaps. Technological Forecasting & Social Change **71**(1), 141–159 (2004)
28. Kriegel, H., Schubert, M., Zimek, A.: Angle-based outlier detection in high-dimensional data. In: Proceedings of the 14th ACM International Conference on Knowledge Discovery & Data Mining (KDD), pp. 444–452 (2008)
29. Kyebambe, M.N., Cheng, G., Huang, Y., He, C., Zhang, Z.: Forecasting emerging technologies: A supervised learning approach through patent analysis. Technological Forecasting & Social Change **125**, 236–244 (2017)
30. Lazarevic, A., Kumar, V.: Feature bagging for outlier detection. In: Proceedings of the 11th ACM International Conference on Knowledge Discovery & Data Mining (KDD), pp. 157–166 (2005)
31. Lee, C., Kwon, O., Kim, M., Kwon, D.: Early identification of emerging technologies: A machine learning approach using multiple patent indicators. Technological Forecasting & Social Change **127**, 291–303 (2018)
32. Liu, F., Ting, K., Zhou, Z.: Isolation forest. 2008 Eighth IEEE International Conference on Data Mining pp. 413–422 (2008)
33. Majja, S., Kaisu, P.: Evaluating technology disruptiveness in a strategic corporate context: A case study. Technological Forecasting and Social Change **74**(8), 1315–1333 (2007)
34. Mariani, M.S., Medo, M., Lafond, F.: Early identification of important patents: Design and validation of citation network metrics. Technological Forecasting & Social Change **146**, 644–654 (2019)
35. Mariani, M.S., Medo, M., Zhang, Y.C.: Identification of milestone papers through time-balanced network centrality. Journal of Informetrics **10**(4), 1207–1223 (2016)
36. Mikolov, T., Chen, K., Corrado, G.S., Dean, J.: Efficient estimation of word representations in vector space. In: Workshop Track Proceedings of the 1st International Conference on Learning Representations (ICLR) (2013)
37. Osborne, F., Mannocci, A., Motta, E.: Forecasting the spreading of technologies in research communities. In: Proceedings of the Knowledge Capture Conference (K-CAP), pp. 1:1–1:8 (2017)
38. Perozzi, B., Al-Rfou, R., Skiena, S.: DeepWalk: Online learning of social representations. Proceedings of the 20th ACM International Conference on Knowledge Discovery & Data Mining (KDD) pp. 701–710 (2014)
39. Porter, A.L., Garner, J., Carley, S.F., Newman, N.C.: Emergence scoring to identify frontier R&D topics and key players. Technological Forecasting & Social Change **146**, 628–643 (2019)
40. Ramaswamy, S., Rastogi, R., Shim, K.: Efficient algorithms for mining outliers from large data sets. In: Proceedings of the ACM International Conference on Management of Data (SIGMOD), pp. 427–438 (2000)
41. Rigatti, S.: Random forest. Journal of Insurance Medicine **47**(1), 31–39 (2017)
42. Rossi, E., Chamberlain, B., Frasca, F., Eynard, D., Monti, F., Bronstein, M.: Temporal graph networks for deep learning on dynamic graphs. In: Proceedings of the ICML Workshop on Graph Representation Learning & Beyond (GRL+) (2020)

43. Rotolo, D., Hicks, D., Martin, B.R.: What is an emerging technology? Research Policy **44**(10), 1827–1843 (2015)
44. Salatino, A.A., Motta, E.: Detection of embryonic research topics by analysing semantic topic networks. In: Proceedings of the WWW International Workshop on "Semantics, Analytics and Visualization: Enhancing Scholarly Data", pp. 131–146 (2016)
45. Salatino, A.A., Osborne, F., Motta, E.: AUGUR: forecasting the emergence of new research topics. In: Proceedings of the 18th ACM/IEEE on Joint Conference on Digital Libraries (JCDL), pp. 303–312 (2018)
46. Schölkopf, B., Platt, J.C., Shawe-Taylor, J., Smola, A., Williamson, R.: Estimating the support of a high-dimensional distribution. Neural Computation **13**, 1443–1471 (2001)
47. Shyu, M., Chen, S., Sarinnapakorn, K., Chang, L.: A novel anomaly detection scheme based on principal component classifier. In: Proceedings of the ICDM International Workshop on Foundations & New Directions of Data Mining, pp. 172–179 (2003)
48. Small, H., Boyack, K.W., Klavans, R.: Identifying emerging topics in science and technology. Research Policy **43**(8), 1450–1467 (2014)
49. de Solla Price, D.J.: Networks of scientific papers. Science **149**(3683), 510–515 (1965)
50. Srinivasan, R., Lilien, G., Rangaswamy, A.: Technological opportunism and radical technology adoption: An application to e-business. Journal of Marketing **66**(3), 47–60 (2002)
51. Sun, J., Gao, J., Yang, B.: Achieving disruptive innovation forecasting potential technologies based upon technical system evolution by TRIZ. In: Proceedings of the 4th IEEE International Conference on Management of Innovation & Technology, pp. 18–22 (2008)
52. Takeda, Y., Kajikawa, Y.: Optics: a bibliometric approach to detect emerging research domains and intellectual bases. Scientometrics **78**(3), 543–558 (2008)
53. Tang, J., Qu, M., Wang, M., Zhang, M., Yan, J., Mei, Q.: Line: Large-scale information network embedding. Proceedings of the 24th International Conference on World Wide Web (WWW) pp. 1067–1077 (2015)
54. Veugelers, R., Wang, J.: Scientific novelty and technological impact. Research Policy **48**(6), 1362–1372 (2019)
55. Vojak, B., Chambers, F.: Roadmapping disruptive technical threats and opportunities in complex, technology-based subsystems: The SAILS methodology. Technological Forecasting & Social Change **71**(1), 121–139 (2004)
56. Weismayer, C., Pezenka, I.: Identifying emerging research fields: A longitudinal latent semantic keyword analysis. Scientometrics **113**(3), 1757–1785 (2017)
57. Xu, S., Hao, L., An, X., Yang, G., Wang, F.: Emerging research topics detection with multiple machine learning models. Journal of Informetrics **13**(4), 100983 (2019)
58. Zhao, Y., Nasrullah, Z., Li, Z.: PyOD: A Python toolbox for scalable outlier detection. Journal of Machine Learning Research **20**(96), 1–7 (2019)

Chapter 13
Link Prediction in Bibliographic Networks

Pantelis Chronis, Dimitrios Skoutas, Spiros Athanasiou, and Spiros Skiadopoulos

Abstract A bibliographic network consists of entities related to scientific publications (e.g., authors, papers and venues) and their relations. Predicting links on a bibliographic network is important for understanding and facilitating the scientific publication process. Bibliographic networks can be represented and analysed as Heterogeneous Information Networks (HINs). In this chapter, we comparatively evaluate three different models for link prediction in HINs, on three instances of a bibliographic network. The selected models represent two main paradigms for link prediction in HINs: path counting models and embedding models. The results indicate that, although being conceptually simpler, path counting models are, overall, slightly more accurate for predicting links in the bibliographic networks of our evaluation.

Pantelis Chronis [1,2]
[1] University of Peloponnese, Tripoli, Greece,
[2] Athena Research Center, Athens, Greece
e-mail: chronis@uop.gr

Dimitrios Skoutas
Athena Research Center, Athens, Greece
e-mail: dskoutas@athenarc.gr

Spiros Athanasiou
Athena Research Center, Athens, Greece
e-mail: spathan@athenarc.gr

Spiros Skiadopoulos
University of Peloponnese, Tripoli, Greece
e-mail: spiros@uop.gr

© The Author(s), under exclusive license to Springer Nature Switzerland AG 2021
Y. Manolopoulos, T. Vergoulis (eds.), *Predicting the Dynamics of Research Impact*,
https://doi.org/10.1007/978-3-030-86668-6_13

13.1 Introduction

Analysing information about scientific publications provides useful insight into the scientific process. Such information is often represented as a network of entities and their connections, e.g., *authors* connected to their *papers*, and *papers* to the *venues* where they were presented or published. This kind of network is referred to as a *bibliographic network*. An example of a bibliographic network, with authors, papers, venues and their connections, is depicted in Fig. 13.1. Link prediction is the task suggesting new edges in a network. These new edges may correspond to connections that actually exist in the real world but have not been captured in the available data or to edges that do not actually exist but may be created in the future. In the first case, which often occurs in automatically harvested bibliographic networks, link prediction methods can be used to automatically retrieve the missing links (e.g., find authors of a paper that are missing or are ambiguous), thus helping to improve the quality of the dataset. In the second case link prediction methods may be used to suggest future links (e.g., future co-authors) or relevant nodes, as recommendations (e.g., relevant papers or conferences to an author). Therefore, an effective link prediction model can be useful in the analysis of bibliographic networks.

A bibliographic network can be represented as a Heterogeneous Information Network (HIN). A HIN is a graph, i.e., a set of *nodes* and *edges*, with the additional property that there are multiple *types* of nodes and edges. The existence of multiple types is important for modelling the graph. In particular, a sequence of node and edge types of a HIN is called a *metapath*. A metapath represents a composite relation between two nodes. For example, in a bibliographic network, the metapath:

$$\text{Author} \xrightarrow{\text{Writes}} \text{Paper} \xrightarrow{\text{isWrittenBy}} \text{Author}$$

connects two authors that have a common paper (e.g., Fig. 13.1, red path), while the metapath

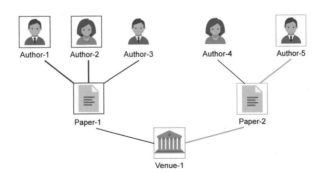

Fig. 13.1: An example of a bibliographic network with authors papers and venues. The red and green paths belong to different metapaths and express different relations

$$\text{Author} \xrightarrow{\text{Writes}} \text{Paper} \xrightarrow{\text{isPublished}} \text{Venue}$$

exist between an author and a venue he has published in (e.g., Fig. 13.1, green path).

The concept of metapath is important for link prediction models in HINs. These models can be categorized in two main categories: (a) *metapath-counting* models, which are based on counting the number of path instances, for each different metapath, between any two nodes [3, 11, 12]; (b) *embedding* models, that use *embeddings*, which are low dimensional vector representations of the nodes of the graph [4, 5, 13].

These two types of models take different approaches for the task of link prediction. Metapath-counting models consider the topological relation between a pair of nodes explicitly, by counting and evaluating the paths that connect them. On the other hand, embedding models focus on the nodes of the network individually and assign to each node a representation that aims to capture properties of the node that affect its links. Each type of model may be more suitable for networks with different characteristics and connectivity patterns. However, it is not straightforward to determine which type of model is more appropriate for link prediction in bibliographic networks.

To answer this question, we comparatively evaluate three models for link prediction in HINs, stemming from the two aforementioned categories: METAPATHCOUNTS [11] that works by counting the metapath instances between any two nodes, METAPATH2VEC [5] which is a embedding model based on random walks and DISTMULT [13] which is an embedding model where each node interacts only with its direct neighbors. We test the three models on three instances of bibliographic networks, obtained from DBLP and AMINER. All networks contain nodes of types Author, Paper and Venue but vary in their edge types. The first instance contains edge types Writes/isWrittenBy, between authors and papers, and Published/isPublished, between papers and venues. The second instance contains the additional edge type Cites/isCitedBy, between papers. The third instance contains the additional edge type isCoauthorOf between authors. The goal of the evaluation is to study the behaviour of each model on networks with different characteristics, and on each different type of edge individually. The results suggest that different models are more accurate on different edge types, but, overall, the metapath-counting model achieves slightly higher accuracy, on average, despite being conceptually simpler.

The rest of the chapter is organized as follows. Sect. 13.2 presents a brief overview of the literature of link prediction in HINs. Sect. 13.3 presents the thee selected models, and Sect. 13.4 presents the experimental evaluation and the results. Sect. 13.6 concludes the chapter.

13.2 Related Work

We study link prediction in bibliographic networks as an instance of link prediction in HINs. Models for link prediction in HINs can be categorized in two categories: (i) those that count the metapaths that exist between any two nodes and (ii) those that use node embeddings. We discuss these categories below.

13.2.1 Metapath-counting Models

Models of this category [3, 11, 12, 14, 15] employ a two-step process. The first step is to extract metapath-based feature vectors, which are used in the second step to train a classifier to predict the probability of a link. In [11], the feature vector for the prediction of links contains the counts of each metapath between the two nodes, as well as normalized versions of these counts. These features are fed into a logistic regression model, which assigns appropriate weights to the counts of each metapath. In [12], the same metapath-based features are used along with various continuous probability distributions, to model the time interval until a link occurs. In [14], the metapath-count based features of [11] are combined with an additional similarity feature based on keywords. To obtain this similarity feature, keywords that correlate with citation edges are identified, papers are grouped based on these keywords, and the similarity between papers is estimated based on their group memberships. This similarity score along with the features obtained by counting the metapaths are used to train a logistic regression classifier for the prediction of links. In a similar fashion, in [15], a similarity measure is constructed by adding two separate measures, one corresponding to metapath-based similarity and another corresponding to similarity based on other node attributes. The resulting similarity measure is trained to predict the existence of edges between nodes. In [3] link prediction is performed collectively, i.e., simultaneously for different types of edges, based on the assumption that predicted edges may provide information for the existence of other edges as well. For each pair of nodes, metapath based similarities are calculated using a metapath-based adjacency matrix, and these similarities are given to a machine learning model which predict the existence of links. The process is repeated, with the predicted links of one iteration provided as input to the next iteration, until convergence.

13.2.2 Embedding Models

Models of the second category use the recently popular approach of node embeddings. A *node embedding* is a vector representation, which captures the topological information of a node in the network. It can be used for various tasks, including link prediction. Metapaths are important for most works in this category as well. In [5], embeddings are obtained through metapath-constrained random walks. For each node, and for each given metapath, a large number of random walks is executed, and the vectors of the nodes are adjusted so that their inner product is reduced whenever two nodes are encountered closely in a random walk. In [9], embeddings are also obtained through metapath constrained random walks. However, in this case, a different embedding is first obtained for each metapath, and then all embeddings are combined to minimize the error of the model on a recommendation task. In this approach, the embeddings are used in combination with an existing matrix factorization algorithm for recommendations. In [4], each node is represented with a vector and each edge type is represented with a matrix. In this case, for each edge, the node vectors are multiplied by the edge type matrix and the euclidean distance is calculated from the resulting vectors, to estimate the likelihood of a link.

Many embedding methods for link prediction have also been developed in the literature of Knowledge Graphs [1, 7, 8, 10, 13]. Due to the similarity between Knowledge Graphs and HINs, these methods can also be applied to HINs. Generally, Knowledge Graphs models do not use random walks. Instead they model single edges individually, similarly to [4]. The difference between these models lies in the specific formulation of the function that combines the node vectors with the edge type vectors or matrices. More specifically, in [1] addition and euclidean distance is while in [10] a series of matrix multiplications and an inner product is employed. In [8] multiplications with a diagonal matrix, addition and an inner product is utilized. Finally, in [13] a simple and effective method that uses the bilinear product between the node vectors and a diagonal matrix assigned to each edge type is presented.

13.3 Description of the Selected Models

In this section we describe the models we selected for the evaluation. We have selected three models, METAPATHCOUNTS, METAPATH2VEC and DISTMULT, that represent different approaches for link prediction in HINs. METAPATHCOUNTS is based on counting the metapaths between any two nodes. METAPATH2VEC is an embedding model based on random walks, from the literature of HINs. DISTMULT is an embedding model that models each single edge individually, from the literature of Knowledge Bases, which can also be applied on HINs.

Formally, a HIN H is defined as $H = (V, E, L_V, L_E)$, where V is the set of nodes, E is the set of edges, L_V is the set of node types, and L_E is the set of edge types. Given two nodes $u, v \in V$ and an edge type $\ell_e \in L$, each model calculates a score function $f_{\ell_e}(u, v)$ for the probability than an edge of type ℓ_e exists between nodes u and v. Node pairs with higher probability scores are considered more likely to have an edge between them. Each model has a different formulation for scoring function f_{ℓ_e}. After training each model, the resulting function f_{ℓ_e} can be used to rank pairs of nodes and find those that are most likely to have an edge (e.g., the top-k). Next we present the exact formulation of each model.

13.3.1 METAPATHCOUNTS

METAPATHCOUNTS [11] models the probability that a link exists between two nodes as a function of the number of metapaths that connect them. The model is developed for each edge type separately. For an edge type $\ell_e \in L_E$, there exists a set $R_m(\ell_e)$, with size $d_{\ell_e} = |R_m(\ell_e)|$ comprising all metapaths up to length m which connect the node types corresponding to ℓ_e. For convenience, we enumerate the metapaths of $R_m(\ell_e)$ as $r_{i,\ell_e}, 1 \leq i \leq d_{\ell_e}$.

For a pair of nodes (u, v) and a metapath r, function $c(r, u, v)$ returns the number of instances of metapath r that exists between u and v (e.g., the number of publications of an author to a venue). The vector of metapath-counts between nodes u and v is defined as:

$$x_{\ell_e} : x_{i,\ell_e} = c(r_{i,\ell_e}, u, v)$$

Then, the probability that an edge of type ℓ_e exists between u and v is modeled as:

$$f_{\ell_e}(u, v) = \sigma(x_{\ell_e} \cdot w + b) \qquad (13.1)$$

where σ is the logistic sigmoid function, $\sigma(x) = \frac{1}{1+e^{-x}}$, and $\cdot$ is the inner product operator. Vector w contains the weights given by the model to the counts of the different metapaths and b is a constant bias on the probability.

To train the model for edge type ℓ_e, we select all pairs (u, v) of H that are connected with an edge of type l_e and form set $P(\ell_e)$. We also define a set of node pairs (z, y), of the node types corresponding to ℓ_e, that do not have an edge between them, which we denote as $N(\ell_e)$. Since bibliographic networks tend to be sparse, which means that the size of $P(\ell_e)$ is $O(n)$, while $N(\ell_e)$ is $O(n^2)$, we do not include all pairs without an edge in $N(\ell_e)$. Instead we select a random sample. The technique of selecting a sample of the *negative* (i.e., non existing) edges is called negative sampling and is widely used in link prediction [6]. We select the negative sample in two steps. First, for each pair $(u, v) \in P(\ell_e)$, we randomly select α nodes y, of the same node type as v, from the m step neighborhood of u, asserting that edge (u,y) does not exist $((u, y) \notin P(\ell_e))$. Then, we randomly select α nodes y, of the same node type as u, again asserting that the edge does not exist $((y, u) \notin P(\ell_e))$. Finally, the parameters of the model are obtained by maximizing the log-likelihood on sets $P(\ell_e)$ and $N(\ell_e)$:

$$L_{\ell_e}(w, b) = \sum_{(u,v)\in P(\ell_e)} \log(f_{\ell_e}(u, v)) + \sum_{(u,v)\in N(\ell_e)} \log(1 - f_{\ell_e}(u, v)) \qquad (13.2)$$

Maximizing this function asserts that the probability that the model assigns to existing edges is as close to 1 as possible, while the probability that it assigns to non-connected pairs is as close to 0 as possible. We note that in [11] a few other variations of the metapath-count features are also described.

13.3.2 METAPATH2VEC

METAPATH2VEC [5] assigns a vector representation for each node of the HIN, based on its proximity to other nodes, and then models the probability of a link as a function of these representations. The proximity is calculated via metapath constrained random walks on the HIN. Formally, each node v_i is associated with a d-dimensional representation $x_i \in R^d$. The probability that a link exists between v_i and v_j is modeled as:

$$f(v_i, v_j) = \frac{e^{x_i \cdot x_j}}{\sum_{v_k \in V} e^{x_i \cdot x_k}} \tag{13.3}$$

The training algorithm, that estimates the representations for each node, receives a set of metapaths R as input and performs the following steps:

- For each metapath $r \in R$ it performs I random walks of length k on the graph. At each step of the random walk, it chooses uniformly at random one of the available neighbors, having the type that is defined by metapath r for the next step.
- For each node v, the algorithm creates a multiset $C(v)$ containing all nodes that were encountered in a distance of w or less steps from v, in all random walks.
- Vectors x_i are randomly initialized and optimized via gradient descent so that they maximize the following log likelihood:

$$L = \sum_{v_i \in V} \sum_{v_j \in C(v)} \log(f(v_i, v_j)) \tag{13.4}$$

Eq. 13.4 require $O(n)$ evaluations of Eq. 13.3 which itself has $O(n)$ complexity, for calculating the denominator. To avoid this complexity, negative sampling is used. Specifically α nodes v_k (Eq. 13.3) are sampled uniformly at random from all nodes in V with the same type as v_i. We note that METAPATH2VEC models all edge types simultaneously, i.e., the score it provides for two nodes is the same for every edge type that may exists between the nodes.

13.3.3 DISTMULT

DISTMULT assigns a vector representation to each node of the network and a diagonal matrix representation for each edge type. It uses the bilinear product between the node vectors and the edge type matrix, as a scoring function for the existence of an edge of a specific type between two nodes. The node and edge type representations are trained on every edge that exists in the network.

Specifically, each node v_i is associated with a d-dimensional vector $x_i \in R^d$. Each edge type $\ell_e \in L_E$ is associated with a diagonal matrix $W_{\ell_e} \in R^{d \times d}$. Given a pair of nodes (v_i, v_j) and an edge type ℓ_e, scoring function $g_r(v_i, v_j)$ is defined as:

$$f_{\ell_e}(v_i, v_j) = x_i^T W_{\ell_e} x_j \tag{13.5}$$

Triples (x_i, ℓ_e, x_j) that result in higher scores are considered more likely to exist. To train the model the algorithm uses the set of existing edges E and a random sample of non existing edges N. The model is trained to minimize the following margin-based loss:

$$L = \sum_{(v_i, \ell_e, v_j) \in E} \left(\sum_{(v_k, \ell_e, v_l) \in N} \max(f_{\ell_e}(v_k, v_l) - f_{\ell_e}(v_i, v_j) + \alpha, 0) \right) \tag{13.6}$$

Minimizing this loss asserts that the scores of existing edges (E) are larger than those of non-existing edges (N) by more than α, which is a hyperparameter of the loss.

13.3.4 Discussion

There are several differences between the three described models. One key difference concerns the way the models treat the topology of the network. METAPATHCOUNTS considers the topology explicitly. When evaluating the existence of a link between u and v, the algorithm finds all paths in the network that connect the nodes and counts them. On the other hand, embedding models consider the topology of the network implicitly. When evaluating the existence of a link between u and v they consider the vector representations of u and v, instead of their location on the network location directly. Therefore, any topological information regarding the nodes must be stored in the node representations, during the model's training. This entails learning many parameters per node during training, typically 64, 128 or 256. This may make the embedding models more demanding, in terms of required sample size, and it may also lead to loss of information. Embedding models also require the selection of several hyper parameters for their training (e.g., embedding dimension, learning-rate, epochs, batch number, regularization type and strength, optimization algorithm and more) which are generally tuned heuristically.

METAPATHCOUNTS is limited by maximum path length m. When evaluating the existence of a link between two nodes, all paths with length greater than m between the nodes are ignored. Also METAPATHCOUNTS ignores the specific nodes of each path, as the paths are aggregated into their metapaths. On opposite, embedding models consider the effects of each node individually. Also the representation of each node is transitively affected by nodes that may be multiple steps away, depending on the number of training iterations. Therefore, they can incorporate information that is not considered by metapath-count based models.

Finally, there are a few differences between the two embedding models as well. The most important is that METAPATH2VEC uses random walks to traverse the graph. The representations of nodes that are more than one steps away on the random walks interact directly during training. On the other hand, in DISTMULT, the embeddings of all nodes that have an edge interact, without any random process, but nodes that are further apart do not interact directly. Also, in DISTMULT the probability of an edge between two nodes also depends on the edge type ℓ, through matrix W_ℓ. Finally, METAPATH2VEC is trained using the log-softmax loss (Eq. 13.4) and DISTMULT uses margin loss (Eq. 13.6).

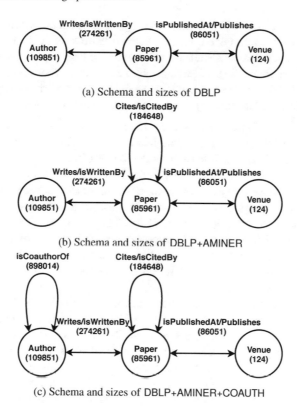

(a) Schema and sizes of DBLP

(b) Schema and sizes of DBLP+AMINER

(c) Schema and sizes of DBLP+AMINER+COAUTH

Fig. 13.2: The network schemas of the datasets

13.4 Evaluation

In this section we describe the process for evaluating the models. The evaluation is performed on three instances of a bibliographic network which we describe in Sect. 13.4.1. The experimental protocol and the metrics used to assess the accuracy of the models are described in Sects. 13.4.2-13.4.3, respectively. In Sect. 13.4.4 we present the hyper-parameter configurations of the models.

13.4.1 Datasets

We apply the algorithms described in Sect. 13.3 on three versions of the DBLP dataset. To limit the running time of the experiments, from the entire DBLP dataset we select only the venues that include the word 'data' (i.e., are related to data), the papers that were published in these venues and their authors. The three versions of the dataset are described next.

DBLP.   The schema of DBLP is shown in Fig. 13.2a It contains node types Author, Paper and Venue, and edge types Writes/isWrittenBy, that exist between authors and papers, and Pulishes/isPublished, that exist between papers and venues. There are 109,851 authors, 85,961 papers and 124 venues, connected with 274,261 writes and 86,051 publishes edges (there exist 90 papers with two reported venues of publication in the dataset). The edges are two-directional, e.g., when edge Author1 $\xrightarrow{\text{Writes}}$ Paper1 exists then edge Paper1 $\xrightarrow{\text{isWrittenBy}}$ Author1 also exists. The density[1] of the DBLP graph is $1.87 \cdot 10^5$. This instance of the network is the most simple and is directly provided by DBLP.

DBLP+AMINER.   The schema of DBLP+AMINER is depicted in Fig. 13.2b. It contains an additional edge type, Cites/isCitedBy, that exists between a paper that cites another paper. These edges are provided by Aminer. Not all citations of all papers are available in the dataset. Overall, 184,648 edges of type Cites/isCitedBy exist in the network. DBLP+AMINER represents a slightly more complex instance of DBLP. The Cites/isCitedBy edge type has the particular property that it connects nodes of the same type Papers. The addition of Cites/isCitedBy edges increases the density of the network to $3.80 \cdot 10^5$.

DBLP+AMINER+COAUTH.   The schema of DBLP+AMINER+COAUTH is depicted in Fig. 13.2c. Edge type isCoauthorOf is added between authors that have written a paper together. We add these edges on AMINER dataset, during a preprocessing step. There are 898,014 edges of this type in the network. Edge type isCoauthorOf has the particular property that it corresponds to a metapath (Author-Paper-Author) in the original HIN. Including a metapath as an edge does not provide additional information for the relations between the nodes, but is a different representation of the available information which may influence the performance of models. For example, in DISTMULT only nodes that are neighbours interact directly during training. With the addition of isCoauthorOf edges, nodes with distance of two steps can also interact directly. Similarly, for METAPATHCOUNTS, 3 edges of type isCoauthorOf correspond to a path of length 6 (each isCoauthorOf edge corresponds to two Writes/isWrittenBy edges), which would be above maximum length m in the original graph. Including isCoauthorOf edge increases the density of the network to $8.45 \cdot 10^5$.

The three datasets contain the same nodes and edges, for the types they share. DBLP is a sub-graph of DBLP+AMINER, which, in turn, is a subgraph of DBLP+AMINER+COAUTH. These three datasets are selected evaluate the performance of the models against several different challenges, i.e., N-M relations with N>>M (Pulishes/isPublished), relations that are closer to 1-1 (Writes/isWrittenBy), relations that start and end at the same type (Cites/isCitedBy) or are composite relations that correspond to a metapath (isCoauthorOf), as well as graphs with varying density.

---

[1] The density of a graph is defined as the number of its actual edges $|E|$ divided by the maximum possible number of edges $|V|^2$ where V is the number of nodes, i.e, $\frac{|E|}{|V|^2}$.

13.4.2 Experimental Setup

The evaluation process requires training the models, applying them and evaluating their results. The evaluation for each edge type is performed separately. For each edge type ℓ we select 20,000 edges (u, v) of type $\tau(u, v) = \ell_e$ (function $\tau(u, v)$ shows the type of edge (u, v)). These edges that exist, in the HIN are called positive edges. For each edge (u, v) we construct 10 edges that do not exist (negative edges) in the following way. First we select 5 nodes y from the l-step neighborhood of u, so that $(u, y) \notin E$ and $\tau_e(u, y) = \ell_e$ hold. We select these nodes that are close because such nodes are more likely to create false positive responses from the models. This is because all three models implicitly assume that nodes that are close on the graph (i.e., neighbors, likely to co-occur on a random walk, joined by at least one path of length m) are more likely to form an edge. We select an additional 5 nodes y, from the entire network, so that $(u, y) \notin E$ and $\tau_e(u, y) = \ell_e$ hold. Both selections are performed uniformly at random without replacement. Overall, we select 220,000 edges (20,000 positive and 200,000 negative). Subsequently, we select 20% (44,000) of the edges, at random, to construct the test set of the models. Due to the large sample size, the proportions of positive and negative edges in both the training and testing sets will be effectively equal to 10%. METAPATHCOUNTS is trained using the remaining 80% (176,000) of the edges. We train the model on this sample to limit the computational complexity of finding all paths between all pairs of the network. The model can be effectively trained on this (or on an even smaller) sample, since it has few parameters. The embedding models are trained, using all the edges of the graph, after removing the positive edges of the test set. All models are trained for each edge type separately. After training the models we use them to obtain a score for each edge of the test set, which is common for all models.

13.4.3 Metrics

We are interested in the ability of each model to assign higher scores to positive than negative edges. We evaluate this ability using two metrics: Area Under Curve (AUC) and Mean Reciprocal Rank (MRR).

- AUC is the area under the ROC curve [2]. Each model assigns a score to each pair of nodes, for the probability that an edge exists. By setting a threshold on this score and considering pairs above the threshold as positive a trade-off occurs, regarding true positives (i.e., pairs with edges that are retrieved) and false positives (i.e., pairs without edges that are wrongly labeled as positive). The ROC curve shows the true positive rate (recall) of the model relative to the false positive rate (1-specificity), for all threshold selections. AUC is the area under the ROC curve, and is an estimate of the probability that the model will rank a randomly chosen positive instance higher than a randomly chosen negative one. The optimal value of AUC is 1.

- MRR is calculated using the rank of the positive edge with the highest score. Specifically, for each node of the test set, we select all edges that contain this specific node and sort them according to their scores. MRR is the inverse of the rank of the first positive edge. The reported numbers are the averages from all the nodes of the test set. The optimal value for MRR is 1.

MRR express the ability to distinguish a single edge, that is the most likely to exist, from all others, i.e., it focuses on the tail of the score distribution. On opposite, AUC expresses the ability of the model to distinguish positive from negative edges across the entire range of the scores.

13.4.4 Hyperparameter Configuration

Parameter Configuration is an important part of model training, especially for the embedding methods that have several hyperparameters. METAPATHCOUNTS has a single hyperparameter, maximum path length m, which we set at $m = 3$. Simple hyperparameter tuning is an important advantage of METAPATHCOUNTS.

For METAPATH2VEC, we used the following configuration: 20 random walks per node, 5 steps per random walk, 3 step window, 128 embedding dimension, 100 training epochs and learning rate from 0.1 to 0.001 (changing linearly with the epochs). All other hyperparameters were left to their default values, provided by the implementation of the gensim library, that we used.

For DISTMULT, we used the following configuration: 128 embedding dimension, 1000 epochs with 5 batches in each epoch, 0.01 learning rate, L2 regularization with $\lambda = 10^{-5}$ and 0.5 loss margin. All other parameters were left at their default values. For DISTMULT we used the implementation of the Ampligraph library.

We note that due to the significant running time for training the embedding models and the number of hyperparameters, a grid search of the hyperparameter space is not feasible. Instead the hyperparameters were selected heuristically with a few tuning experiments. Our tuning experiments as well as the results presented in Fig. 13.7, regarding the effect of different embedding dimensions, suggest that further hyper parameter tuning would not lead to significant improvements.

13.5 Results

Fig. 13.3 shows the average scores on the DBLP dataset. METAPATHCOUNTS has 8% higher AUC (0.81) than both METAPATH2VEC and DISTMULT (0.75 and 0.75). METAPATHCOUNTS also has the highest MRR (0.85), 2.4% higher than the second best, METAPATH2VEC (0.83), and 3.6% higher than DISTMULT (0.82) Comparing the two metrics, this result suggest that the difference between the models is larger at identifying positive edges of all scores, than finding a single highly likely positive edge (MRR).

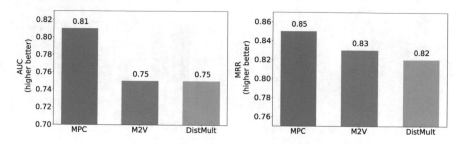

Fig. 13.3: Average results on DBLP

Table 13.1: Link Prediction Scores on DBLP

	Writes		Publishes	
	AUC	MRR	AUC	MRR
METAPATHCOUNTS	0.84	0.86	**0.78**	**0.83**
METAPATH2VEC	0.81	**0.91**	0.68	0.75
DISTMULT	**0.86**	0.88	0.64	0.76

The scores for each edge type of DBLP separately are presented in Table 13.1. A different model has the best performance on each edge type. For Writes edge type, DISTMULT has the highest AUC (0.86) and METAPATH2VEC has the highest MRR (0.91). Overall, embedding models have better performance than METAPATHCOUNTS in this edge type. For Publishes edge type the results are the opposite. METAPATH-COUNTS has significantly higher scores (more than 10%) than both embedding models on both metrics. The poorer performance of the embedding models in this edge type may be due to the small number of venues relative to papers (Fig. 13.3). Due to this asymmetry, each venue must be connected to many papers, possibly making it difficult for the embedding models to assign an appropriate representation that models all the connections. The cause of this problem could be that the embedding of a venue must have a short distance to the embeddings of all papers published in the venue (tens of thousands in our dataset) while retaining a long distance from the embeddings of all other papers. This constitutes a problem with many constraints that may be difficult to solve effectively in the relatively low dimensional embedding space. Moreover, the problem can not be easily avoided because hub-like nodes, such as venues, exist in most bibliographic networks.

Fig. 13.4 shows the average scores on the DBLP+AMINER dataset. METAPATH-COUNTS has slightly better AUC (0.89 to 0.88) and equivalent MRR (0.91) with DISTMULT. METAPATH2VEC has significantly worse performance on both metrics (0.79 AUC, 0.86 MRR) (Fig. 13.4).

The results on each edge types of the DBLP+AMINER dataset are shown separately in Table 13.2. In Cites edge type, DISTMULT achieves the highest scores (0.99 AUC, 0.99 MRR), followed by METAPATHCOUNTS with (0.96 AUC, 0.96 MRR) and METAPATH2VEC with (0.89 AUC, 0.91 MRR). The results suggest that citation edges

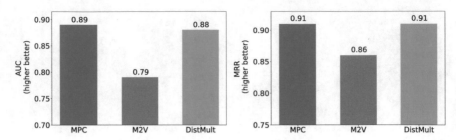

Fig. 13.4: Average results on DBLP+AMINER

Table 13.2: Link Prediction Scores on DBLP+AMINER

	Writes		Publishes		Cites	
	AUC	MRR	AUC	MRR	AUC	MRR
METAPATHCOUNTS	0.87	0.88	**0.85**	**0.89**	0.96	0.96
METAPATH2VEC	0.81	**0.91**	0.68	0.75	0.89	0.91
DISTMULT	**0.88**	0.90	0.78	0.83	**0.99**	**0.99**

are easier to predict. This may be due to the fact that citations are given in groups of related papers, therefore the existence of one citation may be very informative, regarding other possible citations. The introduction of Cites edge type affects the results on Writes and Publishes edge types as well. Specifically, the performance of all models has improved on both edge types, but more significantly on Publishes edge type. It seems that the additional information for the connection between papers is useful both for predicting the venue of publication (e.g., by considering the venues of cited papers) and, to a smaller degree, for predicting authors (e.g., by considering the authors of cited papers). The model that benefited less from the introduction of the new edge is METAPATH2VEC. METAPATH2VEC models all edge types without performing any edge related transformation to the embeddings. By introducing an additional edge type, the obtained embeddings may not be sufficient to model all the existing edge types. On the other hand, DISTMULT transforms the embeddings using the edge type matrix (Eq.13.5), therefore it is less hindered by the existence of more edge types.

Fig. 13.5 shows the average scores on the DBLP+AMINER+COAUTH dataset. In this dataset METAPATHCOUNTS has the highest scores on both metrics (0.86 AUC, 0.87 MRR), followed by DISTMULT (0.84 AUC, 0.85 MRR), and then followed by METAPATH2VEC (0.76 AUC, 0.83 MRR). In DBLP+AMINER+COAUTH, as in DBLP+AMINER, METAPATHCOUNTS performs slightly better than DISTMULT, and both achieve higher scores than METAPATH2VEC by a more significant margin, especially in AUC.

The results on each edge type of the DBLP+AMINER+COAUTH dataset are presented separately in Table 13.3. METAPATHCOUNTS and DISTMULT have similar performance in isCoauthorOf edge type (0.77 AUC, 0.76 MRR for METAPATH-

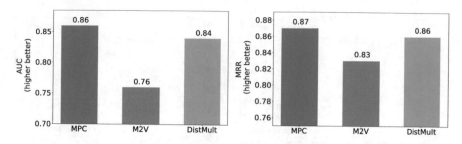

Fig. 13.5: Average results on DBLP+AMINER+COAUTH

Table 13.3: Link Prediction Scores on DBLP+AMINER+COAUTH

	Writes		Publishes		Cites		isCoauthorOf	
	AUC	MRR	AUC	MRR	AUC	MRR	AUC	MRR
METAPATHCOUNTS	0.87	0.87	**0.86**	**0.89**	**0.96**	**0.96**	**0.77**	0.76
METAPATH2VEC	0.81	**0.92**	0.69	0.76	0.90	0.92	0.63	0.72
DISTMULT	**0.89**	0.90	0.75	0.82	0.95	0.96	0.75	**0.77**

COUNTS, 0.75 AUC 0.77 MRR for DISTMULT). It seems that, overall, the models did not benefit from the introduction of edge type isCoauthorOf. In fact, the scores of DISTMULT on Cites edge type decreased noticeably.

AUC is defined as the area under the ROC curve. Observing the ROC curve directly can provide additional information for the performance of the models. The ROC curve (Sect. 13.4.3) shows the true positive rate (recall) of the model relative to the false positive rate (1-specificity), for all threshold selections. For example, if the ROC curve of a model passes from point $(0.2, 0.8)$, it means that there is a threshold for the score above which 0.8 of all positive edges lie and 0.2 of the edge are negative. Fig. 13.6 shows the ROC curves of METAPATHCOUNTS, METAPATH2VEC and DISTMULT for each edge type of the DBLP+AMINER+COAUTH dataset. For all edge types, the curve of METAPATHCOUNTS rises faster than the curves of METAP-ATH2VEC and DISTMULT. This means that METAPATHCOUNTS correctly identifies a larger number of positive edges with relatively few false positives, i.e., it has higher specificity in retrieving the most probable edges. However, the curves of the other two models converge to and eventually overcome the curve of METAPATHCOUNTS, at higher values of false positive rates. This effect is more significant in Writes edge type. This means that that the embedding models are better at identifying positive edges that are harder to distinguish from negatives. This may be due to the fact that METAPATHCOUNTS has a hard threshold for the length of the paths that it considers, therefore pairs that are connected only with longer paths are treated by METAPATH-COUNTS as completely unconnected, while for embedding models the characteristics of nodes change more smoothly and similarities between nodes at larger distances may be detectable.

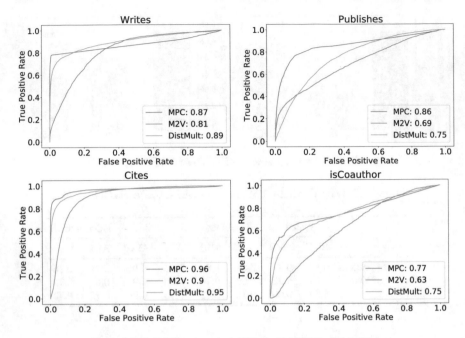

Fig. 13.6: ROC curves on DBLP+AMINER+COAUTH

To evaluate the effects of the path length threshold of METAPATHCOUNTS, we split the test set of the DBLP dataset into two parts: one consisting of pairs that have distance ≤ 3 and another consisting of pairs with distance > 3, and apply the models on the two parts separately. The results are presented on Tables 13.4-13.5. As expected all models have higher scores on pairs with distance ≤ 3 (illustrated in Table 13.4). On these pairs, METAPATHCOUNTS has the highest score for both edge types and metrics. On the opposite, on pairs that are further away (illustrated in Table 13.2), embedding models have higher scores on both edges and metrics (except AUC in Publishes). These results suggest that METAPATHCOUNTS is more accurate for pairs that are close on the graph (below its path length threshold) but embedding models are more accurate on nodes that are further apart.

Table 13.4: Individual Results on DBLP with ≥ 1 paths

	Writes		Publishes	
	AUC	MRR	AUC	MRR
METAPATHCOUNTS	**0.89**	**0.96**	**0.80**	**0.93**
METAPATH2VEC	0.74	0.95	0.79	0.86
DISTMULT	0.88	0.94	0.65	0.87

Table 13.5: Individual Results on DBLP with 0 paths

	Writes		Publishes	
	AUC	MRR	AUC	MRR
METAPATHCOUNTS	0.50	0.55	**0.50**	0.61
METAPATH2VEC	**0.63**	**0.79**	0.34	0.76
DISTMULT	0.59	0.77	0.48	**0.77**

Another important parameter is the embedding dimension of the embedding models. Increasing the embedding dimension allows the model to capture more complex relations between the nodes but also increases the capacity of overfitting. Fig. 13.7 shows the performance of DISTMULT on the DBLP for various embedding dimension sizes. On edge type Writes both metrics increase with the embedding dimension, however on Publishes both metrics decrease as the embedding dimension increases. This suggests that overfitting occurs in the Publishes relation, which can explain the poor performance of the embedding models on this edge type.

A possibly useful product of METAPATHCOUNTS are the weights/coefficients it assigns to the metapaths of the HIN. These weights quantify how important is the existence of an instance of each metapath between two nodes, for the creation of an edge. Table 13.6 presents the coefficients obtained for the metapaths of WRITES edge type, on DBLP+AMINER. The metapath with the highest coefficient is:

$$\text{Author} \xrightarrow{\text{Writes}} \text{Paper} \xrightarrow{\text{isWrittenBy}} \text{Author} \xrightarrow{\text{Writes}} \text{Paper}$$

This metapath exists between an author and a paper when the author has been coauthor, on another paper, with one of the authors of the paper. When this relation occurs between an author and a paper it is more likely that the author and the paper are connected with a WRITES edge. Two other metapaths with positive, and approximately equal, coefficients are:

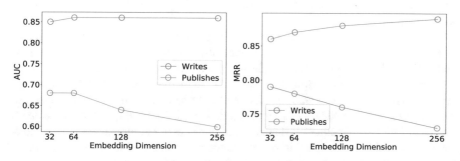

Fig. 13.7: Effect of embedding dimension of DISTMULT on DBLP

Table 13.6: Metapath coefficients for Writes edge type

Metapath	coefficient
Author $\xrightarrow{\text{Writes}}$ Paper $\xrightarrow{\text{isWrittenBy}}$ Author $\xrightarrow{\text{Writes}}$ Paper	4.05
Author $\xrightarrow{\text{writes}}$ Paper $\xrightarrow{\text{Cites}}$ Paper	1.01
Author $\xrightarrow{\text{writes}}$ Paper $\xrightarrow{\text{isCitedBy}}$ Paper	0.97
Author $\xrightarrow{\text{writes}}$ Paper $\xrightarrow{\text{Cites}}$ Paper $\xrightarrow{\text{isCitedBy}}$ Paper	−0.07
Author $\xrightarrow{\text{writes}}$ Paper $\xrightarrow{\text{isCitedBy}}$ Paper $\xrightarrow{\text{Cites}}$ Paper	−0.09
Author $\xrightarrow{\text{Writes}}$ Paper $\xrightarrow{\text{isPublishedAt}}$ Venue $\xrightarrow{\text{Publishes}}$ Paper	−0.21
Author $\xrightarrow{\text{writes}}$ Paper $\xrightarrow{\text{isCitedBy}}$ Paper $\xrightarrow{\text{isCitedBy}}$ Paper	−0.35
Author $\xrightarrow{\text{writes}}$ Paper $\xrightarrow{\text{Cites}}$ Paper $\xrightarrow{\text{Cites}}$ Paper	−0.49

$$\text{Author} \xrightarrow{\text{writes}} \text{Paper} \xrightarrow{\text{Cites}} \text{Paper}$$

and

$$\text{Author} \xrightarrow{\text{writes}} \text{Paper} \xrightarrow{\text{isCitedBy}} \text{Paper}$$

These metapaths occur between an author and a paper when the author has written another paper that cites or is cited by the paper in question. These relations also increase the probability that an edge is formed between an author and a paper, but not as much as:

$$\text{Author} \xrightarrow{\text{Writes}} \text{Paper} \xrightarrow{\text{isWrittenBy}} \text{Author} \xrightarrow{\text{Writes}} \text{Paper}$$

The other metapaths are less important and in fact have negative coefficients. This is paradoxical because it would mean that when these metapaths exist between two nodes it is less likely for an edge to be formed. The negative coefficients are explained by the negative sampling steps employed in the training of METAPATHCOUNTS (Sect. 13.3.1) and in the Evaluation (Sect. 13.4.2). During the negative sampling steps a large number of negative pairs with distance < 3 (i.e., with at least one of the above metapaths) are sampled. this results to the negative coefficients observed for these metapaths.

Overall, METAPATHCOUNTS model had slightly better performance than the embedding models on the datasets of our evaluation. However, for certain edge types, such as Cites in DBLP+AMINER (Table 13.2), DistMult model, had significantly better performance. Including additional information in the network, by adding Cites edge type, improved the performance of all models, but benefited METAPATHCOUNTS and DISTMULT more than METAPATH2VEC. Including the Author-Paper-Author metapath as an additional isCoauthorOf edge type did not benefit the models. Finally, regarding the embedding models, DISTMULT achieved better performance than METAPATH2VEC. For most edge types the performance of the METAPATHCOUNTS and DISTMULT was similar, with the exception of Publishes edge type, particularly in DBLP, where METAPATHCOUNTS was better by a significant margin.

The results suggest that the additional complexity of the embedding models did not benefit their accuracy, overall. On the opposite, the explicit topological approach of METAPATHCOUNTS was able to better capture the patterns that exist in the bibliographic networks we evaluated. Regarding the embedding models, the approach of DISTMULT, which uses single edges for training and includes a transformation for edge types, was more effective than METAPATH2VEC, which is based on random walks and does not include an edge type transformation.

13.6 Conclusion

In this chapter we performed a comparative evaluation of three link prediction models, on three instances of a bibliographic network. The models represent the two main paradigms for link prediction in HINs. The datasets used in the evaluation constitute typical bibliographic networks that may be available in practice. The evaluation was performed on the various different edge types of the datasets individually, and on average for each dataset. The results showed that a different model achieved the best performance on different edge types. Overall, the metapath-count based model (METAPATHCOUNTS) achieved slightly better performance, on average, than the embedding based models (DISTMULT, METAPATH2VEC) on all three instances of the bibliographic network. The evaluation suggests that each type of model performs better in different types of edges, and in different locations of the ROC curve. Therefore, it may be possible to combine the two types of methods, by training a model that combines explicit (metapath-counts) and implicit (embedding vector) topological features, to form a more effective model for link prediction in bibliographic networks. Such a model could potentially benefit from including additional features of the nodes and edges of the network (e.g., dates, keywords). These are possible directions of improvement for future work.

References

1. Bordes, A., Usunier, N., Garcia-Durán, A., Weston, J., Yakhnenko, O.: Translating embeddings for modeling multi-relational data. In: Proceedings of the 26th International Conference on Neural Information Processing Systems (NeurIPS), pp. 2787–2795 (2013)
2. Bradley, A.P.: The use of the area under the ROC curve in the evaluation of machine learning algorithms. Pattern Recognition **30**(7), 1145–1159 (1997)
3. Cao, B., Kong, X., Yu, P.S.: Collective prediction of multiple types of links in heterogeneous information networks. In: Proceedings of the IEEE International Conference on Data Mining (ICDM), pp. 50–59 (2014)
4. Chen, H., Yin, H., Wang, W., Wang, H., Nguyen, Q.V.H., Li, X.: PME: Projected metric embedding on heterogeneous networks for link prediction. In: Proceedings of the 24th ACM International Conference on Knowledge Discovery & Data Mining (KDD), pp. 1177–1186 (2018)

5. Dong, Y., Chawla, N.V., Swami, A.: Metapath2vec: Scalable representation learning for hetero-geneous networks. In: Proceedings of the 23rd ACM International Conference on Knowledge Discovery & Data Mining (KDD), pp. 135–144 (2017)
6. Kotnis, B., Nastase, V.: Analysis of the impact of negative sampling on link prediction in knowledge graphs. ArXiv **abs/1708.06816** (2017)
7. Nickel, M., Rosasco, L., Poggio, T.: Holographic embeddings of knowledge graphs. In: Proceedings of the 30th AAAI Conference on Artificial Intelligence, pp. 1955–1961 (2016)
8. Shi, B., Weninger, T.: ProjE: Embedding projection for knowledge graph completion. In: Proceedings of the 21st AAAI Conference on Artificial Intelligence, pp. 1236–1242 (2017)
9. Shi, C., Hu, B., Zhao, W., Yu, P.S.: Heterogeneous information network embedding for recom-mendation. IEEE Transactions on Knowledge & Data Engineering **31**(02), 357–370 (2019)
10. Socher, R., Chen, D., Manning, C.D., Ng, A.Y.: Reasoning with neural tensor networks for knowledge base completion. In: Proceedings of the 26th International Conference on Neural Information Processing Systems (NeurIPS), pp. 926–934 (2013)
11. Sun, Y., Barber, R., Gupta, M., Aggarwal, C.C., Han, J.: Co-author relationship prediction in heterogeneous bibliographic networks. In: Proceedings of the International Conference on Advances in Social Networks Analysis & Mining (ASONAM), pp. 121–128 (2011)
12. Sun, Y., Han, J., Aggarwal, C.C., Chawla, N.V.: When will it happen? Relationship prediction in heterogeneous information networks. In: Proceedings of the 5th ACM International Conference on Web Search & Data Mining (WSDM), pp. 663–672 (2012)
13. Yang, B., Yih, W., He, X., Gao, J., Deng, L.: Embedding entities and relations for learning and inference in knowledge bases. In: Proceedings of the International Conference on Learning Representations (ICLR) (2015)
14. Yu, X., Gu, Q., Zhou, M., Han, J.: Citation prediction in heterogeneous bibliographic networks. In: Proceedings of the 12th SIAM International Conference on Data Mining (SDM), pp. 1119–1130 (2012)
15. Yu, Z., Feng, L., Kening, G., Ge, Y.: A method of link prediction using meta path and attribute information. In: Proceedings of the 16th International Conference on Web Information Systems & Applications (WISA), pp. 449–454 (2019)

Printed in the United States
by Baker & Taylor Publisher Services